Fundamental Theories of Physics

Volume 223

The international monograph series "Fundamental Theories of Physics" aims to stretch the boundaries of mainstream physics by clarifying and developing the theoretical and conceptual framework of physics and by applying it to a wide range of interdisciplinary scientific fields. Original contributions in well-established fields such as Quantum Physics, Relativity Theory, Cosmology, Quantum Field Theory, Statistical Mechanics and Nonlinear Dynamics are welcome. The series also provides a forum for non-conventional approaches to these fields. Publications should present new and promising ideas, with prospects for their further development, and carefully show how they connect to conventional views of the topic. Although the aim of this series is to go beyond established mainstream physics, a high profile and open-minded Editorial Board will evaluate all contributions carefully to ensure a high scientific standard.

Ken-ji Hamada

Trans-Planckian Physics and Inflation

An Introduction to Renormalizable and Background-Free Quantum Gravity

Springer

Ken-ji Hamada
Institute of Particle and Nuclear Studies
High Energy Accelerator Research
Organization (KEK)
Tsukuba, Japan

ISSN 0168-1222 ISSN 2365-6425 (electronic)
Fundamental Theories of Physics
ISBN 978-981-96-3475-0 ISBN 978-981-96-3476-7 (eBook)
https://doi.org/10.1007/978-981-96-3476-7

This Springer imprint is published by the registered company Springer Nature Singapore Pte Ltd.
The registered company address is: 152 Beach Road, #21-01/04 Gateway East, Singapore 189721, Singapore

Preface

The quantum gravity theory described in this book is defined based on the guiding principles of gravity and quantum field theory, namely diffeomorphism invariance and renormalizability, and has been developed with the aim of revealing the trans-Planckian physics. Traditional approaches to quantizing gravity have been to deal with so-called gravitons as quanta, but such a particle-like description is not applicable at the Planck scale where spacetime is thought to fluctuate violently. Ultimately, it is necessary to quantize spacetime itself. This book presents a concrete method for implementing such a spacetime state and discusses a number of the trans-Planckian physics, including inflation. First and foremost, we concisely mention the historical background and recent developments leading up to this perspective.

The various Planck scales proposed by Planck at the end of the 19th century are fundamental units that characterize a tiny world where quantization of gravity is necessary. The Planck length $l_{\rm pl} = 1.616 \times 10^{-35}$ m is known as the smallest unit of length. Many physicists believe that there is no world shorter than it, or that the world is ruled by a physical law different from diffeomorphism invariance. That is reflected in the approach of putting a cutoff in the Planck length and considering it as an entity of quantized spacetime. However, this way of thinking is exactly the cause of many problems in the theory of gravity. It is simply to escape from the problem of spacetime singularity. Essential problems with gravity such as renormalizability, unitarity, cosmological constant, origin of primordial fluctuations, time, and so on ultimately come down to the problem of how to describe the trans-Planckian world as a continuum quantum field theory.

The root of the problem lies in the basic idea that gravity is distortion of spacetime caused by the presence of matter. The Einstein equation is a concrete manifestation of this view. It is a fundamental equation of the universe that holds in a very wide range from a macroscopic scale beyond the Hubble distance, which is the largest known scale in the universe, to the vicinity of the Planck length. No other equation has such a wide range of applications. Despite being such an excellent equation, the existence of spacetime singularities cannot be denied, and the laws of physics break down there. The cause is not in diffeomorphism invariance introduced as a guiding principle, but in the Einstein equation. The reason why we have to introduce a wall at the Planck scale is that the theory has not been defined properly so that diffeomorphism invariance holds in the world beyond the Planck scale.

In order to resolve the problem of singularities, it is necessary to reconsider the relationship between matter and gravity. The two are by no means standing on the same ground. We need to remember that matter is defined in spacetime, while gravity defines spacetime itself. Matter is normally described as a particle that propagates in space. Gravity is also described as a gravitational wave, or graviton, if its fluctuation amplitude is so small. However, if the fluctuation becomes so large that time and distance become uncertain, the particle picture propagating and scattering in space is no longer justified.

In fact, gravitational waves discovered in recent years by directly observing expansion and contraction of spatial distance are extremely weak, with an amplitude (strain) of less than 10^{-20}. Now, please imagine how the world would change if this amplitude were on the order of unity. If spacetime itself fluctuates greatly to this extent, the traditional particle description does not work.

As we approach the Planck scale, we will encounter such a world in which the substance of time and distance is lost. The words such as "background independence" or "background freedom" often used in the field of quantum gravity research represent such a state. Equations describing general gravitational states, including such quantum states, are what is called the Hamiltonian and momentum constraints, representing that Hamiltonian and momentum of the whole system vanish as a consequence of diffeomorphism invariance.

In order for diffeomorphism invariance to hold beyond the Planck scale, the theory has to be renormalizable, in which the field can be defined everywhere continuously. In addition, we need a quantization of gravity involving background freedom. This cannot be described by conventional quantum gravity theories that consider gravitons as fundamental objects. Some nonperturbative quantum field theory technique is required. In this book, we introduce a renormalizable quantum theory of gravity, in which the extreme world with background freedom is expressed as a special conformal field theory and deviations from it are formulated by perturbations. This is a theory built centered on conformally invariant fourth-derivative gravitational actions.

The conformal invariance that defines the extreme gravitational world is part of diffeomorphism invariance and is a hidden gauge symmetry that arises only when gravity is quantized. Here we call it the BRST conformal invariance to distinguish it from normal one. This is nothing but an algebraic representation of the Hamiltonian and momentum constraints. Even though the Hamiltonian vanishes exactly, there are an infinite number of physical states representing the world beyond the Planck scale, and they can be responsible for entropy of the universe. Moreover, despite describing in terms of the gravitational field, physical states are all given by scalar quantities, no tensor ones, thus it can successfully explain the results of cosmological observations. These facts are closely related to the ghost issue mentioned later.

Quantum spacetime with background freedom is far from the spacetime that we normally perceive. In other words, if such a quantum spacetime exists, there must be a moment when we move from such a quantum spacetime to the current classical spacetime. This moment is called "spacetime phase transitions". The renormalizable quantum gravity suggests the existence of a physical energy scale Λ_{QG} at which such a phase transition occurs. We believe that the moment of this change thought to have occurred at the beginning of the universe is the true identity of the Big Bang. Einstein's theory of gravity does not consider such phase changes in spacetime, and accordingly the source of the current matter is sought in another, usually unknown, matter. In contrast, matter created during the spacetime phase transition is born from gravitational fluctuations. In other words, gravity is the source of everything.

If the new gravitational scale Λ_{QG} is taken to be lower than the Planck energy, a stable inflationary solution exists. Choosing a value that can successfully explain the subsequent cosmic evolution results in about two orders of magnitude lower, 10^{17} GeV. The reciprocal of this value, $\xi_\Lambda = 1/\Lambda_{QG}$, represents the correlation length of quantum gravity, which becomes approximately 100 times larger than l_{pl}. That is, as we approach the center of mass, quantum gravity becomes effective before reaching the Planck scale, and the inside the correlation length is a world with background freedom where space and time fluctuate greatly. This implies that spacetime is practically quantized by ξ_Λ, and suggests the existence of localized excited states with a diameter of ξ_Λ, consisting only of quantum gravitational fluctuations.

As a consequence of diffeomorphism invariance, gravitational systems can be expressed as states in which the Hamiltonian vanishes, regardless of quantum theory or classical theory. Nevertheless, non-trivial states exist. This fact has implications for the existence of ghosts. If all modes were physical, the only such a state would be a trivial vacuum, and there would be no entropy. Thus, it turns out that the No-Go theorem for the ghost problem, which shows the unboundedness of the Hamiltonian, cannot be applied to gravitational systems where it is constrained to zero.

In fact, the existence of the Friedmann spacetime and various gravitational objects essentially depend on a ghost mode that makes the Einstein-Hilbert action indefinite. The large-scale structure of the current universe is also formed by the instability caused by it. Thus, the ghost mode plays an essential role in gravitational systems, but it has to be an unphysical mode that never observed directly. Similarly, in renormalizable quantum gravity, ghost modes are indispensable to ensure diffeomorphism invariance. The difference from Einstein's theory of gravity is that ghost modes exist even though the action is positive definite, i.e., stable. In other words, the gravitational field itself is not a ghost, but the ghost arises as a sub-mode hidden within the field. At this time, a constraint is required to prevent it from appearing as an observable in the real world. The BRST conformal invariance is responsible for that.

On the other hand, in the traditional perturbation theory in which the gravitational system is handled by reducing to a system of special relativity, or Lorentz-invariant system, a quantum field of graviton is dealt with as a Hamiltonian eigenstate, and no constraint that makes the whole Hamiltonian of the system vanishing is imposed. This is based on the premise that influence of gravitons is local and not strong enough to change the entire structure of spacetime, i.e., that back-reactions can be neglected. This approximation method is applicable to Einstein's theory of gravity at low energies, but in the trans-Planckian region where quantization of the entire spacetime should be considered, we cannot ignore back-reactions. In that case, the gravitational field itself must be quantized to obey the Hamiltonian and momentum constraints. Hence, the trans-Planckian physics begins with discarding the particle picture. Diffeomorphism invariance then requires the existence of ghosts, unlike Lorentz invariance.

This book is an easy-to-understand summary of the latest research results on renormalizable quantum gravity that exhibits background freedom asymptotically. Why do we need to quantize gravity? What physical phenomena does quantum gravity have to explain? Even though the Hamiltonian vanishes, why do non-trivial solutions exist, that is, why do gravitational states with entropy exist? We would like to reveal the essence of the problem by stepping into the realm of physics that has been avoided until now. The theory is constructed based on diffeomorphism invariance and renormalizability, which are the basic principles of gravity and quantum field theory, but many of the concepts and techniques described here are new, developed in this century.

This book is organized as follows. Chapter 1 summarizes various problems with gravity that this book aims to reveal. In Chap. 2, we describe the basic structure of renormalizable quantum gravity and present the effective action necessary for discussion in later chapters. The BRST conformal invariance representing background freedom, closely related to the ghost issue, is described in Chap. 3. Chapter 4 discusses physical meaning of the Hamiltonian and momentum constraints, which are significant when considering physics in gravity. The concrete renormalization calculation using dimensional regularization to derive the effective action is described in Chap. 5. The strict derivation and classification of physical states based on the BRST conformal algebra are summarized in Chap. 6. Chapters 7–10 describe the world beyond the Planck scale suggested by this theory. We first show that there is a stable inflationary solution eventually terminated at the spacetime phase transition, and then entropy of the universe can be explained using this solution in Chap. 7. In Chap. 8, we specifically construct a localized massive excitation of quantum gravity with a Schwarzschild tail outside. Chapter 9 presents a solution to the cosmological constant problem from the perspective of quantum gravity. In Chap. 10, we set a system of nonlinear equations of motion that describes the evolution of scalar fluctuations during inflation. Solving it numerically, the fluctuation amplitude is shown to be actually attenuated, and the primordial spectrum is derived. Chapter 11 is devoted to describing a gravitational instanton that represents a spacetime with non-trivial topology and considering its physical implications. Chapter 12 describes a fully nonperturbative lattice formulation of quantum gravity using the dynamical triangulation method and clarifies the relationship to the renormalizable continuum theory. The last is closed with Finale.

Tsukuba, Japan Ken-ji Hamada

The original version of the book has been revised. A correction to this book can be found at https://doi.org/10.1007/978-981-96-3476-7_13

Competing Interests The author has no competing interests to declare that are relevant to the content of this manuscript.

Contents

Chapter 1
What Quantum Gravity Should Reveal

The first place where quantum gravity is needed is near the center of black holes. It is believed that quantization of gravity is necessary to solve the singularity problem. The same is true of the early days of the universe. The universe is not static; it is thought to have been expanding ever since its birth. In other words, if we go back in time, it is inferred that the universe was in a high energy and high density state, and it is thought that we eventually reach a region where quantum gravity is required. Here, we reveal the causes of the various problems that occur in these regions and argue that quantization of gravity is necessary to overcome them.

First of all, we would like to emphasize that the image of elementary particles described by ideal points with no spread is a concept that contradicts Einstein's theory of gravity [Ein1916]. This is because from the perspective of gravity theory, it is nothing but a black hole. Of course, it can be regarded as a point-like particle if mass of the particle is sufficiently smaller than the Planck mass $m_{\rm pl}\,(= 1/\sqrt{G}) = 1.221 \times 10^{19}$ GeV [Pla1899],[1] then the Compton wavelength (reciprocal of mass), which is a measure of position uncertainty, is larger than the size of its event horizon caused by mass. However, if a particle with mass exceeding the Planck mass exists, its horizon size exceeds the position uncertainty, and information about such a particle is no longer visible from the outside. It is exactly like a black hole, as shown in Fig. 1.1.

Since masses of all known elementary particles are much smaller than the Planck mass, there is no need to consider quantum effects of gravity for them. However, in a world where quantum gravity becomes effective, it is natural to think that elementary excitations with the Planck mass as a unit will occur. How will it be described? This is not only the problem with elementary excitations but also the problem with black holes.

---

[1] The Planck mass and the Planck length are defined by $m_{\rm pl} = \sqrt{\hbar c/G}$ and $l_{\rm pl} = \sqrt{\hbar G/c^3}$, respectively, where G is Newton's constant. In this book, we basically use the natural units $c = \hbar = 1$, but $\hbar$ is sometimes reinstated for explanation purposes.

K. Hamada, *Trans-Planckian Physics and Inflation*, Fundamental Theories of Physics 223, https://doi.org/10.1007/978-981-96-3476-7_1

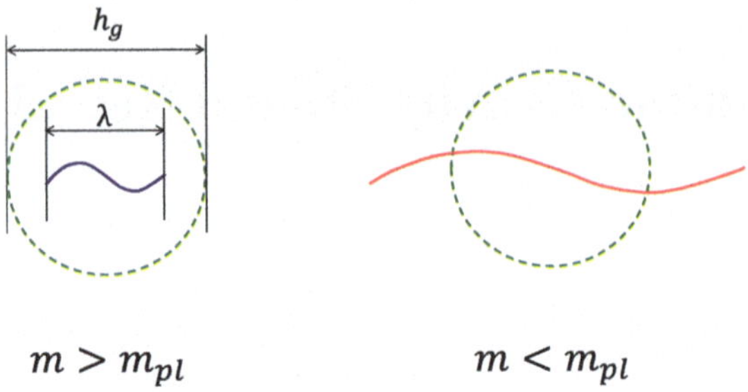

Fig. 1.1 The Compton wavelength of mass m is given by $\lambda \sim 1/m$. On the other hand, the particle horizon size (dotted line) is given by $h_g \sim m/m_{\mathrm{pl}}^2$. Therefore, when $m > m_{\mathrm{pl}}$, $\lambda < h_g$ as shown in the left, information about the elementary excitation is trapped inside the horizon and is lost. In this way, in a world that exceeds the Planck scale, normal particle picture breaks down

Now, as we all known, the fundamental equation that defines Einstein's theory of gravity is the Einstein equation. It is usually expressed as

$$R_{\mu\nu} - \frac{1}{2} g_{\mu\nu} R = 8\pi G \, T^{\mathrm{M}}_{\mu\nu}, \tag{1.1}$$

by placing a matter energy-momentum tensor $T^{\mathrm{M}}_{\mu\nu}$ on the right-hand side. Here, $g_{\mu\nu}$ is the gravitational field, or metric field, $R_{\mu\nu}$ is the Ricci tensor, and R is the scalar curvature. This equation expresses that matter is a source that causes distortions in space and time. From this perspective, in order to eliminate spacetime singularity, matter itself would have to disappear.

The problem with the Einstein equation is that it does not contain the Riemann tensor $R_{\mu\nu\lambda\sigma}$. Since $R_{\mu\nu\lambda\sigma}$ measures the degree of curvature of spacetime, it is nothing but a quantity that represents the strength of the gravitational field, corresponding to the strength of gauge fields such as an electromagnetic field, $F_{\mu\nu}$. The fact that the Riemann tensor is not included in the equation is in stark contrast to that the gauge-field equation of motion is described as $\nabla^\mu F_{\mu\nu} = J_\nu$ using the field strength and a current J_μ. This indicates that the Einstein equation does not have sufficient ability to control curvature and cannot prevent singularities.

Examining the problem of singularities from the action point of view, it comes down to the fact that the Einstein-Hilbert action, given by the scalar curvature R that can take values from $-\infty$ to ∞, is not bounded below, and thus the bottom of energy is missing. Therefore, as long as we are considering small gravitational fluctuations around a regular background spacetime,[2] there would be no problem, but if the fluctuations become large, a gravitational system controlled by this action is doomed to collapse.

Furthermore, physical solutions are those that become finite when substituted them into the action. Famous examples are soliton and instanton solutions in gauge

[2] Recently, the gravitational wave was directly observed in ground-based experiments [Abb+16], and this achievement was awarded the Nobel Prize in Physics in 2017.

theories. According to this way of thinking, since the Einstein-Hilbert action disappears for the Schwarzschild solution [Sch1916] with vanishing scalar curvature, the singularity ends up existing as a physical solution in Einstein's theory of gravity.

In order to solve this problem, adding the square of the Riemann tensor to the action in a positive-definite way is inevitable from a quantum field theory standpoint. This also solves the instability problem. In addition, since the action of curvature squared is dimensionless, renormalizability that is the guiding principle of quantum field theory established by Tomonaga, Schwinger, Feynman, and Dyson [Tom46, Sch48, Fey48, Fey49, Dys49a] is guaranteed. Unfortunately, however, quantum gravity is not a simple matter in which everything works fine even by adding it to the action.

In the late 1970s, from the perspective of renormalizability, a theory of quantum gravity whose main part is given by the square of curvatures was actively researched [Ste77, Tom77, Tom80, FT82a, AB85]. A new difficulty that was discussed at that time was the ghost problem. In general, ghost is a term that refers to something unphysical, whose action has an opposite sign from normal particles and is not bounded below. If a ghost particle were to contribute to scattering as a real field, unitarity will be broken, therefore it should not be observed. The problem here is that even though the action is introduced in a positive-definite manner, such ghost modes arise as a feature of higher-derivative theories [Ost1850, PU50, Woo15]. However, the essence of the problem is not that ghost modes exist, but whether they appear as observables.

In the first place, the existence of ghosts is not limited to quantum theory. In Einstein's theory of gravity, there is a negative-metric ghost mode that causes the indefiniteness of the Einstein-Hilbert action. Locally, it can be removed and the theory can be described using only physical gravitational wave modes. However, when considering the structure of the entire spacetime such as the current universe and gravitational objects, the negative-metric mode plays an essential role. This fact indicates that the ghost mode is necessary to describe gravitational systems, i.e., to preserve diffeomorphism invariance.

Ultimately, the cause of the problem with quantum gravity lies in the weak-field approximation where the gravitational field is mode-expanded as $g_{\mu\nu} = \eta_{\mu\nu} + H_{\mu\nu}$.[3] What we are looking for is a theory that describes an extremely small world at a distance shorter than the Planck length, which is a world of extremely high energy. Nevertheless, the weak-field approximation assumes that the gravitational field becomes weak near the center of the black hole, creating a flat spacetime where gravitons move around. That is quite strange. We think that we are crushed by gravity as we approach the center, but this approximation says that when we get there, we actually feel a calm and flat environment. Alternatively, if we go back to the early

[3] According to the convention in this research field, it is stopped at the first order of $H_{\mu\nu}$, but the inverse of the metric tensor defined by $g^{\mu\lambda}g_{\lambda\nu} = \delta^{\mu}{}_{\nu}$ is expanded by the multiple product of the field the lasts infinitely like $g^{\mu\nu} = \eta^{\mu\nu} - \kappa H^{\mu\nu} + \kappa^2 H^{\mu}{}_{\lambda} H^{\lambda\nu} + \cdots$. The same is true for $\sqrt{-g}$ and R. In general, unlike gauge fields with dimensions, interaction terms of the dimensionless gravitational field become multi-point functions that continue to infinite order.

universe, we believe that there was a state of strong gravity in which all matter was crushed, but this approximation assumes that this is not the case.

This approximation using gravitons as quanta is nothing more than reducing the gravitational system to a system of special relativity, or a Lorentz-invariant system. At that time, the inherent property of the gravitational system, that is, the vanishing of the whole Hamiltonian, is not taken into account strictly, but only approximately. In other words, the description by gravitons is only applicable at low energies sufficiently below the Planck scale where back-reactions to the spacetime structure due to them can be ignored.

Something is missing to overcome the physical and theoretical problems in quantizing gravity. What exactly is that? The image of spacetime in which gravity is quantized is a state in which spacetime fluctuates so much that time and distance become uncertain. It expresses that even if we adopt a particular background spacetime, it no longer has any physical meaning in itself. Such a property is called "background-metric independence", or simply "background independence" or "background freedom". That means that the traditional picture of particles propagating in a specific background spacetime no longer holds true. We need to go back to the basis of gravity and construct a theory that represents such quantum states exactly.

In order to express this, it is necessary to introduce some kind of nonperturbative method. The key of the new formulation introduced in the next chapter is that the core part is described by a conformal field theory, which is a representative of nonperturbative quantum field theories. Here, it is given by a special one in which conformal invariance appears as a gauge symmetry, unlike normal one. In other words, all of worlds with different scales connected by conformal transformations become gauge equivalent. In this way, the background freedom is algebraically expressed, and the condition for the whole Hamiltonian to vanish is also included. Furthermore, it can be shown that ghosts become unphysical due to the presence of this gauge symmetry.

Turning our attention to the early universe, the Einstein Eq. (1.1) also raises a question of how the current universe was created. It is believed that matter making up us was generated during the Big Bang. To do that, we need something that is the source of all matter. Within the framework of Einstein's theory of gravity, it must also be matter. In other words, we have to introduce an unknown field that creates everything. Although the Einstein equation is an excellent equation that describes the magnificent evolution of the universe over several tens of orders of magnitude from the time of the Big Bang to the present day, this equation also shows the limits of Einstein's theory of gravity.

The deeper we prove into the early stage of the universe, the more obvious the need for a theory of quantum gravity becomes. The major advances in cosmology that we are witnessing today are because it has become possible to measure temperature fluctuations (anisotropies) of cosmic microwave background radiation (CMB) with high precision by the Cosmic Background Explorer (COBE), an astronomical satellite launched from NASA's Kennedy Space Center in 1989 [COBE], its successor the Wilkinson Microwave Anisotropies Probe (WMAP) launched in 2001 [WMAP], and more recently by the Planck astronomical satellite launched by the European Space Agency (ESA) in 2009 [PLANCK]. The parameters of standard cosmology have been determined from these experiments, and this provided strong support for the inflation theory proposed by Guth [Gut81], Sato [Sat81], and Starobinsky [Sta80], which states that there was a period of rapid exponential expansion of space like the de Sitter spacetime [Des1917] before the Big Bang.

The idea of inflation was introduced primarily to solve the "horizon problem" of why correlations much larger than the horizon size exist in the early universe and the "flatness problem" of why the curvature of space is so small. If we interpret this idea honestly, the universe has expanded approximately 10^{60} times since its birth. This implies that the Hubble distance, the largest scale in the universe, and the Planck length $l_{\rm pl}$, the smallest scale, are connected. From this, it is expected that the observed CMB spectrum contains traces of quantum fluctuations of gravity at the time of the birth of the universe.

Of course, this fact alone does not allow us to believe that there are any traces of quantum gravity left. Thinking naively, much of information from the early universe would have been lost over time, and even less any trace of quantum gravity effects would remain. However, if we go back in time through the history of the expanding universe, we can answer this question. For that, we need to have a good understanding of the expanding universe solution discovered by Friedmann [Fri1922] (also by Lemaître [Lem1927] independently), which is the basis of standard cosmology.

The Friedmann solution is a dynamical solution of the Einstein equation obtained by assuming the universe to be homogeneous and isotropic. It must be emphasized here that this solution is unstable because it follows the ghost mode mentioned above. Usually, such a solution is not chosen for physics. This is because if we consider small spacetime fluctuations (perturbations) around this solution, they grow over time and deviate greatly from the solution. Nevertheless, the universe can still be well approximated by the Friedmann solution. The results of the Sloan Digital Sky Survey (SDSS) [SDSS], which observed the distribution of galaxies over such a wide area where galaxies can be regarded as mass points, clearly demonstrate this fact (see Fig. 1.2). The fact that the unstable Friedmann universe has continued for over 10 billion years suggests that the amplitude of the initial fluctuations immediately after the Big Bang, the primordial fluctuations, was unnaturally small.

After the Big Bang, these tiny spacetime fluctuations grow due to the instability and form structures such as stars, galaxies, and galaxy clusters. The formation of these structures involves nonlinear effects and thus cannot be described by simple perturbation theory, but until the universe becomes neutral, the fluctuations are still small and perturbation theory can be applied. The study that describes the evolution of the universe perturbatively based on this fact is called cosmological perturbation theory [Bar80, MC81, KS84, HS95, Dur01]. The spacetime fluctuations also cause fluctuations in light and other matters created at the Big Bang. After the neutralization of the universe, light becomes free from interactions with matter (decoupling), thus its fluctuation no longer grows. Therefore, we can find the spectrum of fluctuations

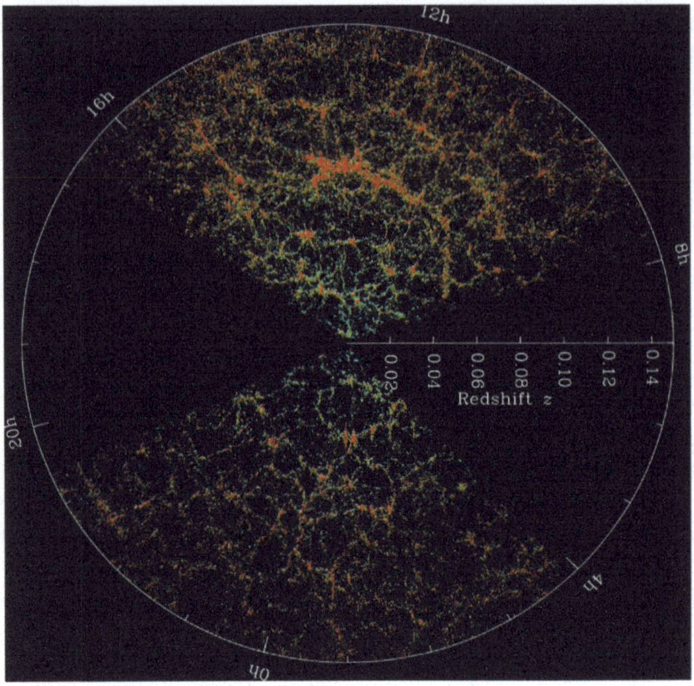

Fig. 1.2 Observation results of galaxy distribution in the entire sky by SDSS. Each dot represents a galaxy. The area with no data is the Milky Way Galaxy plane. Reprinted from https://www.sdss4.org/science/ under the Creative Commons Attribution (CC-BY) license

at the decoupling point from the current CMB temperature fluctuation spectrum,[4] and further trace it back to the time of the Big Bang using cosmological perturbation theory.

The CMB, discovered in 1964 by Penzias and Wilson [PW65], has a Planck distribution with a temperature T of 3 degrees K. The COBE astronomical satellite was the first to observe that the temperature fluctuation $\delta T/T$, the magnitude of deviation from the Planck distribution, was on the order of 10^{-5}. Its successor, WMAP, revealed the detailed distribution of CMB temperature fluctuations across the sky, and this observation was taken over by the Planck astronomical satellite. Fig. 1.3 shows the distribution of temperature fluctuations obtained by the Planck satellite, and Fig. 1.4 shows the temperature fluctuation spectrum obtained by statistically processing this distribution.

The CMB spectrum records the history of the universe from just after the Big Bang to the present. The early spectrum of the current universe, i.e. the primordial spectrum, is thought to have a scale-invariant form with almost no features, and when this spectrum, called the Harrison-Zel'dovich type [Har70, Zel72, Pee93], is drawn

[4] The relationship linking these fluctuations at different times is known as the Sachs-Wolfe effect [SW67].

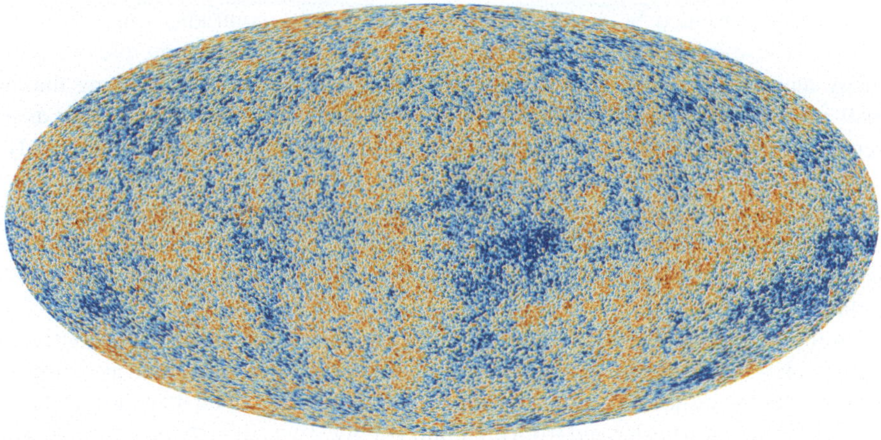

Fig. 1.3 Planck all-sky CMB temperature fluctuation distribution. Reprint from https://www.sci.esa.int/web/planck/ with permission ©ESA and the Planck Collaboration

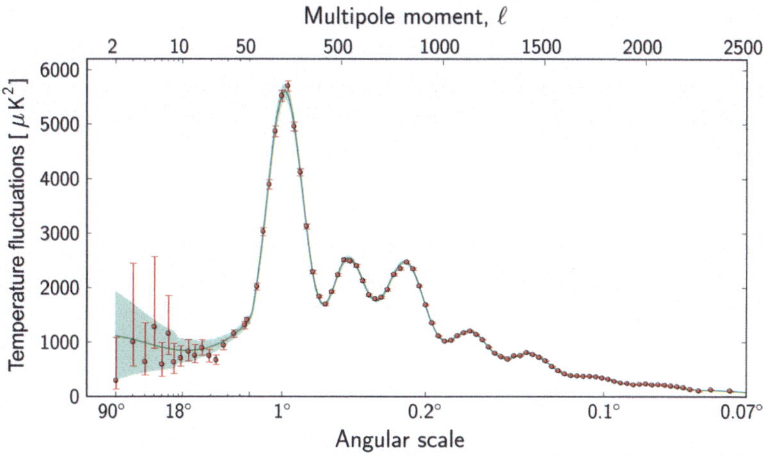

Fig. 1.4 Angular power spectrum of CMB temperature anisotropies observed by Planck. Reprinted from https://www.sci.esa.int/web/planck/ with permission ©ESA and the Planck Collaboration

within Fig. 1.4, it becomes a horizontal straight line passing through the bottom of the concavity where the multipole moment l is small. The deformation from the straight line thus represents dynamics that the fluctuations underwent in the process of cosmic evolution. Most of this deformation occurs during the period from the radiation-dominated era until the universe becomes neutral. Since we can go back in time to anywhere in the past as long as Einstein's theory of gravity is correct, this implies that from the current CMB spectrum, we can extract information that the spectrum just after the Big Bang was almost scale-invariant.

This fact presented by standard cosmology seems to contradict our intuition. This is because it is natural to think that the early universe was a crucible of high-energy reactions and had large fluctuations. Therefore, rather than thinking that a small value was chosen by chance as the initial value, it is more acceptable as a scenario to think that there was some mechanism to make it smaller in the early stage. Inflation is believed to be the key to solving this subject. The story becomes even simpler if we consider that the force causing inflation is given by quantum effects of gravity and the fluctuations become smaller during this period. In addition, the attenuation of quantum fluctuations in spacetime is nothing but the emergence of classical spacetime.

One of the goals of this book is to elucidate such inflationary dynamics and derive a nearly scale-invariant and scalar-like primordial spectrum with very small amplitudes that provides the initial conditions of the Friedmann universe. Well then, how can the inflationary solution be derived from quantum gravity? Even though the gravitational field is described by tensor fields, why are scalar fluctuations dominant and tensor fluctuations absent in the early universe? How is the process by which gravitational fluctuations reduce significantly described? Simple perturbation methods based on the weak-field approximation cannot answer these questions. To address these issues, we introduce a new nonperturbative method and construct a scenario for the evolution of the universe from the trans-Planckian world to the present. In future, we expect that this study will lead to a resolution of the Hubble constant tension that has been reported recently [Rie+21].

Chapter 2
Renormalizable and Background-Free Quantum Gravity

Let us define quantum gravity that describes the world beyond the Planck scale. Diffeomorphism invariance is recognized as a gauge invariance [Uti56], together with ordinary gauge invariance for gauge fields including Yang-Mills fields [YM54]. The guiding principles adopted here are diffeomorphism invariance and renormalizability that continuum quantum field theories should have. Performing the path integral over the gravitational field so as to satisfy these principles is the definition of quantization. Moreover, by adding physical requirements such as initial scalar-fluctuation dominance suggested from cosmological observations, we construct a theory of quantum gravity. This theory has a property called background freedom in the high-energy limit, which allows it to overcome the Planck scale wall. In that region, it can be shown that tensor fluctuations, which impede unitarity and the construction of realistic cosmology, are prohibited as unphysical states.

This chapter begins by sorting out various problems with gravity again, and then introduces an action to overcome them. Next, we describe the quantization method to achieve background freedom, and present the effective action of quantum gravity that will be used in later discussions. Quantized spacetime is significantly different from the common image where gravitons propagate through flat spacetime.

2.1 Quantum Gravity Action and Effective Action

Attempts to quantize Einstein's theory of gravity have been going on for a long time [Dew67a, Dew67b, Dew67c, tHV74, DN74, Vel75]. The biggest problems of this theory are that it is not renormalizable and that there are singularities. The root cause of these problems is that the Einstein-Hilbert action is not bounded below, and furthermore it vanishes for singular solutions. In general, it is believed that a field configuration with finite action can exist physically. Therefore, in order to avoid inescapable singularities, an ultraviolet cutoff is usually introduced in the Planck

K. Hamada, *Trans-Planckian Physics and Inflation*, Fundamental Theories of Physics 223, https://doi.org/10.1007/978-981-96-3476-7_2

scale. This is equivalent to considering spacetime discretized to the Planck length, and is sometimes called spacetime quantization, but the introduction of the cutoff violates diffeomorphism invariance, the guiding principle of gravity.

Returning to the origin of considering diffeomorphism invariance as a first principle, we have to be able to bring the cutoff to infinity, that is, to take the continuum limit. This means that the theory must be renormalizable. At the same time, this property requiring the finiteness of the theory implies that there is no singularity.

The reason why singularities exist in Einstein's theory of gravity is that its action does not contain the Riemann tensor $R_{\mu\nu\lambda\sigma}$, which determines the strength of curvature. Therefore, to construct a renormalizable quantum theory of gravity, we have to consider an action involving its square with the correct sign.[1] Since the action then diverges for singularities, they are eliminated as unphysical objects.[2] Furthermore, since the action is dimensionless, the corresponding coupling constant is also dimensionless, thus the theory becomes renormalizable [Ste77, Tom77, Tom80, FT82a, AB85].

Hence, considering the square of the curvature as the center is necessary for correctly defining the path integral. Meanwhile, the existence of ghosts (negative-metric modes), which are inherent in higher-derivative theory, has been pointed out as a new problem. However, in gravitational systems, the existence of ghost modes is never a problem, as stated below.

The equation of motion for the gravitational field is generally expressed as vanishing of the energy-momentum tensor of the whole system both in classical and quantum theory. In quantum theory, if the gravitational action is I, then it is expressed as an identity of the functional integral like

$$i\frac{1}{2}\langle\sqrt{-g}\,T^{\mu\nu}(x)\rangle = \int [dg]\frac{\delta}{\delta g_{\mu\nu}(x)}e^{iI} = 0, \tag{2.1}$$

where $T^{\mu\nu}$ is the whole energy-momentum tensor defined by a variation of the action with respect to the gravitational field. This is nothing but a Schwinger-Dyson equation [Dys49b, Sch51] for the gravitational field. Strictly speaking, we need to define the path integral measure, as shown later, thus it is still a symbolic formula at this point, but captures the essence.

This is a general conclusion derived from diffeomorphism invariance and renormalizability, and is a quantum-level representation of widely known Hamiltonian and momentum constraints, as described in detail in Chap. 4. Since this includes a condition that the whole Hamiltonian vanishes, if all modes were physical, only a vacuum solution would be allowed, and there would be no gravitational objects

---

1 The correctness is easier to understand if we consider it in a Wick-rotated Euclidean space [Wic54]. The rotation operation is generally applicable to renormalizable quantum field theory, and the correct sign is one in which the Euclidean action is positive definite. For its statistical mechanical description, see Sect. 2.2.

2 This is a general argument without premising finiteness of the theory. Under a setting where there is no singularity obviously, such as in perturbation expansions around regular spacetimes, we often eliminate the $R^2_{\mu\nu\lambda\sigma}$ term using the Euler (Gauss-Bonnet) combination (2.7).

or even the current universe. Thus, the ghost mode is an essential element for the Hamiltonian to vanish, that is, to preserve diffeomorphism invariance. However, it causes a problem if it appears locally as an entity. A problem with early renormalizable quantum gravity was that it could not prevent the ghost mode from appearing as a physical particle state in the ultraviolet limit.

Why does the ghost problem occur? In the end, it turns out that there is a problem with the way of perturbation expansion. The commonly used method is based on the picture of gravitons propagating as a weak quantum field $H_{\mu\nu}$ in a specific background spacetime $\hat{g}_{\mu\nu}$, thus the metric field is expanded as $g_{\mu\nu} = \hat{g}_{\mu\nu} + H_{\mu\nu}$. In this weak-field approximation, a solution of the Einstein equation is employed for the background $\hat{g}_{\mu\nu}$, whereas the quantum fields $H_{\mu\nu}$ representing small perturbations around it are treated as Hamiltonian eigenstates with non-zero eigenvalues. In other words, the condition for the Hamiltonian to vanish is imposed on the background determining the entire spacetime structure, but not on the local gravitons propagating through it. This approach is only applicable to the local particle world, where gravitons do not have a major effect that changes the structure of spacetime, i.e., where back-reactions by them can be ignored. If we choose flat background, it is no longer a gravitational system but a Lorentz-invariant system.

In Einstein's theory of gravity, under the assumption of the weak-field approximation described above, it is possible to construct a Lorentz-invariant system consisting only of physical gravitons, because ghosts can be removed using the gauge degrees of freedom in this case. However, when applied this approximation to fourth-derivative gravity, such a task is not possible. This approximation only deals with local fluctuations in spacetime and cannot be applied to the trans-Planckian world where the entire spacetime is thought to fluctuate significantly. In this case, a nonperturbative method is needed in which the quantized gravitational field itself satisfies the condition for the Hamiltonian to vanish. In addition, if this is not the case, background freedom will not be realized.

Now, we should consider what kind of perturbation theory correctly captures the characteristics of the world beyond the Planck scale. In the first place, assuming that flat spacetime of $R_{\mu\nu\lambda\sigma} = 0$ appears in the ultraviolet limit is an extremely unnatural setting, because it means that the closer we get to the center of a black hole, the weaker the gravitational field becomes. A hint on how to describe it correctly can be found in the early days of the universe with inflation.

Inflation is given by a spacetime configuration in which the Weyl tensor [Wey1918]

$$C_{\mu\nu\lambda\sigma} = R_{\mu\nu\lambda\sigma} - \frac{1}{2} g_{\mu[\lambda} R_{\sigma]\nu} + \frac{1}{2} g_{\nu[\lambda} R_{\sigma]\mu} + \frac{1}{6} g_{\mu[\lambda} g_{\sigma]\nu} R \tag{2.2}$$

vanishes, where $a_{[\mu} b_{\nu]} = a_\mu b_\nu - a_\nu b_\mu$. This fact shows that it is appropriate to perform perturbation expansion around a conformally flat spacetime with $C_{\mu\nu\lambda\sigma} = 0$. Since the Weyl tensor represents the strength of the traceless tensor mode in the gravitational field, this condition requires that this mode is small in the ultraviolet limit. On the other hand, a scalar-like conformal mode primarily for determining distance is not subject to any restriction. In other words, we can successfully explain that scalar fluctuations are predominant than tensor fluctuations in the early universe.

Let ϕ be the field representing the conformal mode and the conformal factor part in the gravitational field is pulled out in the form

$$g_{\mu\nu} = e^{2\phi} \bar{g}_{\mu\nu}, \tag{2.3}$$

so that it is positive-definite regardless of its possible range $(-\infty, \infty)$. Hereinafter, ϕ is called the conformal-factor field. The remaining metric field $\bar{g}_{\mu\nu}$ is expanded as

$$\bar{g}_{\mu\nu} = \hat{g}_{\mu\lambda} \left(e^h\right)^\lambda{}_\nu = \hat{g}_{\mu\lambda} \left(\delta^\lambda{}_\nu + h^\lambda{}_\nu + \frac{1}{2} h^\lambda{}_\sigma h^\sigma{}_\nu + \cdots \right), \tag{2.4}$$

using the traceless tensor field $h_{\mu\nu}$ satisfying $h^\mu{}_\mu = 0$ and the non-dynamical background metric $\hat{g}_{\mu\nu}$. The raising and lowering of indices of the tensor field is done using the background metric as $h_{\mu\nu} = \hat{g}_{\mu\lambda} h^\lambda{}_\nu$. As shown below, ϕ and $h_{\mu\nu}$ each have a positive-definite fourth-derivative action that governs their dynamics. The ghost modes in question arise when these fields are further expanded into sub-modes (see Chap. 6).[3]

Here, introduce the gravitational coupling constant t as a dimensionless parameter that controls the above perturbation expansion. Noting that the field strength of the traceless tensor field is the Weyl tensor, perturbation theory is defined by introducing the reciprocal of t^2 before the Weyl action given by its square with the correct sign. The way of entering this coupling constant is the same as in gauge field theory. Letting the gauge coupling constant be $\mathfrak{g}$ and introducing it as the reciprocal of $\mathfrak{g}^2$ before the action given by the square of the gauge field strength $F_{\mu\nu}$ represents perturbation expansion around $F_{\mu\nu} = 0$.

The action of the entire system including the Weyl action with the coupling constant t is given by[4]

$$I = \int d^4x \sqrt{-g} \left[-\frac{1}{t^2} C^2_{\mu\nu\lambda\sigma} - bG_4 + \frac{1}{\hbar} \left(\frac{1}{16\pi G} R + \mathcal{L}_\mathrm{M} - \Lambda \right) \right]. \tag{2.5}$$

The first Weyl action is conformally invariant as $\sqrt{-g}\, C^2_{\mu\nu\lambda\sigma} = \sqrt{-\bar{g}}\, \bar{C}^2_{\mu\nu\lambda\sigma}$, where the quantities with bars denote those constructed by the metric field $\bar{g}_{\mu\nu}$ (2.4). From (2.2), it can be expressed as

$$C^2_{\mu\nu\lambda\sigma} = R^2_{\mu\nu\lambda\sigma} - 2R^2_{\mu\nu} + \frac{1}{3} R^2. \tag{2.6}$$

[3] In Einstein's theory of gravity, ϕ itself becomes a ghost mode that causes the indefiniteness of the Einstein-Hilbert action.

[4] This action should be expressed in terms of bare quantities, but for simplicity we here express it using renormalized quantities, and renormalization of ultraviolet divergences will be described in Chap. 5.

The second term is a topological term that does not contain a kinetic term, where

$$G_4 = R^2_{\mu\nu\lambda\sigma} - 4R^2_{\mu\nu} + R^2 \tag{2.7}$$

is the Euler density (Gauss-Bonnet combination). Its volume integral is also conformally invariant. The third term is the Einstein-Hilbert action, and $\mathcal{L}_{\rm M}$ represents matter fields coupled conformally to gravity.[5] The last is the cosmological term. The Planck constant $\hbar$ appears only in lower-derivative terms, and its meaning will be discussed later. In the following, unless otherwise specified, $\hbar = 1$.

The coupling constant t represents the degree of deviation from the conformally flat $C_{\mu\nu\lambda\sigma} = 0$ spacetime, and the traceless tensor field becomes smaller in the ultraviolet limit of $t \to 0$. On the other hand, the conformal-factor field ϕ remains largely and nonperturbatively fluctuating. This shows that the picture of particles propagating in a specific spacetime no longer holds true.

When performing the perturbation calculation, the traceless tensor field is usually redefined as $h_{\mu\nu} \to th_{\mu\nu}$ in expansion formula (2.4), but here we proceed without redefining it because we will consider regions where the coupling constant becomes large. The Minkowski metric $\eta_{\mu\nu} = (-1, 1, 1, 1)$ is usually employed as a background metric $\hat{g}_{\mu\nu}$,[6] and then the coordinates are written as $x^\mu = (\eta, \mathbf{x})$, and called the comoving coordinates, following cosmology.

2.1.1 Diffeomorphism Invariance Demands Conformal Anomaly

Unlike the perturbation expansion around flat spacetime, when the gravitational field is decomposed as in (2.3), a quantity called the Wess-Zumino action associated with conformal anomaly is required to preserve diffeomorphism invariance. The existence of this action is the key to the quantization of gravity.

To understand the necessity of the Wess-Zumino action, let us first consider quantum field theory in curved spacetime [UD62], which is the preliminary stage before quantizing gravity. If we write $I_{\rm M}(f, g)$ as an action of a conformally invariant field f contained in the matter sector $\mathcal{L}_{\rm M}$, it satisfies $I_{\rm M}(f, g) = I_{\rm M}(f, \bar{g})$.[7] Here, conformal anomaly refers to the fact that the classically existing conformal invariance is broken when quantized the field, and the ϕ-dependence arises [CD74, DDI76, Duf77, DS93, Duf94].

---

5 Terms with mass parameters such as matter mass terms, the Einstein-Hilbert term, and the cosmological term do not break renormalizability.

6 To be precise, any metric field that satisfies the equation of motion derived from the Weyl action, such as a conformally flat metric, can be used as a background metric.

7 Depending on spacetime dimension and type of field, it is necessary to transform the field f at the same time to remove the ϕ-dependence.

To see this specifically, let us express the partition function in terms of the path integral as

$$Z_{CS}(g) = \int [df]_g e^{iI_M(f,g)}, \tag{2.8}$$

where $[df]_g$ is a diffeomorphism-invariant measure defined on $g_{\mu\nu}$. The measure generally depends on the conformal-factor field ϕ, and its dependence can be extracted as $[df]_g = [df]_{\bar{g}}\, e^{iS(\phi,\bar{g})}$. The partition function is then expressed as

$$Z_{CS}(g) = e^{iS(\phi,\bar{g})} \int [df]_{\bar{g}}\, e^{iI_M(f,\bar{g})} = e^{iS(\phi,\bar{g})} Z_{CS}(\bar{g}). \tag{2.9}$$

Here consider a simultaneous transformation

$$\phi \to \phi - \omega, \qquad \bar{g}_{\mu\nu} \to e^{2\omega} \bar{g}_{\mu\nu}, \tag{2.10}$$

so that the original metric field $g_{\mu\nu}$ is saved unchanged. When applying it to (2.9), the left-hand side is obviously invariant, while the right-hand side changes to

$$e^{iS(\phi-\omega,e^{2\omega}\bar{g})} Z_{CS}(e^{2\omega}\bar{g}) = e^{iS(\phi-\omega,e^{2\omega}\bar{g})} e^{iS(\omega,\bar{g})} Z_{CS}(\bar{g}).$$

In order for this to return to the original $Z_{CS}(g)$, S must satisfy

$$S(\phi - \omega, e^{2\omega}\bar{g}) + S(\omega, \bar{g}) = S(\phi, \bar{g}). \tag{2.11}$$

This is the Wess-Zumino consistency condition [WZ71, BCR83]. Since the invariance under (2.10) guarantees diffeomorphism invariance, the Wess-Zumino action S arises exactly to guarantee that.

The effective action is the original action I_M plus quantum corrections, which corresponds to the logarithm of the partition function. The concrete calculation method will be given in Chap. 5. Denoting the effective action by Γ, conformal anomaly is a quantity obtained by varying Γ with respect to the conformal-factor field ϕ. Despite its name, it is a physically necessary component that obeys diffeomorphism invariance. In general, in renormalizable quantum field theory, even if the action has no scale parameter, performing renormalization generates a new scale that breaks scale invariance, such as the quantum chromodynamics (QCD) scale. The Wess-Zumino action for conformal anomaly plays the role of preserving the effective action in a diffeomorphism-invariant form like

$$\Gamma(g) = S(\phi, \bar{g}) + \Gamma(\bar{g}), \tag{2.12}$$

where $\Gamma(\bar{g})$ is the effective action defined on the metric $\bar{g}_{\mu\nu}$, which consists of the original conformally invariant action I and nonlocal quantum correction terms depending on the new scale.

Since the ϕ-dependence of the effective action comes from the Wess-Zumino action S, conformal anomaly is expressed by $\delta\Gamma/\delta\phi(x) = \delta S/\delta\phi(x) = \sqrt{-g}\,\mathrm{CA}(x)$. There are two types of CA associated with the gravitational field. From studies using Hathrell's renormalization group equation [Hat82a, Hat82b, Fre84, Ham14a], it is known that one is the Weyl tensor squared $C^2_{\mu\nu\lambda\sigma}$, and the other is given by the modified Euler density introduced by Riegert as [Rie84]

$$E_4 = G_4 - \frac{2}{3}\nabla^2 R. \tag{2.13}$$

In addition, the conformal anomaly that appears in gauge theory is given by the square of the field strength of the gauge field, $\mathrm{Tr}(F^2_{\mu\nu})$, regardless of whether it is Abelian or non-Abelian Yang-Mills field.

Conversely, this implies that if we integrate the conformal anomaly over ϕ, we can get the Wess-Zumino action as

$$S(\phi, \bar{g}) = \int d^4x \int_0^\phi d\phi \sqrt{-g}\,\mathrm{CA}(g). \tag{2.14}$$

It can be easily verified that this satisfies the consistency condition (2.11) by decomposing the integral region $[0, \phi]$ into $[0, \omega]$ and $[\omega, \phi]$. Then first, if we apply (2.14) to $C^2_{\mu\nu\lambda\sigma}$ and $\mathrm{Tr}(F^2_{\mu\nu})$ which become conformally invariant when multiplied by $\sqrt{-g}$, we can immediately perform the integration of ϕ and obtain the following types of the Wess-Zumino actions:

$$\int d^4x \sqrt{-g}\,\phi\, C^2_{\mu\nu\lambda\sigma}, \qquad \int d^4x \sqrt{-g}\,\phi\, \mathrm{Tr}\left(F^2_{\mu\nu}\right). \tag{2.15}$$

The Wess-Zumino action obtained by integrating the modified Euler density E_4 is particularly important and is called the Riegert action [Rie84], which is given by

$$\int d^4x \int_0^\phi d\phi \sqrt{-g}\, E_4 = \int d^4x \sqrt{-\bar{g}}\left(2\phi\bar{\Delta}_4\phi + \bar{E}_4\phi\right). \tag{2.16}$$

The ϕ-integration can be easily performed by using

$$\sqrt{-g}E_4 = \sqrt{-\bar{g}}\left(4\bar{\Delta}_4\phi + \bar{E}_4\right),$$

where Δ_4 is a fourth-order conformally invariant differential operator [Pan83, FT82b]

$$\Delta_4 = \nabla^4 + 2R^{\mu\nu}\nabla_\mu\nabla_\nu - \frac{2}{3}R\nabla^2 + \frac{1}{3}\nabla^\mu R\nabla_\mu \tag{2.17}$$

such that $\sqrt{-g}\,\Delta_4 A = \sqrt{-\bar{g}}\,\bar{\Delta}_4 A$ and the self-adjoint condition $\int d^4x \sqrt{-g}\, A\Delta_4 B = \int d^4x \sqrt{-g}\,(\Delta_4 A)B$ are satisfied for scalars A and B.

2.1.2 Quantization and Effective Action

The subject of this section is that we concisely describe quantization of the gravity theory defined by the action I (2.5), and then present the effective action that will be important in later discussions. It is defined by a functional integral over the metric field $g_{\mu\nu}$, and the partition function is symbolically written as

$$Z = \int [dg]_g [df]_g \, e^{iI(f,g)} \tag{2.18}$$

where $[dg]_g$ is the path integral measure for $g_{\mu\nu}$, and the subscript g denotes that it is defined to be diffeomorphism invariant using the dynamical field $g_{\mu\nu}$ itself. For simplicity, matter fields are collectively written as f.

A new quantization problem arises here. The integral measure $[dg]_g$ indicates that if we try to integrate $g_{\mu\nu}$, we also have to integrate the g-dependence existing in the measure. Usually, for practical reasons, path integration is performed by replacing it with a measure defined on a specific spacetime. In flat spacetime, it should be written as $[dg]_\eta$ with the subscript of the Minkowski metric $\eta_{\mu\nu}$. The use of this measure seems to be justified if we adopt the weak-field approximation in which all gravitational modes are expanded as $g_{\mu\nu} = \eta_{\mu\nu} + H_{\mu\nu}$, and then it is sufficient to perform the path integral over $H_{\mu\nu}$ according to the field-theory textbook. However, if we decompose the metric field as (2.3), this simple replacement violates diffeomorphism invariance.

First of all, we have to resolve the nested structure of the measure $[dg]_g$. To do this, as in Sect. 2.1.1, we introduce the Wess-Zumino action S satisfying the consistency condition (2.11) and rewrite the partition function as [HS99, Ham02]

$$\int [dg]_g [df]_g \, e^{iI(f,g)} = \int [d\phi dh]_{\hat{g}} [df]_{\hat{g}} \, e^{iS(\phi,\bar{g})+iI(f,g)}, \tag{2.19}$$

where the measure of the gravitational field defined on the background is orthogonally decomposed into the product of $[d\phi]_{\hat{g}}$ and $[dh]_{\hat{g}}$. The exponential part e^{iS} represents the Jacobian required to ensure diffeomorphism invariance. In this way, we rewrite the quantum gravity theory into a theory defined on the background where the action is given by

$$S_{\mathrm{QG}} = S + I, \tag{2.20}$$

so that ordinary quantum field theory techniques can be applied. In this case, although the action S_{QG} itself is not in a diffeomorphism-invariant form, the effective action with quantum corrections becomes diffeomorphism invariant.

The Wess-Zumino action S is responsible for fourth-derivative dynamics of the conformal-factor field ϕ. The most important term among them is the Riegert action of type (2.16), which exists even in the ultraviolet limit of $t \to 0$. Paying attention to the overall sign and adding the proportionality coefficient, it is given by

$$S_{\rm R} = \int d^4x\sqrt{-\bar{g}}\left[-\frac{b_1}{(4\pi)^2}(2\phi\bar{\Delta}_4\phi + \bar{E}_4\phi)\right]. \tag{2.21}$$

The coefficient generally depends on the coupling constant and is expanded as

$$b_1 = b_c B, \qquad B = 1 - \gamma_1\frac{t^2}{4\pi} + o(t^4), \tag{2.22}$$

whose lowest order is calculated as

$$b_c = \frac{1}{360}\left(N_X + \frac{11}{2}N_W + 62N_A\right) + \frac{769}{180}, \tag{2.23}$$

where N_X, N_W, and N_A are the numbers of scalars, Weyl fermions, and gauge fields coupled to gravity, respectively. Their matter contributions can be calculated within the scope of quantum field theory in curved spacetime [CD74, DDI76, Duf77]. The last term is a correction term that appears when gravity is quantized, and its breakdown is $-7/90$ from quantization of the conformal-factor field ϕ using the Riegert action [AMM92] and $87/20$ from quantization of the traceless tensor field $h_{\mu\nu}$ using the Weyl action [FT82a]. Since b_c is positive regardless of the number or type of matter fields, we find that the Riegert action is always positive-definite.

From the Riegert action (2.21), the two-point function of the quantized conformal-factor field ϕ in the ultraviolet limit is given by a logarithmic function of distance as [AM92]

$$\langle\phi(x)\phi(y)\rangle = -\frac{1}{4b_c}\log(x-y)^2. \tag{2.24}$$

Taking a spatial Fourier transform of this function gives a scale-invariant spectrum with positive amplitude proportional to $1/b_c$, as shown in Chap. 10. This is a consequence of the gravitational field being a dimensionless field with the positive-definite action.

Once again, it should be noted that the metric field with physical meaning is (2.3), while the background $\hat{g}_{\mu\nu}$ was introduced for convenience. If we strictly perform the path integral over the metric field, the integration result will no longer depend on the choice of $\hat{g}_{\mu\nu}$ adopted as a reference, i.e., background freedom appears. Since the conformal-factor field is treated strictly here, the background freedom appears as an invariance to conformal transformations. This is a representation of diffeomorphism invariance that comes out for the first time when quantum gravity is rewritten in terms of the quantum field theory (2.19) on $\hat{g}_{\mu\nu}$. This implies that all different conformally flat spacetimes that can be connected to each other by conformal transformations become gauge-equivalent. To emphasize that it is a gauge symmetry, we specially refer to it as the BRST conformal invariance.[8] This symmetry serves to prevent the

[8] In the 1970s, Becchi, Rouet, Stora, and Tyutin discovered that gauge invariance can be extended to include gauge-fixing terms and associated ghost actions [BRS75, BRS76, Tyu75]. BRST is an

field from being independently gauge invariant, so that tensor states are prohibited as unphysical. The details will be described in Chaps. 3 and 6.

The specific value of the coefficient b_c becomes 7.0 ($N_A = 12$, $N_W = 45$) for the Standard Model [Gla61, Wei67, Sal68]. For grand unified theories (GUT) [Lan81], 9.1 ($N_A = 24$, $N_W = 45$) for $SU(5)$ GUT, 12.0 ($N_A = 45$) for $SO(10)$ GUT, and 17.7 ($N_A = 78$) for E_6 GUT. In this book, $b_c = 7$ is employed.

Letting μ be an arbitrary mass scale associated with renormalization and defining the beta function for the coupling constant t as $\mu dt/d\mu = -\beta_0 t^3 + o(t^5)$, one-loop quantum corrections from both matter fields and the gravitational field itself are given by

$$\beta_0 = \frac{1}{(4\pi)^2}\left[\frac{1}{240}(N_X + 3N_W + 12N_A) + \frac{197}{60}\right], \tag{2.25}$$

where the first three terms in the square brackets come from matter fields [CD74, DDI76, Duf77] and the last term is the sum of the contributions from the conformal-factor field ϕ and the traceless tensor field $h_{\mu\nu}$, given by $-1/15$ [AMM92] and $199/30$ [FT82a], respectively. Since the beta function is negative, it is justified to treat the traceless tensor field perturbatively as in (2.4) in the ultraviolet region.

It should be noted here that this fact does not mean that free-particle states appear asymptotically. Although fluctuations of the traceless tensor field become smaller, fluctuations of the conformal-factor field determining distance remain large and non-perturbative. This represents that a background-free world where the concept of time and distance is lost is realized in the high-energy limit.[9] Therefore, this property is called "asymptotic background freedom" to distinguish it from ordinary asymptotic freedom.

Next, let us describe structures of the diffeomorphism-invariant effective action Γ involving quantum corrections. The coupling constant t is a parameter representing the degree of deviation from conformal invariance, and the running coupling constant, or effective coupling constant, becomes large in the infrared region because the beta function is negative. To see its specific expression, examine the effective action for the Weyl sector. In momentum space, it is calculated as [HY05, HHY06a]

$$\begin{aligned}\Gamma^{\mathrm{W}} &= -\left[\frac{1}{t^2} - 2\beta_0\phi + \beta_0 \log\left(\frac{q^2}{\mu^2}\right)\right]\sqrt{-g}\, C^2_{\mu\nu\lambda\sigma} \\ &= -\frac{1}{\bar{t}^2(Q)}\sqrt{-g}\, C^2_{\mu\nu\lambda\sigma},\end{aligned} \tag{2.26}$$

where the second term in the first line is exactly the Wess-Zumino action of type (2.15). The third term is a nonlocal one-loop quantum correction, and q^2 is the

---

abbreviation of the initial letters of the four names. The BRST quantization method was systematized into the present form by Kugo and Ojima [KO79] soon after the discovery.

[9] Since the traceless tensor field is treated with perturbations, the background freedom for this field is not perfect, but the asymptotic behavior indicates that it is not significant in the ultraviolet limit.

momentum squared defined on the metric $\bar{g}_{\mu\nu}$. The $\bar{t}(Q)$ in the second line is the running coupling constant. Expressing the momentum squared defined in the full metric (2.3) as

$$Q^2 = g^{\mu\nu} q_\mu q_\nu = \frac{q^2}{e^{2\phi}}, \tag{2.27}$$

and letting $\Lambda_{\rm QG} = \mu\, e^{-1/2\beta_0 t^2}$, we find that the inside of square brackets forms into the reciprocal of

$$\bar{t}^2(Q) = \frac{1}{\beta_0 \log(Q^2/\Lambda_{\rm QG}^2)}. \tag{2.28}$$

In cosmology, Q is called physical momentum and q is comoving momentum. The dynamical energy scale $\Lambda_{\rm QG}$ is a renormalization group invariant that satisfies $d\Lambda_{\rm QG}/d\mu = 0$, that is, one of physical constants that do not change during the process of cosmic evolution.

This is exactly the same as in QCD. In the expression (2.26), replace the field strength from $C_{\mu\nu\lambda\sigma}$ to $F_{\mu\nu}$ and the coupling constant from t to $\mathfrak{g}$, and the associated energy scale is written as $\Lambda_{\rm QCD}$, then the running coupling constant is given by $\bar{\mathfrak{g}}^2(Q) = \left[\beta_0 \log(Q^2/\Lambda_{\rm QCD}^2)\right]^{-1}$. At this time, the Wess-Zumino action of type $\phi\, {\rm Tr}(F_{\mu\nu}^2)$ contributes.

The diffeomorphism-invariant effective action associated with the Riegert action (2.21) is given by [Rie84]

$$\Gamma^{\rm R} = \int d^4x \sqrt{-g} \left[-\frac{b_1}{(4\pi)^2} \frac{1}{8} E_4 \frac{1}{\Delta_4} E_4 \right], \tag{2.29}$$

adding a quantum correction term that does not depend on ϕ as in (2.12), where $\Delta_4^{-1} E_4(x) \equiv \int d^4y\, G(x,y) \sqrt{-g} E_4(y)$ and $\Delta_4 G(x,y) = \delta^4(x-y)/\sqrt{-g}$. In fact, expanding $\Gamma^{\rm R}$, we get

$$\int d^4x \sqrt{-\bar{g}} \left[-\frac{b_1}{(4\pi)^2} \left(2\phi \bar{\Delta}_4 \phi + \bar{E}_4 \phi + \frac{1}{18} \bar{R}^2 \right) \right] + W(\bar{g}), \tag{2.30}$$

where $W(\bar{g}) = \Gamma^{\rm R}(\bar{g}) + b_1 \int d^4x \sqrt{-\bar{g}} \bar{R}^2/18(4\pi)^2$ is a nonlocal term starting from $o(h^3)$ when expanded in the traceless tensor field $h_{\mu\nu}$, and the $\bar{R}^2/18$ terms are allocated to do so.

The various Wess-Zumino actions mentioned here are typical ones that appear in one-loop corrections. In higher-loop corrections, there arise general Wess-Zumino actions obtained by further integrating them by ϕ [Ham20b]. For structure of the effective action involving higher-loop corrections, see Appendix Sect. 2.5. When discussing dynamics of quantum gravity below, we will use the effective action (2.26) for the Weyl sector and (2.29) for the Riegert sector in which t^2 in the coefficient b_1 (2.22) is replaced with $\bar{t}^2(Q)$.

Here, we present only the part related to the fourth-derivative gravitational action in the effective action. It is derived from the quantization of the Wess-Zumino action and the ϕ-independent part of the original action I (2.5), i.e., the part that does not have mass parameters. In contrast, the Einstein-Hilbert term and the cosmological term containing mass parameters depend on ϕ in the form of an exponential, and the effective action involving them will be described separately in Chap. 5, along with the definitions of physical constants.

Finally, so far proceeding the argument without paying attention to the Planck constant $\hbar$, but it is meaningful to clarify where it appears when reinstated. Considering the action I by dividing it into the fourth-derivative gravitational part $I^{(4)}$ and the lower-derivative term $I_{\rm EG}$ consisting of the Einstein-Hilbert action and so on, the reciprocal of $\hbar$ is entered only in front of $I_{\rm EG}$. Since the gravitational field is essentially dimensionless by definition, $I^{(4)}$ becomes completely dimensionless, thus $\hbar$ does not enter into this action. Similarly, the Wess-Zumino action S derived from the path integral measure does not involve $\hbar$. This means that all fourth-derivative gravitational actions are quantities that describe purely quantum dynamics, and the measure including them as weights could be regarded as a true measure of the path integral. In other words, the path integral of quantum gravity is often symbolically expressed using only $I_{\rm EG}$ as $\int \mathcal{D}g_{\mu\nu}\, e^{iI_{\rm EG}/\hbar}$, but here it is shown that the measure is expressed exactly to be

$$\mathcal{D}g_{\mu\nu} = [d\phi dh]_\eta\, e^{iS+iI^{(4)}}. \tag{2.31}$$

This is also the reason why the Weyl action and the Wess-Zumino action resulting from the measure are treated on the same footing in the following discussion. Moreover, this consideration of $\hbar$ shows that the ghost mode involved in the fourth-derivative gravity is essentially a quantum entity and not a classical entity like a particle.

2.2 Reality of Fields and Unitarity

Quantum spacetime where quantum gravity is fully activated is described in terms of conformal field theory so that the particle picture no longer holds true and the scattering matrix is not defined. In this case, unitarity is represented as the field being real, i.e., two-point functions are positive-definite and structure constants (three-point functions) are real [FGG74, Mac77, EPP+12].

In such a case, it is better to consider quantum field theory in Euclidean space.[10] Letting $\mathcal{I}$ be the Euclidean action, the partition function can be discussed in terms of statistical mechanics without annoying imaginary units as

$$Z_E = \int e^{-\mathcal{I}},$$

thus the unitarity issue becomes easier to understand.

If the action $\mathcal{I}$ is real and bounded below, the Boltzmann weight $e^{-\mathcal{I}}$ is positive and finite, and thus the path integral is correctly defined. In this case, the reality of the field is not lost and unitarity holds. In addition, field configurations in which $\mathcal{I}$ positively diverges can be removed as unphysical states that do not exist stochastically because the Boltzmann weight vanishes.

On the other hand, in the case of an indefinite action where $\mathcal{I}$ can be negative infinity so that the bottom of energy is missing, the path integral diverges, and thus the reality of the field is sacrificed in order to regularize it. As can be seen from this fact, the path integral of quantum gravity cannot be correctly defined by the indefinite Einstein-Hilbert action alone, and the square of curvatures is required to ensure positive-definiteness.[11]

As mentioned in the first half of Sect. 2.1, the cause of the conventional ghost problem lies in the fact that we have not employed a perturbation expansion method suitable for describing quantum spacetime. Usually, gravity is quantized using the weak-field approximation in which gravitons propagate through a specific spacetime, but when this approach in which quantized gravitons are treated as the Hamiltonian eigenstates with non-zero eigenvalues is applied to fourth-derivative quantum gravity, modes with positive and negative metrics appear as independent physical eigenstates. Therefore, a problem arises in which ill-defined correlation functions among negative-metric modes have physical meaning.

Conversely, if we consider physical quantities using the gravitational field as a "fundamental field" without dividing it into the positive- and negative-metric modes, correlation functions of the field are correctly defined because the fourth-derivative actions of the field, the Riegert and Weyl actions, are positive-definite. Therefore, the reality of the field is not lost.

Hence, fourth-derivative quantum gravity requires some kind of constraints that do not allow the separation of the gravitational field into positive- and negative-metric modes when dealing with physical quantities. After all, the problem in the weak-field approximation is that there is no constraint on quantum fields that imposes the Hamiltonian to vanish. This approximation is applicable only when influence of gravitons is local and not strong enough to change the spacetime structure. In the

[10] Specifically, the Wick rotation can be defined for the background metric $\hat{g}_{\mu\nu}$ (selecting flat metric) in the metric decomposition (2.3) and (2.4). In Chap. 5, we will perform renormalization calculations after rotating to Euclidean space in this way.

[11] Similarly, a positive cosmological constant is necessary, even if small, to ensure positive definiteness at low energies.

trans-Planckian region where we need to consider the entire spacetime, the gravitational field has to be quantized to obey the Hamiltonian and momentum constraints. The BRST conformal invariance plays exactly the role of these constraints.

In gravitational systems, since the Hamiltonian vanishes identically, we cannot apply No-Go theorems proving that energy spectrum is not bounded below such as the Ostrogradsky theorem [Ost1850, Woo15]. Reconciling quantum mechanics and gravity requires a deep understanding of what it means for the Hamiltonian to vanish. The ghost mode is reminiscent of Bohm's "hidden variable", in the sense that it is "invisible but being" [Gil09]. Note, however, that no such variable exists in a system of special relativity, and ghosts are necessary only when gravity makes a substantial contribution.

2.3 Physical Constants and Spacetime Phase Transition

In general, a measurable physical constant $A_{\rm phy}$ in renormalizable quantum field theory must be a renormalization group invariant. It is expressed as a quantity independent of the renormalization scale μ [Col84], satisfying

$$\frac{d}{d\mu}A_{\rm phy} = 0. \tag{2.32}$$

This means that it is a constant that does not change during evolution of the universe.

The effective action is originally defined as a bare quantity, and is expressed in terms of renormalized ones, thus it is by definition a renormalization group invariant (see Chap. 5). The Planck mass $m_{\rm pl}$ and the cosmological constant $\Lambda_{\rm cos}$ that appear in the effective action are also renormalization group invariants, i.e., physical constants, unlike the bare constants in the action I (2.5). The three constants, including the dynamical scale $\Lambda_{\rm QG}$, have to be determined from observations. Two of them, $m_{\rm pl}$ and $\Lambda_{\rm QG}$, determine dynamics of the early universe. On the other hand, the observed $\Lambda_{\rm cos}$ is so small and can be ignored there.

The energy-momentum tensor is an equation obtained by performing a variation of the effective action with respect to the gravitational field. This is also a renormalization group invariant and is a normal product that gives a finite value. The equation of motion of quantum gravity is given as an equation in which the total energy-momentum tensor vanishes. This is a general conclusion that follows from diffeomorphism invariance and renormalizability, and means that there is no correction due to zero-point energy, which is also consistent with the absence of an ultraviolet cutoff.

Now, what does the existence of the novel gravitational energy scale $\Lambda_{\rm QG}$ indicate? The coupling constant t is a parameter introduced to control conformal dynamics due to the Weyl action. Considering similarly to the case of QCD, the fact that the running coupling constant (2.28) diverges at $\Lambda_{\rm QG}$ means that the conformal invariance completely disappear at this energy scale. In other words, all the fourth-derivative actions responsible for conformal gravity dynamics disappear, and spacetime shifts to a new phase [HY05, HHY06a].

Then, how is the low-energy phase of spacetime is described? In QCD, when the gluon field action given by at most second derivatives disappears, the gluon dynamics completely disappear. In contrast, in quantum gravity, even if all the fourth-derivative gravitational actions disappear, the second-derivative Einstein-Hilbert action remains, which provides the kinetic term of the gravitational field at low energies. In other words, Einstein's theory of gravity appears as a low-energy effective theory of gravity. The gravitational field can no longer be considered separately into the conformal-factor field and the traceless tensor field as in (2.3), and the gravitational field as a composite field that they are tightly coupled becomes a new field variable.

The value of Λ_{QG} at which the spacetime phase transition occurs must be determined from observations. From the scenario of quantum gravity inflation described in Chap. 7, it has been suggested that the value is about 10^{17} GeV. Since this value is about two orders of magnitude smaller than the Planck mass m_{pl}, the quantum gravity correlation length

$$\xi_\Lambda = \frac{1}{\Lambda_{QG}}$$

is about 100 times the Planck length. This implies that if we increase the energy higher and higher and explore shorter distances, Einstein's theory of gravity will break down before reaching the Planck scale, and spacetime will enter a new phase. Hence, it can solve many of the problems that Einstein's theory of gravity has.

The correlation length ξ_Λ gives a minimum measurable distance. Physics at longer distances than this is described by Einstein's theory of gravity, thus distances can be measured classically, but at shorter distances, we enter the quantum gravity phase and the gravitational field begins to fluctuate significantly. That is, a background-free world in which distances fluctuate largely is realized, and distances shorter than ξ_Λ become virtually impossible to measure. Theoretically, this is expressed as all spacetimes with different scales being gauge-equivalent. In this way, even though the renormalizable quantum gravity deals with spacetime continuously without introducing an ultraviolet cutoff, we end up with a spacetime in which distance is practically quantized by ξ_Λ.

2.4 Several Comments on Renormalization

One of the features of renormalizable quantum field theory is that the highest-derivative kinetic term does not contain a scale. Therefore, the higher energy, the more dominant kinetic terms become, where terms with dimensionful parameters such as mass terms can be ignored. In contrast, in Einstein's theory of gravity, in which the Einstein-Hilbert action is treated as the highest derivative term of the theory, no high-energy limit is defined where Newton's constant can be ignored. In other words, it can be said to be an unrenormalizable theory with an upper limit on energy that cannot be transferred to a scale-invariant world. This problem is solved by adding the square of curvatures to the action.

The condition that determines the fourth-derivative quantum gravity action is, of course, diffeomorphism invariance. However, this condition alone does not determine the action uniquely. We here determine it by adding conformal invariance as a new condition. The conformally invariant action does not contain the kinetic and interaction terms of the conformal-factor field. The key to the formulation of the theory is that the Wess-Zumino action of conformal anomalies derived from the path integral measure is responsible for these terms. See Chap. 5 for a more precise definition using dimensional regularization.

Conformal invariance also appears in the ultraviolet limit of ordinary gauge theories, but the big difference is that in quantum gravity it appears as a gauge symmetry, representing background freedom. It is a quantum diffeomorphism invariance that arises for the first time by quantizing the gravitational field, and is distinguished by using the term, BRST conformal invariance, from normal one. It is important that all ghost modes become unphysical under this gauge symmetry.

In the following, in order to better understand the quantization method described above, we will briefly explain how it differs from traditional methods.

2.4.1 Historical Background of Weak-Field Approximation

One of the problems with the quantization of forth-derivative gravity using the weak-field approximation $g_{\mu\nu} = \eta_{\mu\nu} + H_{\mu\nu}$, which was actively researched in the 1970s, is that the setting does not match physical intuition, as mentioned in Chap. 1. This setting also cannot explain the fact that the early universe is dominated by scalar fluctuations. Leaving aside the physical strangeness of this setting, here we will discuss theoretical problems that arise when quantization of gravity is performed using the weak-field approximation.

In those days, the weak-field expansion was adopted widely because it was the only methodology that could perform concrete calculations, but the success of QCD was also a major factor. This was the period when QCD was established as a theory to describe hadrons with the discovery of asymptotic freedom [Poli73, GW73]. Quantum gravity was also studied under the assumption that it had similar properties as one of non-Abelian gauge theories.

At that time, it was known that several theoretical problems arose in the weak-field perturbation theory. One is the following. In this approach, the path integral measure is defined in the same way as in ordinary quantum field theory, thus the Wess-Zumino action introduced above does not appear from the measure. Therefore, we have to introduce R^2 in addition to the Weyl action as a kinetic term for the conformal mode of $H_{\mu\nu}$. However, when introducing a new coupling constant with the correct sign for the R^2 action, a problem arises that "tachyon mode" appears.[12]

The other is the ghost problem. The work of Lee and Wick [LW69, Nak72] was often cited as a solution to this problem at the time. They showed that in a quantum field theory of fourth-derivative fields with asymptotically-free interactions, negative-metric ghost modes do not appear in the real world due to quantum corrections [Tom77, Tom80, FT82a, BG84]. However, although it shows that ghosts do not appear in the real world at low energies, when the interaction disappears at high energies, we end up with a unitary-broken world in which ghost particles propagate as real entities. Since information on scattering matrices there will be transmitted to low energies, like hadron jets in QCD, simple perturbation theory using the weak-field approximation has not been able to solve the problem essentially.

The key to the Lee-Wick ideas is that two modes with different metrics are connected through interactions and are no longer independent of each other. Therefore, if a gauge symmetry exists in the ultraviolet limit that connects the two modes, the story of the ghost problem changes. A ghost appears in the real world when it becomes gauge invariant by itself, but if there is a gauge symmetry that does not allow this, then individual modes can be made unphysical. Such a situation cannot be realized by the weak-field perturbation theory, which assumes the existence of asymptotic particle states. The BRST conformal invariance is a new idea to solve such problems.

2.4.2 *Renormalization in Different Perspectives*

If we try to quantize Einstein's theory of gravity with perturbation theory, we have to use Newton's constant, which has a dimension of length squared, as a coupling constant. Therefore, it is easy to understand from dimensional analysis that higher-derivatives divergences occur, so that gravitational counterterms consisting of multiple products of curvatures are necessary to eliminate them. It indicates that the theory is not renormalizable because as increasing the order of perturbation, we eventually have to introduce an infinite number of new counterterms. Still, many people try to formulate the theory around Einstein's theory of gravity. In the following, we comment on such attempts.

First, consider a system with only the gravitational field, and then an attempt was made to eliminate gravitational counterterms, or divergences, by imposing the Einstein equation expressed as the Ricci flatness condition on external gravitational fields in Feynman diagrams. With this method, although it is possible to eliminate gravitational counterterms for about one loop, some higher-derivative gravitational counterterms required for higher loops never go away, and it soon stops working.

This led to the idea of supergravity theory with supersymmetry that imposes stronger theoretical constraints [Nie81, WB83]. Supersymmetry is a symmetry that

[12] Unlike the case of (2.3), in the weak-field perturbation, the Einstein-Hilbert action provides mass terms to the gravitational field, but in this case its sign for the conformal mode becomes opposite.

connects bosons and fermions that was discovered in the 1970s, and it was hoped that by combining it with gravitational equations of motion, the number of gravitational counterterms would be limited to a finite number. This attempt was actively researched after Hawking predicted that everything would be fixed if increasing the level of supersymmetry to the highest level, but we cannot say that it has been successful. This can be said to be a last resort devised for quantum gravity, which is different from normal renormalization. This approach was eventually absorbed into superstring theory [GSW87, Pol98], and supergravity remains as its effective theory.[13]

A different perspective is the idea of "asymptotic safety" proposed by Weinberg in the 1970s [Wei79]. This was proposed as a property that a correct quantum theory of gravity should exhibit when bringing the ultraviolet cutoff to infinity. It allows the existence of an infinite number of gravitational counterterms, but requires the existence of an ultraviolet fixed point in renormalization group flows of many new coupling constants introduced with them. This represents the expectation that the theory should converge to the fixed point, and diffeomorphism invariance lost by the introduction of the cutoff should be restored there.

Researches to demonstrate the existence of such fixed points are still actively being conducted [Reu98, RS18]. A method often used in this case is the functional renormalization group method [Wil75, Wet93, Mor94], which was devised to investigate responses of the effective field theory to changes in the cutoff. This is a broader concept than the renormalization group presented in renormalizable quantum field theory, and is also applied to unrenormalizable theories such as Einstein's theory of gravity.

If such a fixed point exists, it is reasonable to consider that some kind of conformal invariance is realized there. The asymptoticaly background-free quantum gravity is a renormalizable quantum field theory whose core part is given by a certain conformal field theory and the perturbation expansion is defined as deviations from it. Therefore, it can be said that asymptotic background freedom is a representation of asymptotic safety.

2.4.3 QED and Landau Singularity

The Landau singularity is a problem that is thought to occur generally in asymptotically non-free quantum field theory whose beta function is positive, and refers to having a singular point where the running coupling constant diverges on the high-energy side. Quantum electrodynamics (QED) is known as a representative of field theories with the Landau singularity. Since such a theory is not defined in regions beyond energy at which the singularity appears, it has to be replaced with a new

[13] Unlike renormalizable quantum field theories, superstring theory is a manifestly finite theory, and thus its effective action is in local form. The only energy scale that exists in the theory is basically string tension, the coefficient of string action. Due to its structure, like Einstein's theory of gravity, there is no defined high-energy region where the scale can be ignored, thus it gives a limit energy scale (Hagedron temperature) that cannot be exceeded [HW70, KT90].

theory there. Hence, a theory with the Landau singularity is not appropriate for a fundamental theory that describes a microscopic world.

Such a bad thing is usually thought to occur near the Planck scale. Therefore, we have to consider a scenario that the theory is integrated into another particle model that shows the asymptotic freedom such as GUT models before it breaks down, otherwise restrict parameters of the theory so that it works up to the Planck scale. The latter is a passive way of thinking that once we reach the Planck scale, there is no point in thinking about the world beyond it, thus we do not need to worry about it any more.

However, the dynamical scale Λ_{QG} is set to 10^{17} GeV here, which is two orders of magnitude lower than the Planck energy. Therefore, the theoretical situation changes because quantum gravity corrections are added before reaching the Planck scale.

Now, consider whether the Weinberg's asymptotic safety requirement mentioned above can be applied not only to quantum gravity but also to QED. That of quantum gravity has already been revealed as the BRST conformal invariance, and here we expect that QED with quantum gravity corrections also has a non-trivial ultraviolet fixed-point where its beta function β_e vanishes. Although nonperturbative calculations to confirm this are so difficult, it is known that quantum gravity loop corrections described in Chap. 5 give a negative contribution to the QED beta function. Therefore, it can be expected that β_e will decrease and reach an ultraviolet fixed-point when energy exceeds Λ_{QG}.

During the inflation period, the conformally invariant dynamics in which the Weyl tensor disappears are well retained and the running coupling constant $\bar{t}(Q)$ does not increase until near the point where the spacetime phase transition occurs. Let us then assume that the QED beta function β_e is also mostly stuck at a fixed point, maintaining a zero value which indicates the presence of conformal invariance. Approaching the phase transition point, the running coupling constant increases rapidly, and eventually the conformally invariant dynamics of gravity disappear completely, shifting to a low-energy effective field theory based on Einstein's theory of gravity. At that time, if β_e also becomes non-zero and the Wess-Zumino interaction $\sqrt{-g}\,\beta_e \phi F_{\mu\nu}^2$ opens, we can consider a scenario that photons are generated from the scalar fluctuation of quantum gravity through this interaction.

The mechanism of matter creation during the spacetime phase transition also applies to the QCD and GUT models if the photon field strength $F_{\mu\nu}$ is replaced with that of the Yang-Mills field. In these cases, the Landau singularity does not occur because the beta functions disappear in the high energy limit.

2.5 Appendix: More About Effective Action

The ϕ-dependence of the physical momentum Q (2.27) that appears in the effective action originates from the Wess-Zumino action S. When higher-order corrections are taken into account, the following more general multipoint interactions arise [Ham20b]:

$$\int d^4x\sqrt{-g}\,\phi^n\,\mathrm{Tr}\big(F_{\mu\nu}^2\big), \qquad \int d^4x\sqrt{-g}\,\phi^n\,C_{\mu\nu\lambda\sigma}^2, \tag{2.33}$$

and

$$\int d^4x\sqrt{-\bar{g}}\,\phi^n\big(2\bar{\Delta}_4\phi+\bar{E}_4\big), \tag{2.34}$$

The $n=1$ type is the Wess-Zumino action discussed before. On the other hand, $n\geq 2$ obtained by integrating over ϕ repeatedly appear at higher loops, which give local parts of the Wess-Zumino actions containing ϕ^n as their top terms.[14] If (2.33) and (2.34) are collectively denoted as $S_{\rm top}^{(n)}$, then

$$\int d^4x\frac{\delta S_{\rm top}^{(n)}}{\delta\phi(x)}=n\,S_{\rm top}^{(n-1)}$$

holds, where we use the self-adjointness of the differential operator $\bar{\Delta}_4$ (2.17) and $\int d^4x\sqrt{-\bar{g}}\,\bar{\Delta}_4 A=0$ derived from it when we show that (2.34) satisfies this relation. If we employ dimensional regularization that preserves diffeomorphism invariance manifestly, these $S_{\rm top}^{(n)}$ are systematically derived from the Laurent expansions (5.17), (5.18), and (5.19) of the action, as shown in Chap. 5.

The Wess-Zumino action of type (2.33) plays the role of converting a nonlocal loop correction $\log^n(q^2/\mu^2)$ into a diffeomorphism invariant form $\log^n(Q^2/\mu^2)$ in the effective action, and it is eventually incorporated into the higher-order correction term of each running coupling constant. In the following, we first see the structure of the effective action in the case of gauge theory specifically, and next describe the structure of the effective action in quantum gravity, including contributions from the type (2.34).

2.5.1 QCD Effective Action in Curved Space

The basic structure of the Wess-Zumino actions for conformal anomalies in gauge theory does not depend on gauge groups. If we use dimensional regularization, then a non-Abelian gauge field action is also expanded as in (5.17) of the QED action, and its effective action can be calculated in the same way. Since diffeomorphism invariance is guaranteed, it can be expressed using the physical momentum squared $Q^2=q^2/e^{2\phi}$ (2.27).

---

[14] Here note that the expressions for $n\geq 2$ do not satisfy the Wess- Zumino consistency condition (2.11) because their integrand, $n=1$, is not a quantity composed by the full metric field $g_{\mu\nu}$. For these to satisfy this condition, it is necessary to add proper nonlocal terms, as shown later in an example.

Specifically, letting the QCD coupling constant be $\mathfrak{g}$ and the beta function be

$$\beta_{\mathfrak{g}} = \mu \frac{d\mathfrak{g}}{d\mu} = -\beta_0 \mathfrak{g}^3 - \beta_1 \mathfrak{g}^5 - \beta_2 \mathfrak{g}^7, \tag{2.35}$$

the effective action is given by[15]

$$\begin{aligned}
\Gamma_{\rm QCD} = &-\frac{1}{2}\int \frac{d^4q}{(2\pi)^4}\bigg\{\frac{1}{\mathfrak{g}^2} + (\beta_0 + \beta_1\mathfrak{g}^2 + \beta_2\mathfrak{g}^4)\log\left(\frac{q^2}{\mu^2}\right) \\
&- \frac{1}{2}\beta_1\beta_0\mathfrak{g}^4\log^2\left(\frac{q^2}{\mu^2}\right) - 2(\beta_0 + \beta_1\mathfrak{g}^2 + \beta_2\mathfrak{g}^4)\phi \\
&- 2\beta_1\beta_0\mathfrak{g}^4\left[\phi^2 - \phi\log\left(\frac{q^2}{\mu^2}\right)\right]\bigg\}\,{\rm Tr}(\bar{F}^2_{\mu\nu})(q) \\
= &-\frac{1}{2}\int \frac{d^4q}{(2\pi)^4}\bigg\{\frac{1}{\mathfrak{g}^2} + (\beta_0 + \beta_1\mathfrak{g}^2 + \beta_2\mathfrak{g}^4)\log\left(\frac{Q^2}{\mu^2}\right) \\
&- \frac{1}{2}\beta_1\beta_0\mathfrak{g}^4\log^2\left(\frac{Q^2}{\mu^2}\right)\bigg\}\,{\rm Tr}(\bar{F}^2_{\mu\nu})(q),
\end{aligned} \tag{2.36}$$

where the linear term of ϕ in the second line is the famous Wess-Zumino action of $n = 1$. The term containing ϕ^2 in the third line is the $n = 2$ Wess-Zumino action, and the nonlocal term containing ϕ appears as a loop correction so that the combination in square brackets satisfies the Wess-Zumino consistency condition (2.11) (See also Footnote 14).

Since the beta function has a negative value, the running coupling constant becomes a function that decreases at high energies. We here consider how the effective action can be written using it. Expressing (2.35) as an integral yields

$$\log\mu + C = \int_{\bar{\mathfrak{g}}(C)}^{\mathfrak{g}} \frac{d\mathfrak{g}}{\beta_{\mathfrak{g}}(\mathfrak{g})},$$

where C is an integration constant and the lower limit $\bar{\mathfrak{g}}(C)$ of the integral is also a constant, i.e., a renormalization group invariant satisfying $\mu d\bar{\mathfrak{g}}(C)/d\mu = 0$. Letting C be $-\log Q$ and denoting the lower limit as a function $\bar{\mathfrak{g}}(Q)$ satisfying the condition $\bar{\mathfrak{g}}(\mu) = \mathfrak{g}$, we get

$$-\frac{1}{2}\log\frac{Q^2}{\mu^2} = \int_{\bar{\mathfrak{g}}(Q)}^{\mathfrak{g}} \frac{d\mathfrak{g}}{\beta_{\mathfrak{g}}(\mathfrak{g})} = \frac{1}{2}\int_{\mathfrak{g}^2}^{\bar{\mathfrak{g}}^2(Q)} \frac{d\lambda}{\lambda^2}\frac{1}{\beta_0 + \beta_1\lambda + \beta_2\lambda^2}.$$

---

[15] It is convenient to calculate the gauge-invariant effective action using the background field method [Abb81]. The difference from QED is that the beta function is negative and that the gauge field is normalized so that $1/\mathfrak{g}^2$ is extracted in front of the lowest action.

This $\bar{g}(Q)$ is the running coupling constant, and performing the integration yields

$$\begin{aligned}&\frac{1}{\bar{g}^2(Q)}+\frac{\beta_1}{\beta_0}\log\bigl(\beta_0\bar{g}^2(Q)\bigr)+\frac{\beta_2\beta_0-\beta_1^2}{\beta_0^2}\bar{g}^2(Q)\\&=\beta_0\log\frac{Q^2}{\mu^2}+\frac{1}{g^2}+\frac{\beta_1}{\beta_0}\log\bigl(\beta_0 g^2\bigr)+\frac{\beta_2\beta_0-\beta_1^2}{\beta_0^2}g^2.\end{aligned} \tag{2.37}$$

Since the left-hand side is renormalization group invariant, the right-hand side is also so. Therefore, by introducing a renormalization group invariant scale Λ_{QCD} and writing the right-hand side as $\beta_0\log(Q^2/\Lambda_{\mathrm{QCD}}^2)$, we obtain the following expression:

$$\Lambda_{\mathrm{QCD}}^2=\mu^2\exp\Bigl(-\frac{1}{\beta_0 g^2}\Bigr)\bigl(\beta_0 g^2\bigr)^{-\frac{\beta_1}{\beta_0^2}}\exp\Bigl(-\frac{\beta_2\beta_0-\beta_1^2}{\beta_0^3}g^2\Bigr).$$

This is known as the QCD physical scale.

Now, letting $L(Q)=\log(Q^2/\Lambda_{\mathrm{QCD}}^2)$ and solving (2.37) with the right-hand side replaced by $\beta_0 L(Q)$ iteratively under $Q\gg\Lambda_{\mathrm{QCD}}$, we find that the running coupling constant can be expressed as

$$\begin{aligned}\bar{g}^2(Q)=&\frac{1}{\beta_0 L(Q)}\biggl\{1-\frac{\beta_1\log L(Q)}{\beta_0^2\,L(Q)}\\&+\frac{\beta_1^2}{\beta_0^4\,L^2(Q)}\biggl[\Bigl(\log L(Q)-\frac{1}{2}\Bigr)^2+\frac{\beta_2\beta_0}{\beta_1^2}-\frac{5}{4}\biggr]\biggr\}.\end{aligned}$$

If we rewrite the effective action (2.36) using this, we can find that it is summarized in the following form:[16]

$$\Gamma_{\mathrm{QCD}}=-\frac{1}{2}\int\frac{d^4q}{(2\pi)^4}\frac{1}{\bar{g}^2(Q)}\mathrm{Tr}\bigl(\bar{F}_{\mu\nu}^2\bigr)(q).$$

This expression is derived perturbatively, but since it is expressed in terms of renormalization group invariants, or physical quantities, it is considered to be valid over a wider momentum range. It indicates that when the physical momentum Q

[16] In fact, the reciprocal of the running coupling constant squared can be expanded as

$$\begin{aligned}\frac{1}{\bar{g}^2(Q)}=&\beta_0 L(Q)+\frac{\beta_1}{\beta_0}\log L(Q)+\frac{\beta_1^2}{\beta_0^3\,L(Q)}\Bigl(\log L(Q)-\frac{\beta_2\beta_0}{\beta_1^2}+1\Bigr)\\&+\frac{\beta_1^3}{\beta_0^5\,L^2(Q)}\Bigl(-\frac{1}{2}\log^2 L(Q)+\frac{\beta_2\beta_0}{\beta_1^2}\log L(Q)+\frac{1}{2}\Bigr)\\=&\frac{1}{g^2}+\bigl(\beta_0+\beta_1 g^2+\beta_2 g^4\bigr)\log\Bigl(\frac{Q^2}{\mu^2}\Bigr)-\frac{1}{2}\beta_1\beta_0 g^4\log^2\Bigl(\frac{Q^2}{\mu^2}\Bigr).\end{aligned}$$

becomes less than $\Lambda_{\rm QCD}$ and $\bar{g}(Q)$ diverges, the kinetic term of the gauge field, namely, the dynamics, disappears, suggesting that confinement occurs.

2.5.2 *Quantum Gravity Effective Action*

The Weyl sector of quantum gravity has a structure similar to gauge theory. Since the beta function of the coupling constant t is negative, the logarithmic part can be expressed using the running coupling constant $\bar{t}(Q)$ as in the case of QCD. That is to say that replacing the field strength from $F_{\mu\nu}$ to $C_{\mu\nu\lambda\sigma}$ and the coupling constant from g to t, and accordingly rewriting the name of the physical scale from $\Lambda_{\rm QCD}$ to $\Lambda_{\rm QG}$, the effective action can be summarized in the form of

$$\Gamma^{\rm W} = -\int \frac{d^4q}{(2\pi)^4} \frac{1}{\bar{t}^2(Q)} \bar{C}^2_{\mu\nu\lambda\sigma}(q),$$

even at higher orders.

This suggests that when $\bar{t}(Q)$ becomes large at the new physical scale $\Lambda_{\rm QG}$, the fourth-derivative kinetic term disappears, and spacetime shifts to a world described by the Einstein-Hilbert action.

The Wess-Zumino action of type (2.34) will arise to make higher-order correction terms in the Riegert effective action $\Gamma^{\rm R}$ (2.29) diffeomorphism invariant. The zeroth-order of t in $\Gamma^{\rm R}$ does not depend on the renormalization scale μ, but the higher-order correction terms exhibit scale dependence. Similar to the Weyl sector, it will be given by a form that the t^2-dependence in the coefficient b_1 of the Riegert action is replaced with $\bar{t}^2(Q)$. In fact, since expanding (2.28) in t^2 yields

$$\bar{t}^2(Q) = t^2 + 2\beta_0 t^4 \phi - \beta_0 t^4 \log\left(\frac{q^2}{\mu^2}\right) + o(t^6),$$

we can see that the Wess-Zumino action (2.34) is required for the substitution to occur. Furthermore, paying attention to $\log \Delta_4 = -4\phi + \log \bar{\Delta}_4$, it is inferred that the diffeomorphism-invariant effective action involving (2.34) as a top term is given by

$$\frac{1}{8}\frac{(-1)^{n-1}}{4^{n-1}} \int d^4x \sqrt{-g} E_4 \log^{n-1}\left(\frac{\Delta_4}{\mu^4}\right) \frac{1}{\Delta_4} E_4 \tag{2.38}$$

in coordinate space. The logarithmic part will finally be written as the running coupling constant squared, $\bar{t}^2(Q)$.

Chapter 3
Conformal Invariance as Background Freedom

In this chapter, we show that the BRST conformal invariance comes out when quantum gravity is formulated as an ordinary quantum field theory defined on the background spacetime. After clarifying the difference from normal conformal invariance, we derive the BRST conformal transformation from diffeomorphism, and describe its physical meaning.

3.1 BRST Conformal Invariance

The key to the asymptotically background-free quantum gravity lies in special properties that the core part of perturbation theory possesses. In the ordinary weak-field approximation, the zeroth-order of perturbation is represented by free fields (gravitons) propagating in flat spacetime, and perturbation theory describes how they interact and scatter. On the other hand, free fields do not appear in this theory. The core part is described by a special conformal field theory, and the coupling constant t represents the degree of deviation from it.

In normal conformal transformations, when the coordinates are transformed to $x^\mu \to x'^\mu$, the line-element ds representing distance changes only by a conformal factor like

$$ds^2 \to ds'^2 = \Omega^2(x) ds^2, \tag{3.1}$$

where the metric field remains fixed at this time. Conversely, it can also be defined as a transformation called Weyl rescaling that expands the metric field by a conformal factor without changing the coordinates. In this section, we describe it as the coordinate transformation.

In contrast, since the BRST conformal transformation is a manifestation of diffeomorphism in the ultraviolet limit, the metric field is also transformed under the coordinate transformation so that the line-element remains unchanged as

K. Hamada, *Trans-Planckian Physics and Inflation*, Fundamental Theories of Physics 223, https://doi.org/10.1007/978-981-96-3476-7_3

$$ds^2 = ds'^2. \tag{3.2}$$

Moreover, physical quantities must also be conformally invariant in the BRST conformal field theory, which is very different from normal conformal field theory where only the vacuum is conformally invariant.

Hence, normal conformal invariance refers to a theory that is independent of any particular scale, although there is a metric that defines distance. On the other hand, conformal invariance in quantum gravity occurs when the distance itself fluctuates, thus it represents a world without scale in the truest sense.

Specifically, it is expressed as follows. First, diffeomorphism is defined by is a transformation in which the line-element $ds^2 = g_{\mu\nu}dx^\mu dx^\nu$ remains unchanged as (3.2) under the coordinate transformation $x^\mu \to x'^\mu = x^\mu - \xi^\mu$. When the gauge parameter ξ^μ is infinitesimal, diffeomorphism is expressed as

$$\delta_\xi g_{\mu\nu} = g_{\mu\lambda}\nabla_\nu \xi^\lambda + g_{\nu\lambda}\nabla_\mu \xi^\lambda.$$

If the metric field $g_{\mu\nu}$ is expressed as the product of the conformal factor $e^{2\phi}$ and the metric field with the bar $\bar{g}_{\mu\nu}$ (2.4) as in (2.3), then diffeomorphism is decomposed as

$$\delta_\xi \phi = \xi^\lambda \partial_\lambda \phi + \frac{1}{4}\hat{\nabla}_\lambda \xi^\lambda,$$
$$\delta_\xi \bar{g}_{\mu\nu} = \bar{g}_{\mu\lambda}\bar{\nabla}_\nu \xi^\lambda + \bar{g}_{\nu\lambda}\bar{\nabla}_\mu \xi^\lambda - \frac{1}{2}\bar{g}_{\mu\nu}\hat{\nabla}_\lambda \xi^\lambda,$$

where we use the fact that $\bar{\nabla}_\lambda \xi^\lambda = \hat{\nabla}_\lambda \xi^\lambda$ holds from $\sqrt{-\bar{g}} = \sqrt{-\hat{g}}$. Furthermore, by expanding both sides of the second equation,[1] we obtain the transformation law of the traceless tensor field as

$$\delta_\xi h_{\mu\nu} = \hat{\nabla}_\mu \xi_\nu + \hat{\nabla}_\nu \xi_\mu - \frac{1}{2}\hat{g}_{\mu\nu}\hat{\nabla}_\lambda \xi^\lambda + \xi^\lambda \hat{\nabla}_\lambda h_{\mu\nu} + \frac{1}{2}h_{\mu\lambda}\left(\hat{\nabla}_\nu \xi^\lambda - \hat{\nabla}^\lambda \xi_\nu\right)$$
$$+\frac{1}{2}h_{\nu\lambda}\left(\hat{\nabla}_\mu \xi^\lambda - \hat{\nabla}^\lambda \xi_\mu\right) + o(h^2), \tag{3.3}$$

where the covariant vector of the gauge parameter is defined by $\xi_\mu = \hat{g}_{\mu\nu}\xi^\nu$ using the background metric.

The lowest term in (3.3),

$$\delta_\xi h_{\mu\nu} = \hat{\nabla}_\mu \xi_\nu + \hat{\nabla}_\nu \xi_\mu - \frac{1}{2}\hat{g}_{\mu\nu}\hat{\nabla}_\lambda \xi^\lambda, \tag{3.4}$$

is the part that contributes the most in the ultraviolet limit. The kinetic term of the Weyl action given as a quadratic term of the traceless tensor field becomes invariant

---

[1] Paying attention to $\delta_\xi \bar{g}_{\mu\nu} = \delta_\xi(\hat{g}e^h)_{\mu\nu} = \delta_\xi h_{\mu\nu} + h_{\lambda(\mu}\delta_\xi h^\lambda{}_{\nu)} + o(h^3)$, expand both sides and determine each order of h.

under this transformation. This gauge degree of freedom is fixed using the standard gauge-fixing method in gauge theory.

Even if these gauge degrees of freedom are fixed, there still remain 15 gauge degrees of freedom ζ^μ satisfying the conformal Killing equation

$$\hat{\nabla}_\mu \zeta_\nu + \hat{\nabla}_\nu \zeta_\mu - \frac{1}{2}\hat{g}_{\mu\nu}\hat{\nabla}_\lambda \zeta^\lambda = 0. \tag{3.5}$$

For this gauge degree of freedom, since the lowest term in transformation law (3.3) disappears and the next term becomes effective, diffeomorphism is expressed as [HH03, Ham09, Ham12a, Ham12b, Ham12c]

$$\begin{aligned} \delta_\zeta \phi &= \zeta^\lambda \partial_\lambda \phi + \frac{1}{4}\hat{\nabla}_\lambda \zeta^\lambda, \\ \delta_\zeta h_{\mu\nu} &= \zeta^\lambda \hat{\nabla}_\lambda h_{\mu\nu} + \frac{1}{2} h_{\mu\lambda}\left(\hat{\nabla}_\nu \zeta^\lambda - \hat{\nabla}^\lambda \zeta_\nu\right) + \frac{1}{2} h_{\nu\lambda}\left(\hat{\nabla}_\mu \zeta^\lambda - \hat{\nabla}^\lambda \zeta_\mu\right). \end{aligned} \tag{3.6}$$

These are conformal transformations of scalar and tensor with conformal dimension 0, representing the zero-dimensionality of the gravitational field. In the transformation of the conformal-factor field ϕ, a field-independent shift term appears, which is not present in normal conformal transformations. It represents that this transformation is derived from diffeomorphism.

In addition, a gauge field A_μ and a conformally invariant scalar field X are also transformed at the same time, and their transformation laws are defined by

$$\begin{aligned} \delta_\xi A_\mu &= \xi^\lambda \nabla_\lambda A_\mu + A_\lambda \nabla_\mu \xi^\lambda, \\ \delta_\xi X &= \xi^\lambda \partial_\lambda X. \end{aligned}$$

Leaving only ζ^μ as the gauge degree of freedom, the transformation law of the gauge field reduces to

$$\delta_\zeta A_\mu = \zeta^\nu \hat{\nabla}_\nu A_\mu + \frac{1}{4} A_\mu \hat{\nabla}_\nu \zeta^\nu + \frac{1}{2} A_\nu \left(\hat{\nabla}_\mu \zeta^\nu - \hat{\nabla}^\nu \zeta_\mu\right),$$

where the conformal Killing equation (3.5) is used. For the scalar field, if we remove the ϕ-dependence in its action by redefining the field as $X = e^{-\phi} X'$, and denote the redefined field X' as X again, then we obtain

$$\delta_\zeta X = \zeta^\mu \partial_\mu X + \frac{1}{4} X \hat{\nabla}_\mu \zeta^\mu,$$

where the second term on the right-hand side arises to compensate for the change in the ϕ-field. These are the same as the conformal transformations of vector and scalar fields with conformal dimension 1, respectively.

Although the residual gauge degrees of freedom ζ^λ are only 15, this gauge symmetry imposes strong constraints on physical states because the right-hand side of each conformal transformation law is field-dependent. Generator for each transformation is constructed using the energy-momentum tensor derived from the action of each field, and those for (3.6) are derived from the Riegert and Weyl actions, respectively (see Chap. 6). The shift term in the transformation law then comes from the linear term of ϕ in the Riegert action.

The asymptotic background freedom represents that the BRST conformal invariance is realized in the high-energy limit. Conversely, it also has the implication that diffeomorphism gradually deviates from the conformal transformation in the infrared region, as can be seen from the full transformation law (3.3). One of the main themes of this book is exactly to discuss dynamical processes of the deviation.

3.2 Background Freedom and Unitarity

Physical quantities of quantum gravity in the high-energy limit must be invariant under the conformal transformation (3.6). One of the characteristics of this transformation is that the right-hand side depends on the field. In normal gauge transformations, the lowest order is given by a field-independent term consisting only of gauge parameters. Under such a transformation, modes that make up the field do not mix and become independent. On the other hand, under the transformation (3.6), modes mix with each other and do not be gauge-invariant independently.

Thus, the gravitational ghost modes do not themselves become physical states, but they are essential elements for diffeomorphism invariance to hold. Physical states can be expressed as states in which the whole Hamiltonian exactly vanishes, as detailed in later chapters. The reason why they can exist as non-trivial states with entropy rather than an empty vacuum is due to the ghost modes. The BRST conformal invariance truly defines physical states that appear in the ultraviolet limit far beyond the Planck scale.

Here again, let us see the difference between the BRST conformal invariance and normal conformal invariance more directly in terms of the background freedom. Unlike normal conformal transformation (3.1), diffeomorphism is a transformation making the line-element invariant as in (3.2) in which the metric field $g_{\mu\nu}$ is also transformed along with the coordinate transformation. Therefore, it is difficult to intuitively understand that the conformal invariance is included in diffeomorphism invariance. The BRST conformal invariance is a hidden symmetry that becomes visible when the theory is rewritten as a quantum field theory (2.19) in the background spacetime. The conformal transformation can then be expressed by a Weyl rescaling for the background line-element $d\hat{s}^2 = \hat{g}_{\mu\nu}dx^\mu dx^\nu$ as

$$d\hat{s}^2 \to d\hat{s}'^2 = e^{2\omega} d\hat{s}^2.$$

On the other hand, since the full metric field $g_{\mu\nu}$ is given by (2.3), this Weyl rescaling can be absorbed by shifting the conformal-factor field as $\phi \to \phi' = \phi - \omega$. In quantum gravity, the shift change is simply a rewriting of the integral variable. Since the integration interval of the conformal-factor field ϕ is defined by $(-\infty, \infty)$, the path integral measure becomes invariant as $[d\phi']_{\hat{g}} = [d\phi]_{\hat{g}}$. After all, it indicates that the theory does not change even if the background is Weyl-rescaled.

Now, using the expression of the partition function (2.19), let us demonstrate that the background freedom holds when the coefficient of the Riegert action S_R is given by b_c (2.23). At the time, note that the value of b_c is calculated using the Riegert action itself that possesses it as a coefficient. Thus, it is first necessary to resolve such a nested structure to determine the coefficient b_c without self-contradiction.

For that, for the time being, let the coefficient be an arbitrary value b. Denoting this action S to distinguish it from S_R, we perform the path integral (2.19) at $t \to 0$. Here it is significant that the coefficient of the Euler-type conformal anomaly for this theory is determined to be b_c, independent of b. Therefore, if the background field $\hat{g}_{\mu\nu}$ is Weyl-rescaled based on this fact, we find that the patition function transforms as follows:

$$\begin{aligned} Z(e^{2\omega}\hat{g}) &= \int [d\phi dh]_{e^{2\omega}\hat{g}} [df]_{e^{2\omega}\hat{g}}\, e^{iS(\phi, e^{2\omega}\bar{g}) + iI(e^{2\omega}g)} \\ &= \int [d\phi dh]_{\hat{g}} [df]_{\hat{g}}\, e^{iS_R(\omega, \bar{g})} e^{iS(\phi, e^{2\omega}\bar{g}) + iI(e^{2\omega}g)} \\ &= \int [d\phi dh]_{\hat{g}} [df]_{\hat{g}}\, e^{iS_R(\omega, \bar{g}) + iS(\phi - \omega, e^{2\omega}\bar{g}) + iI(g)}. \end{aligned}$$

In the second equality, the ω-dependence of the measure is pulled out as a conformal anomaly. Since its coefficient is determined to be b_c regardless of b, it is given by S_R. The third equality is obtained by shifting the conformal-factor field to $\phi \to \phi - \omega$, where the fact that the measure is invariant under the shift transformation is used. From this, if we set $b = b_c$ and thus $S = S_R$, then we can show that from the Wess-Zumino consistency condition (2.11), the background freedom holds as [HH03, Ham12a]

$$\begin{aligned} Z(e^{2\omega}\hat{g}) &= \int [d\phi dh]_{\hat{g}} [df]_{\hat{g}}\, e^{iS_R(\omega, \bar{g}) + iS_R(\phi - \omega, e^{2\omega}\bar{g}) + iI(g)} \\ &= \int [d\phi dh]_{\hat{g}} [df]_{\hat{g}}\, e^{iS_R(\phi, \bar{g}) + iI(g)} = Z(\hat{g}). \end{aligned}$$

Thus, in order to realize the BRST conformal invariance, it is important that the path integral is performed over the conformal-factor field ϕ and its measure is invariant under the shift transformation. Even though the Riegert action originally arises as a quantity yielding conformal anomaly, it works to give exact conformal invariance when gravity is quantized. Here, it is shown by using the Weyl rescaling, but the conformal transformation (3.6) expresses it as a coordinate transformation on the background spacetime $d\hat{s}^2$.

This property that all theories with different backgrounds connected to each other by conformal transformations are gauge-equivalent comes from the fact that the theory is originally written in terms of the metric field $g_{\mu\nu}$. In short, it is just another manifestation of diffeomorphism invariance. This is an algebraic expression of the background freedom, or background-metric independence, and represents a situation where the conformal factor of $g_{\mu\nu}$ is greatly fluctuates so much that it is virtually impossible to measure physical distance.

The success or failure of the unitarity in quantum gravity is represented by whether the reality of physical fields is maintained, similar to that of conformal field theory. The ghost modes are necessary for the BRST conformal invariance to hold, but they are not physical entities because they do not themselves become invariant under the BRST conformal transformation. In fact, all BRST conformally invariant quantities are given by real composite scalar fields. In the terminology of conformal field theory, they are real primary states with zero spin, and there are no tensor states. Since the fourth-derivative quantum gravity action $S_{\rm QG}$ (2.20) is positive definite, the reality of them is ensured not to be broken, as described in Sects. 2.1 and 2.2.

What should be noted here is that the ghost mode involved in the Einstein-Hilbert action makes Einstein's theory of gravity unstable, but the ghost modes in the fourth-derivative quantum gravity action does not induce such instability. When quantizing the gravitational field, it is often considered by decomposing it into modes, but the fact that the action is positive-definite when viewed in the original field indicates that mode decomposition is only for convenience. In fact, each mode is not independent, but connected to each other through the gauge transformation.

As emphasized before, even though physical states are states in which the total Hamiltonian vanishes, there are an infinite number of physical states, and therefore entropy exists. This is owing to the ghosts, which are "invisible but being". Furthermore, the absence of tensor states in quantum gravity is consistent with the CMB observation results, which suggest that the early universe was filled with scalar fluctuations.

For detailed discussion of physical states based on the BRST conformal algebra, see Chap. 6.

3.2.1 On Historical Background

Here is mentioning a historical background of the BRST conformal invariance in quantum gravity briefly. This quantization method was started by Polyakov [Poly81] in the first half of the 1980s for the case that spacetime is two-dimensional, and then led to the discovery of an exact solution of two-dimensional quantum gravity by Polyakov, Knizhnik, and Zamolodchikov [KPZ88] in the latter half of the the same decade. Following this discovery, Distler and Kawai [DK89], David [Dav88] formulated two-dimensional quantum gravity as a quantum Liouville field theory. Its action is called the Liouville action, which is a two-dimensional Wess-Zumino action originally discovered by Polyalov, and the Riegert action was found as a four-dimensional counterpart of that.

In two dimensions, the traceless tensor field has two components, which can be eliminated using the gauge degrees of freedom of diffeomorphism, thus quantum gravity can be expressed as a quantum field theory of the conformal-factor field ϕ only. In addition, the Einstein-Hilbert action, which is the action I in this case, becomes a topological invariant (Euler characteristic), and quantization is represented as a path integral that performs over all possible gravitational field-configurations in each topology and then sums up for topology.

See Appendix Sect. 6.4 for quantization of two-dimensional gravity and derivation of its physical states when a cylinder background $\mathbb{R} \times S^1$ is employed.

Chapter 4
Physical Meaning of Hamiltonian/Momentum Constraints

Diffeomorphism invariance, together with gauge invariance for gauge fields, is believed to be a fundamental principle that holds true from classical theory to quantum theory, and from the infinitesimal scale to the cosmic scale. Although it is often pointed out that gravity is incompatible with quantum theory, they are never contradictory. However, we need to be careful about some differences from normal quantization in which Hamiltonian eigenstates are considered. The most important differences is what is commonly called the Hamiltonian and momentum constraints, which states that the Hamiltonian and momentum of the whole system vanish. When considering quantum gravity, we should pay more attention to their existence. The background freedom is an ultimate form shown under the constraints. Here, we briefly summarize the physical meaning of them before describing dynamics of quantum gravity in later chapters.

4.1 About Gravitational Equations of Motion

It is generally thought that the gravity occurs in the presence of matter. To emphasize this point, the Einstein equation is written in the form (1.1) with the matter energy-momentum tensor placed on the right-hand side as a source. However, this convention hides the true nature of the symmetry. The Einstein equation is an equation of motion for the gravitational field, which is derived from the variational principle as an equation in which the variation of the entire action, including the Einstein-Hilbert action, with respect to the gravitational field vanishes. Since the gravitational field is coupled with all other fields, the quantity obtained by varying the action with respect to the gravitational field is nothing but the energy-momentum tensor $T_{\mu\nu}$ of the whole system. In other words, the Einstein equation is originally correct to express it as

K. Hamada, *Trans-Planckian Physics and Inflation*, Fundamental Theories of Physics 223, https://doi.org/10.1007/978-981-96-3476-7_4

$$T_{\mu\nu} = M_{\mathrm{P}}^2 \left(-R_{\mu\nu} + \frac{1}{2} g_{\mu\nu} R \right) + T^{\mathrm{M}}_{\mu\nu} = 0, \tag{4.1}$$

where $M_{\mathrm{P}} = 1/\sqrt{8\pi G}$ is the reduced Planck mass, which is the Planck mass m_{pl} divided by $\sqrt{8\pi}$. In particular, the time-time component of the energy-momentum tensor is the Hamiltonian density, and the integral of it over the spatial volume is the Hamiltonian. That is to say that the Hamiltonian vanishes.

Employing the path integral method, the field equation of motion in renormalizable quantum field theory can be expressed as an identity called the Schwinger-Dyson equation. The formula obtained by intuitively applying this method to the gravitational field is

$$\int [dg] \frac{\delta}{\delta g_{\mu\nu}(x)} e^{iI(g)} = \frac{i}{2} \langle \sqrt{-g}\, T^{\mu\nu}(x) \rangle = 0, \tag{4.2}$$

as shown in (2.1) at the beginning of Chap. 2.[1] That is, the energy-momentum tensor, including quantum corrections, vanishes identically. To define this equation strictly, it is essential that both diffeomorphism invariance and renormalizability hold in the world beyond the Planck scale.

Let us express (4.2) with a more accurate equation according to the description in Chap. 2. The action I (2.5) consists of two fourth-derivative conformally invariant actions, the Weyl tensor squared and the Euler term, to which the Einstein-Hilbert action and lower-derivative matter actions are added. Defining the diffeomorphism-invariant measure on a practical background as in (2.19), the Schwinger-Dyson equation can be expressed as

$$\begin{aligned} \int [d\phi dh]_{\hat{g}} \frac{\delta}{\delta \phi(x)} e^{iS_{\mathrm{QG}}} &= i \big\langle \mathbf{T}^{\lambda}{}_{\lambda}(x) \big\rangle = 0, \\ \int [d\phi dh]_{\hat{g}} \frac{\delta}{\delta h^{\nu}{}_{\mu}(x)} e^{iS_{\mathrm{QG}}} &= \frac{i}{2} \Big\langle \mathbf{T}^{\mu}{}_{\nu}(x) - \frac{1}{4} \delta^{\mu}{}_{\nu} \mathbf{T}^{\lambda}{}_{\lambda}(x) \Big\rangle = 0, \end{aligned} \tag{4.3}$$

where S_{QG} is the action (2.20) involving the Wess-Zumino action. For accuracy, the notation of the energy-momentum tensor is changed to bold (see also Appendix C). These are collectively written as $\langle \mathbf{T}^{\mu}{}_{\nu}(x) \rangle = 0$. From the way of perturbation expansion, (2.3) and (2.4), this equation holds exactly for the conformal-factor field while perturbatively for the traceless tensor field.

The expectation value of the equation of motion (4.3) contains quantum corrections. In other words, it is expressed as an equation in which the variation of the effective action by the background metric vanishes. The equation has the following structure:

---

[1] This is an intuitive formula, and thus the path integral measure of the gravitational field is written as $[dg]$ without the subscript g, instead of $[dg]_g$. However, when using dimensional regularization, we find that this expression is correct, as will be described in Sect. 5.3.

$$\mathbf{T}^{(4)\mu}{}_{\nu} + M_{\mathrm{P}}^2\sqrt{-g}\left(-R^{\mu}{}_{\nu} + \frac{1}{2}\delta^{\mu}{}_{\nu}R\right) + \mathbf{T}^{\mathrm{M}\mu}{}_{\nu} = 0, \tag{4.4}$$

where $\mathbf{T}^{(4)\mu}{}_{\nu}$ is the fourth-derivative gravitational term, in which contributions from not only the Wess-Zumino action and the Weyl action in S_{QG} itself but also loop corrections yielded by quantizing these actions are included. The specific expression will be given when discussing dynamics of quantum gravity in the following chapters.

Here, we make some comments regarding the method using canonical quantization. The terms, the Hamiltonian and momentum constraints, were originally used in canonical quantization of gravity [ADM59, ADM62, Dew67a, BHS83, Ash86, Kie12]. The Wheeler-DeWitt equation which often appears as an equation defining quantum gravity states is an expression of the Hamiltonian constraint in terms of functional differentiation [Whe64, Dew67a, Kie12]. It is the same as the equations of motion (4.2) in that it is derived from diffeomorphism invariance. However, it should be noted that many of such constraints are discussed within the framework of Einstein's theory of gravity. In addition, since canonical quantization itself treats time in a special way when setting commutation relations, a problem arises in that diffeomorphism invariance becomes unclear. Furthermore, from the perspective of the path integral, it can be said that this quantization method does not pay attention to the measure. The canonical quantization method should be applied after rewriting it into a form that can be handled by ordinary quantum field theory techniques as shown in (2.19). Even after the rewrite, the issue of normal ordering still remains, which is discussed below.

4.2 What Diffeomorphism Invariance Shows

Let us now consider what it means physically for the energy-momentum tensor of the entire system to vanish. There are two important properties to discuss [Ham22].

4.2.1 No Absolute Time

One is that since the Hamiltonian vanishes, there is no absolute time defined over the entire system as assumed in normal quantum theory. That is, the usual concept of time evolution, such as the Schrödinger equation, no longer exists.

This is also true in classical gravity, as can be seen by rewriting the Einstein equation as (4.1). Time progresses differently, depending on the strength of the gravitational field. In places where the gravitational field is weak, such as on Earth, the gravitational potential changes monotonically, and it is possible to measure how time progresses depending on location, but it does not progress in the same way throughout the system.

In a fully fluctuating quantum spacetime, the concept of time itself disappears. This property, called the background freedom, states that even if we choose flat spacetime as a reference background, if the gravitational field fluctuates significantly around it, the reference no longer has any meaning. In other words, the physical time and distance defined by the metric field (2.3) fluctuate greatly, making it virtually impossible to measure. In the previous chapter, we showed that this property is realized as the BRST conformal invariance in the high-energy limit.

The approximate time is given as a dynamical change in the gravitational field that results from solving the equation (4.4) where the energy-momentum tensor vanishes. When an average of the conformal factor, the scale factor in cosmology, begins to increase monotonically, it becomes time. This can be expressed as a process in which conformal invariance is gradually violated by physical scales such as the Planck mass. That is to say that the flow of time begins with inflation, as will be shown in Chap. 7. Furthermore, as will be shown in Chap. 10, during the inflationary period, the amplitude of fluctuations involved in the subsequent structure formation of the universe become smaller. As a result, time as an average value becomes substantial, and a uniform flow throughout the universe is created. The reduced fluctuation is then inherited by the Friedmann universe, and time continues as an entity.

What should be noted here is that even though we consider solutions where the Hamiltonian vanishes, there are non-trivial solutions. Time arises due to a change of state while the Hamiltonian remains zero. The reason why such a change occurs in Einstein's theory of gravity is that the Einstein-Hilbert action is not bounded below. If this action were positive definite like those of matter fields, there would only be a trivial vacuum solution, and time would not be created.

In quantum gravity, for the path integral to be well-defined, the action must be positive definite and renormalizable. Furthermore, ghost modes are necessary because there exists a non-trivial physical state in which the Hamiltonian is preserved to zero. Therefore, diffeomorphism invariance naturally requires the action of the curvature squared involving ghosts.

Here, we turn our attention to the fact that due to the presence of the $\mathbf{T}^{(4)\mu}{}_{\nu}$ term, the equation of motion (4.4) holds true even if the $\mathbf{T}^{\mathrm{M}}_{\mu\nu}$ term disappears. This suggests the existence of objects that consist only of gravitational fluctuations without matter. Thus, we can discuss spacetime from a completely different perspective than the Einstein equation which shows that matter creates gravity. Furthermore, the equation (4.4) can also describe the process of phase transition from a world of gravity alone to a world with matter. This transition process plays an important role in constructing inflationary cosmology and localized excited states of quantum gravity later.

Since an absolute time does not exist, the normal meaning of "conservation" no longer holds true. To be conserved means not to change in the process of evolution of the universe, which is here expressed as being renormalization group invariant. The energy-momentum tensor is exactly a renormalization group invariant. In addition, since the total energy is zero, entropy of the universe is given by the effective action itself, and it is also conserved as a renormalization group invariant. This shows that entropy of the universe originates from quantum fluctuations of gravity. The fact that it is conserved indicates that the conservation of entropy, which is an ansatz in Gamow's hot Big Bang model [Gam48, ABG48], holds strictly true.

4.2.2 No Zero-Point Energy

Another important fact indicated by the equation of motion (4.4) is that there is no zero-point energy.

Before understanding this fact, it should be pointed out that the energy-momentum tensor is a "normal product". It is a composite field that behaves as a finite operator in correlation functions. In the case of free fields, it refers to normal ordered operators. In ordinary quantum field theory, if considering a finite correlation function $\langle\cdots\rangle = \int [df]_g \cdots e^{iI(f,g)}$ in curved spacetime, then the variation of it by the classical metric field, $\delta\langle\cdots\rangle/\delta g_{\mu\nu}(x) = i\sqrt{-g}\langle T^{\mu\nu}(x)\cdots\rangle/2$, is clearly finite, thus it can be generally shown that the energy-momentum tensor $T^{\mu\nu}$ involving quantum corrections is a finite operator.[2] This is a conclusion drawn from diffeomorphism invariance, and can be clearly demonstrated as shown here if quantization is performed using the path integral method that preserves the invariance manifestly.

Adler, Collins, and Duncan [ACD77] showed that $T^{\mu\nu}$ is actually represented by a normal product, as part of their study of conformal anomalies using dimensional regularization. Brown and Collins [BC80], and Hathrell [Hat82a, Hat82b] took this research further and derived a special set of renormalization group equations from the fact that the energy-momentum tensor is a normal product and is a renormalization group invariant. It has been shown that solving their renormalization group equations determines the form of the fourth-derivative gravitational action [Ham14a] (see Chap. 5 for its specific form).

Ordinary quantum field theory only shows that the energy-momentum tensor is finite, but quantum theory of gravity shows that it vanishes identically. This means that there is no zero-point energy.

In general, performing normal ordering after canonical quantization in quantum field theory can be said to be an operation to restore diffeomorphism invariance. In fact, as shown in Chap. 6, when finding the generator of the BRST conformal transformation by quantizing (2.19) employing the canonical quantization method, it can be shown that in order to correctly close the conformal algebra, that is, to preserve diffeomorphism invariance, it is necessary to apply normal ordering to the generator.[3]

The fact that zero-point energy does not exist has implications for the origin of fluctuations in the early universe. Within the framework of Einstein's theory of gravity, the source of matter that makes up the current universe must be some kind of matter, and when considering inflation theory, an unknown scalar field called inflaton

---

[2] For more precise descriptions, see Sect. 5.3 and Appendix Sect. 5.7.1.

[3] Two-dimensional quantum gravity with the same numbers of time and space coordinates is special, in which case diffeomorphism invariance is preserved even when canonical quantization is used.

is introduced for this purpose. At that time, the origin of the fluctuation is explained as the zero-point energy of inflaton. However, in quantum theory of gravity, such a zero-point energy does not exist; instead, an excited state of quantum gravity is thought to be the source of everything.

This is also related to what is called the cosmological constant problem [Zel67, Wei89]. This problem arises from the fact that when we calculate the zero-point energy by introducing an ultraviolet cutoff to the Planck scale, a large cosmological constant of the Planck mass to the fourth power is obtained that does not match the observation. However, the quantum gravity discussed here is a theory that describes the world beyond the Planck scale and has no such a cutoff, thus the cause of this problem itself does not exist. This problem is also related to renormalizability, and will be discussed in detail in Chap. 9.

Chapter 5
Renormalization Using Dimensional Regularization

There are several methods to regularize ultraviolet divergences, but among them, dimensional regularization [tHV72, Col84, Mut87] is the only method that allows systematically performing higher-order renormalization calculations while preserving diffeomorphism invariance. This is a method of making the theory finite by reducing spacetime dimension to less than 4, so that ultraviolet divergences are extracted as poles of $D - 4$ and renormalization calculations can be performed relatively easily. After removing ultraviolet divergences, return the dimension to 4 and find physical quantities. In this regularization method, information of the Wess-Zumino action S, which is the contribution from the path integral measure, is included in the D-dimensional bare action. Therefore, when applying this method to gravitational theories, it is necessary to pay close attention to the D-dependence of the action involving the conformal-factor field. On the other hand, it has an excellent property that the results do not depend on how to choose the measure.

The purpose of this chapter is to describe concisely how the effective action presented before was actually derived, but even if you skip this chapter, you will still be able to understand the trans-Plankian physics in the latter half. Further details on computations in this chapter refer to the references cited or the book [Ham18].

5.1 How to Determine Quantum Gravity Action

One of the conditions that determine the quantum gravity action is, of course, diffeomorphism invariance, but a difficulty is that this condition alone does not determine the D-dimensional action uniquely. In fact, there are three candidates for the fourth-derivative actions of the gravitational field: the Riemann tensor squared, the Ricci tensor squared, and the scalar curvature squared. At first glance it seems that any combination of these can be chosen. However, by imposing renormalizability conditions, it can be shown that the action is determined to a certain combination. Here,

K. Hamada, *Trans-Planckian Physics and Inflation*, Fundamental Theories of Physics 223, https://doi.org/10.1007/978-981-96-3476-7_5

we briefly introduce this study, and use only the results. For mathematical details of what follows, see Appendix Sect. 5.7.

In order to determine the gravitational action, first consider renormalizable QED and QCD in curved spacetime. That is, consider a gauge theory in which the gravitational field remains classical and all other fields are quantized. Here, among counterterms necessary to remove ultraviolet divergences, those consisting only of the gravitational field are called gravitational counterterms. The reason for considering such a theory is that by imposing both gauge invariance and diffeomorphism invariance, the (bare) actions of quantum fields, gauge fields and fermions, are uniquely determined. In other words, the only part that has not been determined at this point is gravitational counterterms. However, since interactions between quantum fields and the gravitational field are now fixed, if we derive ultraviolet divergences by calculating Feynman diagrams in which external lines are described by the gravitational field and internal lines by gauge fields and fermions, then gravitational counterterms should be uniquely determined.

Calculations to find gravitational counterterms using dimensional regularization preserve both gauge invariance and diffeomorphism invariance. Renormalizability is expressed in terms of renormalization group equations, which imposes significant restrictions on the form of counterterms. In fact, in the 1980s, Hathrell [Hat82a, Hat82b, Fre84] has derived a set of special renormalization group equations that correlation functions among normal products such as the energy-momentum tensor should satisfy, in the study of conformal anomalies of gauge theory in curved spacetime. Solving the system of renormalization group equations shows that the forms of the gravitational counterterms can be determined for all orders of perturbation in gauge coupling. They are neatly grouped into two types, which become the conformally invariant gravitational actions given before when going back to four dimensions.

Thus, gravitational counterterms are completely determined, including parts that deviate from four dimensions. Information about the Wess-Zumino action S is then contained in them. This is a natural result because quantization is performed so that diffeomorphism invariance is preserved. Furthermore, the results are expected to be independent of gauge group, fermion representation, number of fermions, and so on.

This result originates from the fact that interactions in gauge theory are conformally invariant in four dimensions. If an interaction that is apparently not conformally invariant is added, an additional counterterm of the scalar curvature squared will be added to these two. What we want to emphasize here is that from the renormalizability conditions, the D-dimensional gravitational counterterms are classified into three fixed types, that is, their arbitrariness mentioned at the beginning disappears.

Quantum field theory in curved spacetime can be seen as part of quantum gravity theory, thus if we require gravity to be renormalizable, then its action has to take the same form as the gravitational counterterm described above. Therefore, we have to adopt it as a bare action of quantum gravity.

Once the form of the action is determined, we can find how to introduce the coupling constant. Here, we have assumed in advance that the conformal-factor field becomes dominant in the ultraviolet limit. That is, the reciprocal of the square of the dimensionless coupling constant t is introduced before the Weyl action with the correct sign, so as to be the perturbation expansion around a conformally flat spacetime.

Under this action, we need to calculate and confirm whether there is no contradiction as a quantum theory of gravity and whether the beta function is correctly negative so that conformal invariance is realized in the ultraviolet limit. In fact, it can be shown that the beta function of the coupling constant t takes a negative value regardless of the contents of matter fields. This indicates that, like QCD, the dynamical scale of quantum gravity, $\Lambda_{\rm QG}$, can be defined, and conformal invariance arises at the energy region above that.

5.2 Quantum Gravity Action in D Dimensions

Throughout this chapter, we consider the Wick-rotated D-dimensional Euclidean space, which is convenient for dimensional regularization.[1]

In general, in regularization methods strictly defined in four dimensions, such as the DeWitt-Schwinger method [BD82, BV85, PT09], conformal anomalies are derived by evaluating a divergent quantity, $\delta^{(4)}(0) = \langle x|x'\rangle|_{x'\to x}$.[2] This gives us the contribution from the path integral measure to restore diffeomorphism invariance. On the other hand, dimensional regularization is a regularization method that does not depend on the measure because this quantity vanishes identically due to $\delta^{(D)}(0) = \int d^D k = 0$.

In dimensional regularization, information about conformal anomalies and their Wess-Zumino actions is contained between 4 and D (< 4) dimensions. That is, conformal anomalies are given by finite quantities that remains after poles of ultraviolet divergences due to loop corrections and parts of the counterterms representing deviations from four dimensions cancel each other out, such as

$$\frac{1}{D-4} \times o(D-4) \to \text{finite}.$$

They are also generated from higher poles. Therefore, it is necessary to pay close attention to the D-dependence of the action.

The D-dimensional action of quantum gravity is given by the gravitational counterterm described in Sect. 5.1. Writing the result down, it is expressed as

$$S_{\rm QG} = \int d^D x \sqrt{g} \left\{ \frac{1}{t_0^2} F_D + b_0 G_D - \frac{M_0^2}{2} R + \Lambda_0 + \mathcal{L}_{\rm M} \right\}. \tag{5.1}$$

---

[1] The rotation to Euclidean time is defined by $t_E = i x^0$, where x^0 is time in Minkowski background.

[2] It can be expressed as $\delta^{(4)}(0) = \langle x|e^{-sK}|x\rangle|_{s\to 0}$ using a positive-definite regulator K specific to the theory. This quantity can be found by solving the heat conduction equation $(\partial_s + K)\langle x|e^{-sK}|x\rangle = 0$.

Here, the intention of denoting the action as $S_{\rm QG}$ instead of I is that this action itself also contains information about the Wess-Zumino action, as shown explicitly below. The quantities with subscript 0 denote bare ones before renormalization. The first term is the Weyl action, given by the square of the D-dimensional Weyl tensor,

$$F_D = C_{\mu\nu\lambda\sigma}^2 = R_{\mu\nu\lambda\sigma}^2 - \frac{4}{D-2}R_{\mu\nu}^2 + \frac{2}{(D-1)(D-2)}R^2. \tag{5.2}$$

The second term is a quantity that reduces to the Euler density when returned to four dimensions, and is given by [Ham14a, Ham14b]

$$G_D = G_4 + (D-4)\chi(D)H^2, \tag{5.3}$$

where $H = R/(D-1)$ and G_4 are defined by the combination of (2.7). The coefficient χ is a finite function of only D, which is defined by expanding with non-negative powers of $D-4$ as $\chi = \sum_{n=1}^{\infty} \chi_n (D-4)^{n-1}$. The first three terms are found by sequentially solving the Hathrell's renormalization group equations and are given by

$$\chi(D) = \frac{1}{2} + \frac{3}{4}(D-4) + \frac{1}{3}(D-4)^2 + \cdots. \tag{5.4}$$

This coefficient can in principle be determined to an infinite order.

The two actions F_D (5.2) and G_D (5.3) are given as gravitational counterterms to renormalize ultraviolet divergences generated from interactions that are conformally invariant in four dimensions. The coefficient χ will be then a universal coefficient, regardless of the types of interactions if they are conformal invariant in four dimensions. In fact, the first two terms in (5.4) are universal values that have been confirmed in gauge theories in curved spacetime to be independent of gauge group and fermion representations. The third term is currently confirmed only for QED.

In the following, we take QED as an example of the matter sector $\mathcal{L}_{\rm M}$, and explain the structure of the theory, pointing out the similarities between the gauge field and the traceless tensor field. Letting the number of Dirac fermions be n_F, the QED part of the action is given by

$$\int d^D x \sqrt{g} \left\{ \frac{1}{4} F_{0\mu\nu} F_0^{\mu\nu} + \sum_{j=1}^{n_F} i \bar{\psi}_{0j} \gamma^\mu D_\mu \psi_{0j} \right\}, \tag{5.5}$$

where the gauge field strength is $F_{0\mu\nu} = \nabla_\mu A_{0\nu} - \nabla_\nu A_{0\mu} = \partial_\mu A_{0\nu} - \partial_\nu A_{0\mu}$. The gamma matrix in curved spacetime is defined by $\gamma^\mu = e_a^{\ \mu} \gamma^a$ using the vielbein field $e_a^{\ \mu}$ and the Clifford algebra $\{\gamma^a, \gamma^b\} = -2\delta^{ab}$. The covariant derivative for fermions is defined by $D_\mu \psi_0 = \partial_\mu \psi_0 + \omega_{\mu ab} \Sigma^{ab} \psi_0 / 2 + i e_0 A_{0\mu} \psi_0$, where $\omega_{\mu ab}$ is the spin connection and $\Sigma^{ab} = -[\gamma^a, \gamma^b]/4$.

Perturbation expansion with the coupling constant t_0 means expanding the metric field around a conformally-flat configuration with $C_{\mu\nu\lambda\sigma} = 0$. Thus, the conformal factor is extracted as in $g_{\mu\nu} = e^{2\phi}\bar{g}_{\mu\nu}$ (2.3) and the remaining $\bar{g}_{\mu\nu}$ is expanded as

$$\bar{g}_{\mu\nu} = (\hat{g}e^{t_0 h_0})_{\mu\nu} = \hat{g}_{\mu\lambda}\left(\delta^{\lambda}_{\ \nu} + t_0 h^{\lambda}_{0\ \nu} + \frac{t_0^2}{2} h^{\lambda}_{0\ \sigma} h^{\sigma}_{0\ \nu} + \cdots\right) \tag{5.6}$$

using the traceless tensor field $h^{\mu}_{0\ \nu}$, where $h^{\mu}_{0\ \mu} = 0$, and a conformally-flat background metric $\hat{g}_{\mu\nu}$. Raising and lowering the indices is done by using the background metric as $h_{0\mu\nu} = \hat{g}_{\mu\lambda} h^{\lambda}_{0\ \nu}$. In addition, unlike (2.4), t_0 is introduced into the field-expansion formula (5.6) so that the coupling constant does not appear in the kinetic term of the traceless tensor field.

Letting L be Laurent series with pure poles, renormalization factors usually denoted by Z have the structure of $1 + L$. The renormalization factors in the QED part are introduced as

$$e_0 = \mu^{2-D/2} Z_3^{-1/2} e, \quad A_{0\mu} = Z_3^{1/2} A_\mu, \quad \psi_{0j} = Z_2^{1/2} \psi_j, \tag{5.7}$$

where an arbitrary mass scale μ is introduced in order to compensate for missing dimensions, and then the renormalized coupling e becomes dimensionless. The definition of e uses the Ward-Takahashi identity $Z_e Z_3^{1/2} = 1$.[3] This identity holds for systems coupled to quantum gravity as well [Ham02].

As in gauge theories, the renormalization factors in the gravitational part are introduced for the coupling constant and the traceless tensor field as

$$t_0 = \mu^{2-D/2} Z_t t, \qquad h_{0\mu\nu} = Z_h^{1/2} h_{\mu\nu},$$

respectively, where the renormalized coupling t is dimensionless.

On the other hand, the conformal-factor field ϕ is not subject to renormalization because it is treated strictly without introducing any specific coupling constant for it. Therefore, its renormalization factor becomes [Ham02]

$$Z_\phi = 1. \tag{5.8}$$

This is one of the most important properties in renormalization calculations.

5.2.1 *Renormalization Group Equations and Beta Functions*

Since bare quantities are original ones before renormalization, they should not depend on the mass scale μ introduced arbitrarily. Thus,

[3] It is usually expressed as $Z_1 = Z_2$, where Z_1 is the renormalization factor to the QED interaction term.

$$\mu \frac{d}{d\mu}[\text{bare quantities}] = 0.$$

Rewriting this equation in terms of renormalized quantities yields a renormalization group equation that specifies how they respond to the scale.

In the following, we employ

$$\alpha_e = \frac{e^2}{4\pi}, \qquad \alpha_t = \frac{t^2}{4\pi}, \tag{5.9}$$

as the notation for the coupling constants.

First, consider the famous renormalization group equation for the QED coupling constant α_e, which is given by

$$\mu \frac{d}{d\mu}\left(\frac{e_0^2}{4\pi}\right) = \frac{\mu^{4-D}}{Z_3}\alpha_e \left(4 - D - \mu \frac{d}{d\mu}\log Z_3 + \frac{\mu}{\alpha_e}\frac{d\alpha_e}{d\mu}\right) = 0.$$

Defining the beta function as

$$\beta_e = \frac{1}{\alpha_e}\mu \frac{d\alpha_e}{d\mu} = D - 4 + \bar{\beta}_e, \tag{5.10}$$

we get an expression $\bar{\beta}_e = \mu d(\log Z_3)/d\mu$. Similarly, if we define the beta function of the coupling constant α_t as

$$\beta_t = \frac{\mu}{\alpha_t}\frac{d\alpha_t}{d\mu} = D - 4 + \bar{\beta}_t, \tag{5.11}$$

then we obtain an expression $\bar{\beta}_t = \mu d(\log Z_t^{-2})/d\mu$ from the renormalization group equation for $t_0^2/4\pi$. Since the beta function is finite, we can find that a recurrence relation holds among residues of poles for each renormalization factor, whose specific expression will be presented in Sect. 5.4.

In addition to the usual renormalization factors, a bare constant b_0 is introduced to remove ultraviolet divergences proportional to the G_D term. Since its volume integral becomes topological in four dimensions, it does not contain the tree-level kinetic term of the gravitational field. Thus, b_0 is not an independent coupling constant, and is expanded in a series of pure poles as

$$b_0 = \frac{\mu^{D-4}}{(4\pi)^{D/2}} L_b, \qquad L_b = \sum_{n=1}^{\infty} \frac{b_n}{(D-4)^n}, \tag{5.12}$$

unlike normal renormalization factors. Since the field-dependent part of the volume integral of G_D starts from $o(D-4)$, the bare action defined by multiplying it by b_0 produces a finite term through the cancellation of the pole and this part, as shown later.

In general, residues are functions of coupling constants and do not contain any constant term that does not depend on them. In contrast, although the residues b_n satisfy this for $n \geq 2$, the residue of the simple pole has a constant term, which is decomposed as

$$b_1 = b + b_1', \tag{5.13}$$

where b_1' is the part that depends on the coupling constants, while b is just the number.

The Riegert action is one of the Wess-Zumino actions that remains even in the vanishing coupling limit, which is induced as a part proportional to the constant b and is responsible for the kinetic term of the conformal-factor field ϕ. At this time, in order to incorporate the Riegert action systematically by eliminating the nested structure, described in Sects. 2.1.2 and 3.2, the following procedure is required.

For the time being, regarding b that should be a constant as a new renormalized coupling constant, we calculate the effective action according to the normal renormalization procedure. It then becomes finite except for a topological term as

$$\Gamma = \frac{\mu^{D-4}}{(4\pi)^{D/2}} \frac{b - b_c}{D-4} \int d^D x \sqrt{\hat{g}}\, \hat{G}_4 + \Gamma_{\rm ren}(t, b)$$

where $\Gamma_{\rm ren}$ is a renormalized quantity that depends on t and b. The divergent term only occurs when we choose a curved background spacetime. The constant b_c is the 1-loop quantum correction (2.23), which includes a contribution from the induced Riegert action as well. If the matter sector is (5.5), then it is given by

$$b_c = \frac{11 n_F}{360} + \frac{40}{9}.$$

After completing all renormalization procedures, we set $b = b_c$. In this way, we obtain a finite effective action $\Gamma_{\rm ren}(t, b_c)$ whose dynamics are controlled by a single gravitational coupling constant t.

Now, let us see the renormalization group equation $\mu db_0/d\mu = 0$ [Ham14b]. In this case, reflecting that b is the coefficient (5.13) of the simple pole of b_0, its beta function is defined by

$$\mu \frac{db}{d\mu} = (D-4)\bar{\beta}_b. \tag{5.14}$$

By solving the renormalization group equations while noting that the derivative of an arbitrary scale μ is expressed as

$$\mu \frac{d}{d\mu} = \mu \frac{\partial}{\partial \mu} + \mu \frac{d\alpha_t}{d\mu} \frac{\partial}{\partial \alpha_t} + \mu \frac{d\alpha_e}{d\mu} \frac{\partial}{\partial \alpha_e} + \mu \frac{db}{d\mu} \frac{\partial}{\partial b},$$

we obtain the following recurrence relation between the residues b_n as the condition for poles to disappear:

$$\left(\alpha_t \frac{\partial}{\partial \alpha_t} + \alpha_e \frac{\partial}{\partial \alpha_e} + \bar{\beta}_b \frac{\partial}{\partial b} + 1\right) b_{n+1} + \left(\bar{\beta}_t \alpha_t \frac{\partial}{\partial \alpha_t} + \bar{\beta}_e \alpha_e \frac{\partial}{\partial \alpha_e}\right) b_n = 0.$$

In addition, $\bar{\beta}_b$ is expressed using the simple pole b_1 as

$$\bar{\beta}_b = -\left(\frac{\partial b_1}{\partial b}\right)^{-1} \left(b_1 + \alpha_t \frac{\partial b_1}{\partial \alpha_t} + \alpha_e \frac{\partial b_1}{\partial \alpha_e}\right). \tag{5.15}$$

Thus, (5.14) shows that b is a constant satisfying $\mu db/d\mu = 0$ in four dimensions, which guarantees that we can eventually replace b with the constant b_c.

5.2.2 Kinetic Terms, Interactions, and Counterterms

In the following, the background metric $\hat{g}_{\mu\nu}$ is set to the flat Euclidean metric $\delta_{\mu\nu}$. As a convention, Euclidean indices are expressed by subscripts and the same index denotes doing contraction by $\delta_{\mu\nu}$.

Let us see specifically how the Riegert action and other Wess-Zumino actions that describe dynamics of the conformal-factor field are derived. Performing Laurent expansion of the action (5.1) around four dimensions using renormalized quantities, terms with zeroth and positive powers of $D-4$ give kinetic and interaction terms, and terms with negative powers give counterterms.

First, we describe the gauge field part of the D-dimensional QED action (5.5). Let us Laurent-expand the renormalization factor of the gauge field as

$$Z_3 = 1 + \frac{x_1}{D-4} + \frac{x_2}{(D-4)^2} + \cdots . \tag{5.16}$$

In any D dimension, since the gauge field has a ϕ-dependence as $\sqrt{g} F_{0\mu\nu} F_0^{\mu\nu} = Z_3\, e^{(D-4)\phi}\, \bar{F}_{\mu\nu} \bar{F}^{\mu\nu}$, where $\bar{F}_{\mu\nu} \bar{F}^{\mu\nu} = \bar{g}^{\mu\lambda} \bar{g}^{\nu\sigma} F_{\mu\nu} F_{\lambda\sigma}$, the action is Laurent-expanded as

$$\begin{aligned}
&\frac{1}{4}\int d^D x \sqrt{g} F_{0\mu\nu} F_0^{\mu\nu} \\
&= \frac{1}{4}\int d^D x \biggl\{ \left(1 + \frac{x_1}{D-4} + \frac{x_2}{(D-4)^2} + \cdots\right) \bar{F}_{\mu\nu} \bar{F}^{\mu\nu} \\
&+ \left(D - 4 + x_1 + \frac{x_2}{D-4} + \cdots\right) \phi \bar{F}_{\mu\nu} \bar{F}^{\mu\nu} \\
&+ \frac{1}{2}\left((D-4)^2 + (D-4)x_1 + x_2 + \cdots\right) \phi^2 \bar{F}_{\mu\nu} \bar{F}^{\mu\nu} + \cdots \biggr\}.
\end{aligned} \tag{5.17}$$

The first line on the right-hand side gives the usual kinetic term and counterterms. In the second line, the Wess-Zumino action of type (2.15) and new counterterms associated with it arise. The subsequent lines provide the higher-order Wess-Zumino actions. Furthermore, when $\bar{g}_{\mu\nu}$ is expanded as (5.6), interactions and counterterms with the traceless tensor field arise.

Since the Weyl action is also expressed as $\sqrt{g}\, C^2_{\mu\nu\lambda\sigma} = e^{(D-4)\phi}\, \bar{C}^2_{\mu\nu\lambda\sigma}$ in any dimension, it can be expanded as

$$\frac{1}{t_0^2}\int d^Dx \sqrt{g} F_D = \int d^Dx \left\{ \frac{1}{t_0^2} \bar{C}^2_{\mu\nu\lambda\sigma} + \frac{D-4}{t_0^2} \phi\, \bar{C}^2_{\mu\nu\lambda\sigma} + \cdots \right\}, \tag{5.18}$$

in the same way as the gauge field action. Further expanding the bare quantities with the renormalized quantities yields a kinetic term, interactions, and counterterms.

The second term of the action (5.1), which is topological in four dimensions, is expanded using (5.12) as

$$\begin{aligned} b_0 &\int d^Dx \sqrt{g} G_D \\ &= \frac{\mu^{D-4}}{(4\pi)^{D/2}} \int d^Dx \left\{ \left(\frac{b_1}{D-4} + \frac{b_2}{(D-4)^2} + \cdots \right) \bar{G}_4 \right. \\ &\quad + \left(b_1 + \frac{b_2}{D-4} + \cdots \right) \left(2\phi \bar{\Delta}_4 \phi + \bar{E}_4 \phi + \frac{1}{18} \bar{R}^2 \right) \\ &\quad + \frac{1}{2} \left[(D-4) b_1 + \cdots \right] \left(2\phi^2 \bar{\Delta}_4 \phi + \bar{E}_4 \phi^2 + 6\phi \bar{\nabla}^4 \phi + \cdots \right) \\ &\quad \left. + \frac{1}{6} \left[(D-4)^2 b_1 + \cdots \right] \left(2\phi^3 \bar{\Delta}_4 \phi + \cdots \right) + \cdots \right\}, \end{aligned} \tag{5.19}$$

where Δ_4 and E_4 are defined by (2.17) and (2.13), respectively, even in D dimensions. The Riegert action appears in the second line on the right-hand side. The ones in the third and fourth lines are its higher-order versions. See (A.6) in Appendix A for more detailed information on this expression.

The action of massless fermions is conformally invariant in any D dimension (see Appendix B). That is, there is no Wess-Zumino action associated with fermions. The ϕ-dependence of the fermion action can be removed by appropriately redefining the field. Since dimensional regularization does not depend on the choice of measure, it is convenient to use fermion fields defined such that the ϕ-dependence disappears when calculating the effective action.

5.2.3 Gauge Fixing and Propagators

From the lowest of the expansion formula of the Weyl action (5.18), the kinetic term of the traceless tensor field is given by

$$\int d^D x \left\{ \frac{D-3}{D-2} \left(h_{0\mu\nu} \partial^4 h_{0\mu\nu} + 2\chi_{0\mu} \partial^2 \chi_{0\mu} \right) - \frac{D-3}{D-1} \chi_{0\mu} \partial_\mu \partial_\nu \chi_{0\nu} \right\}, \quad (5.20)$$

where $\chi_{0\mu} = \partial_\nu h_{0\mu\nu}$ and $\partial^2 = \partial_\mu \partial_\mu$ is d'Alembertian on the flat background.

Quantization of the gauge field and the traceless tensor field is performed by applying the familiar BRST gauge-fixing method, and only the latter field case is presented here. According to the Kugo-Ojima procedure [KO79], the gauge-fixing term and the associated ghost action are given by [FT82a, Ham02, HM16]

$$S_{\text{gh+gf}} = \int d^D x \, \delta_{\text{B}} \left\{ -i \, \tilde{c}_{0\mu} N_{\mu\nu} \left(\chi_{0\nu} - \frac{\zeta_0}{2} B_{0\nu} \right) \right\},$$

where $\tilde{c}_{0\mu}$ is an antighost field, $B_{0\mu}$ is an auxiliary field, and ζ_0 is a gauge-fixing parameter. $N_{\mu\nu}$ is a symmetric second-order differential operator, given by

$$N_{\mu\nu} = \frac{2(D-3)}{D-2} \left(-2\partial^2 \delta_{\mu\nu} + \frac{D-2}{D-1} \partial_\mu \partial_\nu \right),$$

so that only the first ∂^4 term of the kinetic term (5.20) remains when the Feynman gauge defined later is adopted. The BRST transformation δ_{B} is defined by introducing a ghost field $c_{0\mu}$ for diffeomorphism as

$$\begin{aligned}
\delta_{\text{B}} h_{0\mu\nu} &= \partial_\mu c_{0\nu} + \partial_\nu c_{0\mu} - \frac{2}{D} \delta_{\mu\nu} \partial_\lambda c_{0\lambda} + t_0 c_{0\lambda} \partial_\lambda h_{0\mu\nu} \\
&\quad + \frac{t_0}{2} h_{0\mu\lambda} \left(\partial_\nu c_{0\lambda} - \partial_\lambda c_{0\nu} \right) + \frac{t_0}{2} h_{0\nu\lambda} \left(\partial_\mu c_{0\lambda} - \partial_\lambda c_{0\mu} \right) + \cdots, \\
\delta_{\text{B}} \phi &= t_0 c_{0\lambda} \partial_\lambda \phi + \frac{t_0}{D} \partial_\lambda c_{0\lambda}, \\
\delta_{\text{B}} c_{0\mu} &= t_0 c_{0\lambda} \partial_\lambda c_{0\mu}, \qquad \delta_{\text{B}} \tilde{c}_{0\mu} = i B_{0\mu}, \\
\delta_{\text{B}} B_{0\mu} &= 0,
\end{aligned}$$

where the ghost field is normalized so that the lowest order of $\delta_{\text{B}} h_{0\mu\nu}$ does not depend on t_0.

Applying δ_{B} yields the gauge-fixing term and the ghost action,

$$S_{\text{gh+gf}} = \int d^D x \left\{ B_{0\mu} N_{\mu\nu} \chi_{0\nu} - \frac{\zeta_0}{2} B_{0\mu} N_{\mu\nu} B_{0\nu} + i \tilde{c}_{0\mu} N_{\mu\nu} \partial_\lambda (\delta_{\text{B}} h_{0\nu\lambda}) \right\}.$$

Moreover, if the auxiliary field $B_{0\mu}$ is integrated out, the gauge-fixing term becomes

$$S_{\text{gf}} = \int d^D x \left\{ \frac{1}{2\zeta_0} \chi_{0\mu} N_{\mu\nu} \chi_{0\nu} \right\}. \quad (5.21)$$

Here note that there is no interaction with the conformal-factor field in the gauge-fixing term and the ghost action.

Rewrite the bare quantities with the renormalized ones to derive a propagator, interactions, and counterterms. At that time, the renormalization factor of the gauge-fixing parameter is defined as $\zeta_0 = Z_h \zeta$. Since (5.21) is then expressed only with the renormalized quantities, the counterterm becomes a gauge-invariant form. It is also necessary to introduce a renormalization factor to the ghost fields.

Expressing the equation of motion of the traceless tensor field involving the gauge-fixing term (5.21) as $K^{(\zeta)}_{\mu\nu,\lambda\sigma}(k) h_{\lambda\sigma}(k) = 0$ in momentum space, we obtain

$$
\begin{aligned}
K^{(\zeta)}_{\mu\nu,\lambda\sigma}(k) = \frac{2(D-3)}{D-2} \Biggl\{ & I^{\mathrm{H}}_{\mu\nu,\lambda\sigma} k^4 + \frac{1-\zeta}{\zeta} \biggl[\frac{1}{2} k^2 (\delta_{\mu\lambda} k_\nu k_\sigma + \delta_{\nu\lambda} k_\mu k_\sigma \\
& + \delta_{\mu\sigma} k_\nu k_\lambda + \delta_{\nu\sigma} k_\mu k_\lambda) - \frac{1}{D-1} k^2 (\delta_{\mu\nu} k_\lambda k_\sigma + \delta_{\lambda\sigma} k_\mu k_\nu) \\
& + \frac{1}{D(D-1)} \delta_{\mu\nu} \delta_{\lambda\sigma} k^4 - \frac{D-2}{D-1} k_\mu k_\nu k_\lambda k_\sigma \biggr] \Biggr\},
\end{aligned}
$$

where

$$
I^{\mathrm{H}}_{\mu\nu,\lambda\sigma} = \frac{1}{2} \left(\delta_{\mu\lambda} \delta_{\nu\sigma} + \delta_{\mu\sigma} \delta_{\nu\lambda} \right) - \frac{1}{D} \delta_{\mu\nu} \delta_{\lambda\sigma}
$$

is a projection tensor satisfying $(I^{\mathrm{H}})^2 = I^{\mathrm{H}}$. The propagator is given by its inverse defined by $K^{(\zeta)}_{\mu\nu,\lambda\sigma}(k) \langle h_{\lambda\sigma}(k) h_{\rho\kappa}(-k) \rangle = I^{\mathrm{H}}_{\mu\nu,\rho\kappa}$. If it is expressed as

$$
\langle h_{\mu\nu}(k) h_{\lambda\sigma}(-k) \rangle = \frac{D-2}{2(D-3)} \frac{1}{k^4} I^{(\zeta)}_{\mu\nu,\lambda\sigma}(k), \tag{5.22}
$$

then the last tensor is given by

$$
\begin{aligned}
I^{(\zeta)}_{\mu\nu,\lambda\sigma}(k) = I^{\mathrm{H}}_{\mu\nu,\lambda\sigma} + (\zeta - 1) \biggl[& \frac{1}{2} \left(\delta_{\mu\lambda} \frac{k_\nu k_\sigma}{k^2} + \delta_{\nu\sigma} \frac{k_\mu k_\lambda}{k^2} + \delta_{\mu\sigma} \frac{k_\nu k_\lambda}{k^2} \right. \\
& \left. + \delta_{\nu\lambda} \frac{k_\mu k_\sigma}{k^2} \right) - \frac{1}{D-1} \left(\delta_{\mu\nu} \frac{k_\lambda k_\sigma}{k^2} + \delta_{\lambda\sigma} \frac{k_\mu k_\nu}{k^2} \right) \\
& + \frac{1}{D(D-1)} \delta_{\mu\nu} \delta_{\lambda\sigma} - \frac{D-2}{D-1} \frac{k_\mu k_\nu k_\lambda k_\sigma}{k^4} \biggr],
\end{aligned} \tag{5.23}
$$

reflecting traceless and symmetric properties of the field.

The gauge choice of $\zeta = 1$ where the tensor (5.23) is only the I^{H} term is called the Feynman gauge. On the other hand, from

$$
k_\mu I^{(\zeta)}_{\mu\nu,\lambda\sigma}(k) = \zeta \left(\frac{1}{2} k_\lambda \delta_{\nu\sigma} + \frac{1}{2} k_\sigma \delta_{\nu\lambda} - \frac{1}{D} k_\nu \delta_{\lambda\sigma} \right),
$$

the choice of $\zeta = 0$ satisfying the transverse condition is called the Landau gauge.

From the Riegert action that arises in the second line on the right-hand side of the expansion formula (5.19), we find that the kinetic term of the conformal-factor field is given by

$$\frac{\mu^{D-4}}{(4\pi)^{D/2}} 2b \int d^D x \phi \, \partial^4 \phi,$$

and thus the propagator is expressed in momentum space as

$$\langle \phi(k) \phi(-k) \rangle = \mu^{4-D} \frac{(4\pi)^{D/2}}{4b} \frac{1}{k^4}. \tag{5.24}$$

This expression indicates that loop corrections due to the conformal-factor field appear in the order of $1/b$.

This implies that as b increases, the conformal factor becomes classical. The $1/b$ expansion corresponds to the so-called large N expansion in which the number of matter fields is large.

5.3 Finiteness of Energy-Momentum Tensors

Now, we describe more precisely what was stated in Chap. 4 based on the path integral method using dimensional regularization. Since this regularization method exactly preserves diffeomorphism invariance, it is suitable for investigating general properties of gravity.

First, we consider renormalizable quantum field theory in curved spacetime. The renormalizable fields other than the gravitational field are collectively denoted as f, and their diffeomorphism-invariant action is denoted as S_M. Since the gravitational field is assumed to be classical, the variation by it acts from outside the partition function $Z_{CS} = \int [df] e^{-S_\mathrm{M}}$ and is given by

$$\frac{\delta}{\delta g_{\mu\nu}} Z_{CS} = \int [df] \frac{\delta}{\delta g_{\mu\nu}} e^{-S_\mathrm{M}} = -\frac{1}{2} \sqrt{g} \, \langle T_\mathrm{M}^{\mu\nu} \rangle, \tag{5.25}$$

where the energy-momentum tensor is defined by $T_\mathrm{M}^{\mu\nu} = (2/\sqrt{g}) \times \delta S_\mathrm{M} / \delta g_{\mu\nu}$. As emphasized at the beginning of this chapter, dimensional regularization does not depend on the choice of the path integral measure, thus the variation operation passes through it. This is why the measure $[df]$ does not have the subscript g unlike the description $[df]_g$ used in the previous chapters. Information on the conformal anomaly, which is considered to be the contribution from the measure, is contained between 4 and D dimensions.

The following can be seen from the formula (5.25) [ACD77, BC80, Hat82a, Hat82b]:

1. Since the theory is renormalizable, the partition function is finite. The right-hand side is also obviously finite, and therefore this formula expresses that the energy-momentum tensor is a finite quantity, namely, a normal product.
2. The action is originally defined in terms of bare quantities. Therefore, the energy-momentum tensor obtained by varying the action is also a bare quantity, namely, a renormalization group invariant. Since the energy-momentum tensor is finite, it can be eventually rewritten in terms of renormalized quantities.

Next, let us see how this expression is changed in renormalizable quantum theory of gravity. Since we perform the path integral over the gravitational field, it is expressed as the Schwinger-Dyson equation [Ham14b]

$$\int [dg] \frac{\delta}{\delta g_{\mu\nu}} \left(\int [df]\, e^{-S_{\rm QG}} \right) = -\frac{1}{2} \langle \sqrt{g}\, T^{\mu\nu} \rangle = 0,$$

where $T^{\mu\nu} = (2/\sqrt{g}) \times \delta S_{\rm QG}/\delta g_{\mu\nu}$ is the energy-momentum tensor of the whole system, including the gravitational part. From this equation, we can see that the energy-momentum tensor, which is finite in curved spacetime, vanishes identically in quantum gravity. Since it is a normal product and a renormalization group invariant, the fact that it is zero means that the whole Hamiltonian vanishes, thus there is no zero-point energy.

5.4 Main Results of Loop Calculations

First, write down the procedure for obtaining the effective action briefly. Strict discussions on the effective action using the Legendre transformation refer to field theory textbooks [IZ80, Col84, Mut87, PS95]. One of the advantages of performing loop calculations in Euclidean space is that troublesome imaginary units do not appear, and thus it is only necessary to pay attention to signs and coefficients, and integrations can be performed immediately. The partition function is then given by the path integral $Z = \int e^{-S}$, and its effective action $\Gamma = -\log Z$ is simply expressed up to the tree part as follows.

$$\Gamma\,|_{\rm Loop} = -e^{-S_{\rm int}}\Big|_{\rm 1PI} = -\sum_{n=0}^{\infty} \frac{1}{n!} (-S_{\rm int})^n \Big|_{\rm 1PI},$$

where $S_{\rm int}$ denotes interaction terms, including counterterms. The 1PI restriction which corresponds to taking the logarithm of the partition function is that for all possible one-particle irreducible diagrams, Wick contractions are taken, i.e., connected by propagators. The 1PI connections are represented by Feynman diagrams.

When performing loop calculations, it is necessary to pay attention not only to ultraviolet divergences but also to infrared divergences, because this is a fourth-derivative theory. In order to regularize infrared divergences, we introduce an infinitesimal mass z ($\ll 1$) into the gravitational field, that is, calculate loop corrections by replacing the denominator of the propagatiors of $h_{\mu\nu}$ and ϕ, (5.22) and (5.24), with

$$\frac{1}{k^4} \to \frac{1}{k_z^4} = \frac{1}{(k^2+z^2)^2}.$$

Infrared divergences then arise in the form of $\log z$. Since this mass term is not diffeomorphism invariant, infrared divergences should eventually cancel out and disappear.

Here it should be pointed out that the Einstein-Hilbert term and the cosmological term cannot be regarded as mass terms of the fourth-derivative gravitational field. The ϕ-dependence in the fourth-derivative gravitational actions appears as polynomials in the expansion by the coupling constant t, whereas the lower-derivative gravitational actions have an exponential factor of ϕ even at the zeroth order of t. Since the lower-derivative actions become diffeomorphism invariant only when this conformal factor is included, mass terms defined by a quardratic term of the field is not gauge-invariant. The renormalization of composite fields with such conformal factors will be discussed later.

As already shown by the renormalization group equations, the beta functions (5.10) and (5.11) for the coupling constants α_e and α_t are given by $\bar{\beta}_e = \mu d(\log Z_3)/d\mu$ and $\bar{\beta}_t = \mu d(\log Z_t^{-2})/d\mu$ using the renormalization factors Z_3 and Z_t, respectively. Here, these renormalization factors are Laurent-expanded as

$$\log Z_3 = \sum_{n=1}^{\infty} \frac{f_n}{(D-4)^n}, \qquad \log Z_t^{-2} = \sum_{n=1}^{\infty} \frac{g_n}{(D-4)^n}. \tag{5.26}$$

The relationship between f_n and the residue x_n in (5.16) is then given by

$$x_1 = f_1, \qquad x_2 = f_2 + \frac{1}{2} f_1^2,$$

and so on.

Since the beta function $\bar{\beta}_e$ is finite, poles must cancel each other out when the above expression is Laurent-expanded. From this condition, we can see that the residue f_n satisfies the following recurrence relation:

$$\left(\alpha_e \frac{\partial}{\partial \alpha_e} + \alpha_t \frac{\partial}{\partial \alpha_t} + \bar{\beta}_b \frac{\partial}{\partial b} \right) f_{n+1} + \left(\bar{\beta}_e \alpha_e \frac{\partial}{\partial \alpha_e} + \bar{\beta}_t \alpha_t \frac{\partial}{\partial \alpha_t} \right) f_n = 0. \tag{5.27}$$

In addition, the beta function of QED can be expressed using the simple pole as

$$\bar{\beta}_e = \alpha_e \frac{\partial f_1}{\partial \alpha_e} + \alpha_t \frac{\partial f_1}{\partial \alpha_t} + \bar{\beta}_b \frac{\partial f_1}{\partial b}. \tag{5.28}$$

A similar recurrence relation holds for the residue g_n, and the beta function of quantum gravity is given by

$$\bar{\beta}_t = \alpha_e \frac{\partial g_1}{\partial \alpha_e} + \alpha_t \frac{\partial g_1}{\partial \alpha_t} + \bar{\beta}_b \frac{\partial g_1}{\partial b}. \tag{5.29}$$

In the following, the calculation results for loop corrections will be shown. In Feynman diagrams, the propagators of the conformal-factor field ϕ, the traceless tensor field $h_{\mu\nu}$, gauge field A_μ, and fermion ψ are represented by solid lines, spiral lines, wavy lines, and dotted lines with arrows, respectively. The ϵ^n factor depicted at some vertices in the diagrams denotes that corresponding interactions have a factor $(D-4)^n$. Loop calculations involving the gravitational field are so complicated that only the main results are written here.

From loop calculations of QED in curved spacetime, the α_e-dependence of the simple pole b_1 and the double pole b_2 in the counterterm proportional to $\bar{G}_4$ is calculated as [Hat82b, Ham14a]

$$b_1 = b - \frac{n_F^2}{6}\left(\frac{\alpha_e}{4\pi}\right)^2, \qquad b_2 = \frac{2n_F^3}{9}\left(\frac{\alpha_e}{4\pi}\right)^3, \tag{5.30}$$

respectively. This result will be used below. Substituting this into (5.15) yields

$$\bar{\beta}_b = -b + \frac{n_F^2}{2}\left(\frac{\alpha_e}{4\pi}\right)^2. \tag{5.31}$$

This beta function has been calculated up to $o(\alpha_e^3)$.

5.4.1 Non-renormalization Theorem $Z_\phi = 1$

Since the conformal-factor field ϕ is treated strictly without introducing a coupling constant specific to this field, $Z_\phi = 1$ (5.8) holds at all orders. Here, we will give some examples and demonstrate that the conformal-factor field does not actually undergo renormalization [Ham02].

First, we show that this holds for the first order of α_t. Loop corrections of $o(\alpha_t)$ for two-point functions of the conformal-factor field are represented by two Feynman diagrams in Fig. 5.1. If we evaluate the diagram (a) by adopting the Feynman gauge ($\zeta = 1$), we obtain

$$\int \frac{d^D k}{(2\pi)^D} \phi(k)\phi(-k) \left\{ -\frac{D-2}{2(D-3)} \frac{b}{6} \frac{t^2}{(4\pi)^{D/2}} \int \frac{d^D l}{(2\pi)^D} \frac{1}{l_z^4 (l+k)_z^4} \right.$$
$$\times \left[6(l^2 k^6 + l^6 k^2) + 24 l^4 k^4 - 16 (l \cdot k)(l^2 k^4 + l^4 k^2) \right.$$

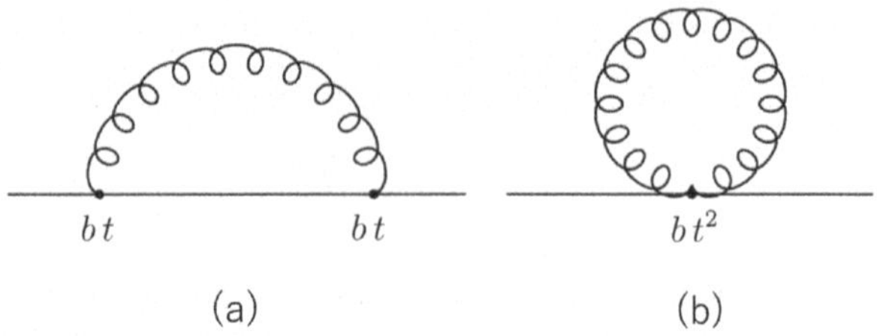

Fig. 5.1 Loop corrections of $o(\alpha_t)$ to the conformal-factor field. The three- and four-point interactions are obtained by expanding the $\phi\bar{\Delta}_4\phi$ term in Laurent expansion (5.19) with $h_{\mu\nu}$

$$
-20(l\cdot k)^2 l^2 k^2 - 2(l\cdot k)^2(l^4+k^4) + 8(l\cdot k)^3(l^2+k^2) + 8(l\cdot k)^4
+\frac{4-D}{3D}\Big(-36 l^4 k^4 + 24(l\cdot k)(l^2k^4+l^4k^2) + 40(l\cdot k)^2 l^2 k^2
-4(l\cdot k)^2(l^4+k^4) - 16(l\cdot k)^3(l^2+k^2) - 16(l\cdot k)^4\Big)\Big]\Big\}.
$$

Integrating the momentum l under the condition $z \ll 1$, the inside of the braces { } yields

$$
\frac{\mu^{D-4}}{(4\pi)^{D/2}} 2bk^4\left[3\frac{\alpha_t}{4\pi}\left(\frac{2}{D-4}+\log\frac{z^2}{\mu^2}-\frac{7}{6}\right)\right].
$$

At this time, nonlocal finite terms of $\log(k^2/\mu^2)$ cancel out and disappear. The contribution from the tadpole diagram (b) can be calculated relatively easily and yields[4]

$$
\frac{\mu^{D-4}}{(4\pi)^{D/2}} 2bk^4\left[-3\frac{\alpha_t}{4\pi}\left(\frac{2}{D-4}+\log\frac{z^2}{\mu^2}-\frac{7}{12}\right)\right].
$$

Adding these two contributions from Feynman diagrams in Fig. 5.1 shows that both the ultraviolet and infrared divergences cancel out. In this way, we can demonstrate that $Z_\phi = 1$ (5.8) holds at $o(\alpha_t)$.

If we calculate the two-point function of ϕ in which the QED fields propagate inside, we can demonstrate that it is finite up to $o(\alpha_e^3)$. First, the two-point function of $o(\alpha_e)$ can be calculated using the Wess-Zumino interactions of type $\phi F_{\mu\nu}^2$ that appears in the Laurent expansion (5.17) of the gauge field action. Since it is given by 1-loop Feynman diagrams containing the ϵ ($\propto D-4$) factor in one of the vertices, it becomes obviously finite.

Next, consider loop corrections of $o(\alpha_e^2)$. Feynman diagrams are give in Fig. 5.2, in which ultraviolet divergence actually occurs in (a) to (d), and the sum of them

---

[4] It is generally known that in dimensional regularization, the contribution from a tadpole diagram in which a massless particle with a lower-derivative kinetic term propagates inside disappears. However, note that here the kinetic term is a fourth-derivative one, thus its contribution cannot be ignored.

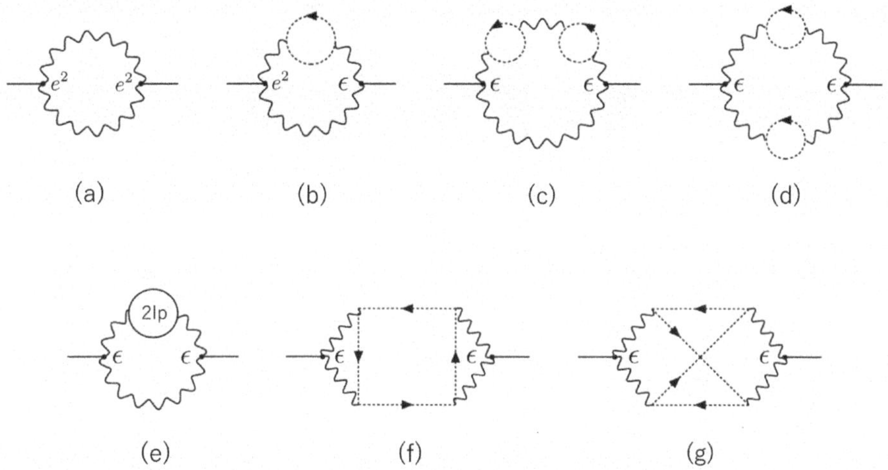

Fig. 5.2 Loop corrections of $o(\alpha_e^2)$ to the conformal-factor field, in which the three-point vertex with e^2 is derived from the residue $x_1 = (8n_F/3)(\alpha_e/4\pi)$ and the part written as "2lp" inside the circle denotes the ordinary two-loop photon self-energy diagram in QED. Diagrams that include counterterms as part in order to eliminate divergences in subdiagrams are not depicted here

results in a finite amount. The diagrams (e) to (g) are each finite due to the ϵ factor in the vertices.

Similarly, we can show $Z_\phi = 1$ in the case of $o(\alpha_e^3)$. In this case, it should be noted that since the residue b_2 has a contribution of $o(\alpha_e^3)$, a simple pole counterterm of this order, which cancels the entire ultraviolet divergence of the two-point function of ϕ, appears.

5.4.2 Beta Function Results

Here we present results for the beta functions including corrections due to quantum gravity.

QED beta function The contribution of $o(\alpha_t)$ to the two-point function of the gauge field is given by Feynman diagrams in Fig. 5.3. The self-energy diagram (a) and the tadpole diagram (b) each give rise to ultraviolet divergences, but it can be shown that they cancel out and the quantum correction becomes finite. Therefore, there is no contribution of $o(\alpha_t)$ to the beta function β_e.

Next, calculate the quantum correction proportional to $1/b$ resulting from Feynman diagrams in which the conformal-factor field propagates inside. Figures 5.4 and 5.5 provides the simple-pole residue f_1 of $o(\alpha_e^2/b)$ and the double-pole residue f_2 of $o(\alpha_e^3/b)$, respectively.

The results for the residues f_1 and f_2 in the renormalization factor Z_3 (5.26) obtained by adding loop corrections from the conformal-factor field of $o(1/b)$ to the

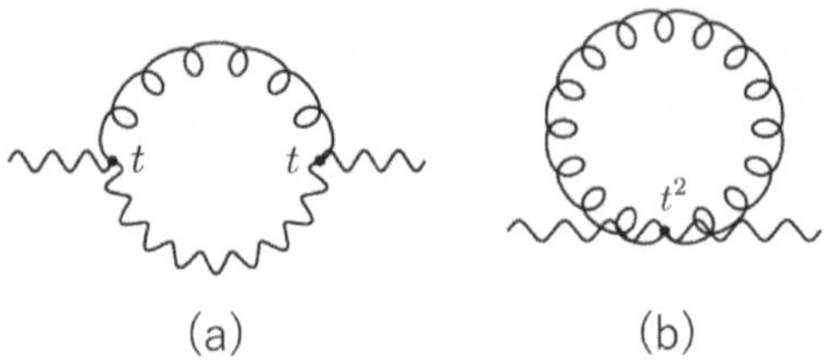

Fig. 5.3 Loop corrections of $o(\alpha_t)$ to the gauge field

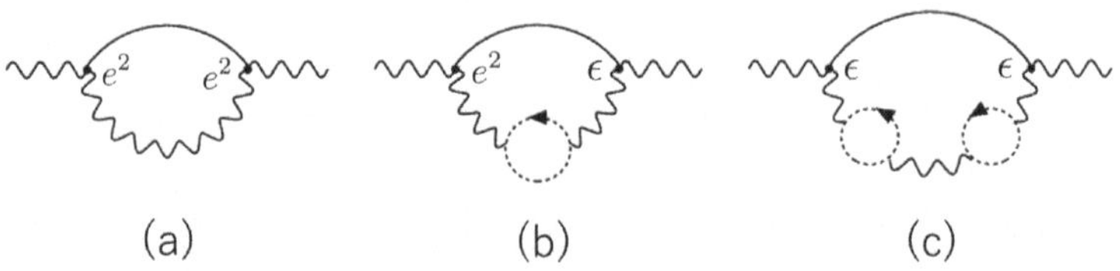

Fig. 5.4 Feynman diagrams providing the simple-pole residue f_1 of $o(\alpha_e^2/b)$

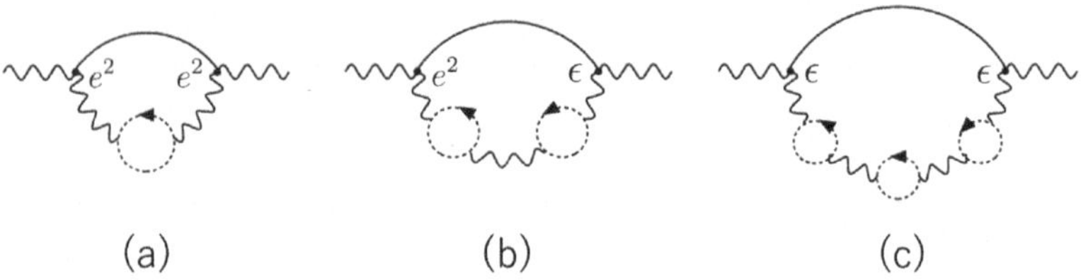

Fig. 5.5 Feynman diagrams providing the double-pole residue f_2 of $o(\alpha_e^3/b)$

ordinary QED corrections are given by [Ham02]

$$f_1 = \frac{8n_F}{3}\frac{\alpha_e}{4\pi} + \left(4n_F - \frac{16n_F^2}{27b}\right)\left(\frac{\alpha_e}{4\pi}\right)^2,$$

$$f_2 = -\frac{32n_F^2}{9}\left(\frac{\alpha_e}{4\pi}\right)^2 - \left(\frac{128n_F^2}{9} - \frac{160n_F^3}{81b}\right)\left(\frac{\alpha_e}{4\pi}\right)^3.$$

Here, it has already been mentioned that there is no 1-loop correction of $o(\alpha_t)$ to f_1, and it can be confirmed that the residues f_1 and f_2 satisfy the recursion relation (5.27) by paying attention to $\bar{\beta}_b = -b + o(\alpha_e^2)$ (5.31).

The beta function (5.28) of QED is determined from the simple-pole residue f_1 as

$$\bar{\beta}_e = \frac{8n_F}{3}\frac{\alpha_e}{4\pi} + \left(8n_F - \frac{16n_F^2}{9}\frac{1}{b}\right)\left(\frac{\alpha_e}{4\pi}\right)^2.$$

At the end, replacing b with the constant b_c yields the final result. Since $b_c > 0$, it can be seen that the quantum gravity correction of $o(1/b_c)$ gives a negative contribution.

Quantum gravity beta function To determine the renormalization factor of the coupling constant t, it is necessary to calculate two- and three-point functions of the traceless tensor field as in Fig. 5.6. In general, renormalization calculations involving the traceless tensor field are hard, and as in the case of non-Abelian gauge field theories, the background field method [Abb81] is often applied.

Letting the background metric be $\hat{g}_{\mu\nu} = (e^{t_0 \hat{h}_0})_{\mu\nu}$, instead of the flat metric, and setting a renormalization factor $Z_{\hat{h}}$ as $\hat{h}_{0\mu\nu} = Z_{\hat{h}}^{1/2} \hat{h}_{\mu\nu}$ for the background field, the background field satisfies gauge invariance, that is, $Z_t Z_{\hat{h}}^{1/2} = 1$ holds as an identity (corresponding to $Z_e Z_3^{1/2} = 1$ in QED). Thus, an advantage of the background field method is that since Z_t can be determined by calculating $Z_{\hat{h}}$, we only calculate the two-point function of $\hat{h}_{\mu\nu}$ to get the beta function.

The one-loop result (2.25) has been calculated using this method. From this and [DH80], the beta function (5.29) is given by

$$\bar{\beta}_t = -\left(\frac{n_F}{20} + \frac{20}{3} \right) \frac{\alpha_t}{4\pi} - \frac{7n_F}{36} \frac{\alpha_e \alpha_t}{(4\pi)^2}.$$

Since the value is negative, it can be seen that the traceless tensor field exhibits the asymptotic freedom, that is, the perturbation expansion around a conformally flat spacetime is justified.

5.5 Cosmological Term and Einstein-Hilbert Term

Since we strictly handle the conformal-factor field ϕ, the cosmological term is expressed by its exponential as

$$S_\Lambda = \Lambda_0 \int d^D x \sqrt{g} = \Lambda_0 \int d^D x e^{D\phi}.$$

The Einstein-Hilbert action is also expressed in terms with the exponential as

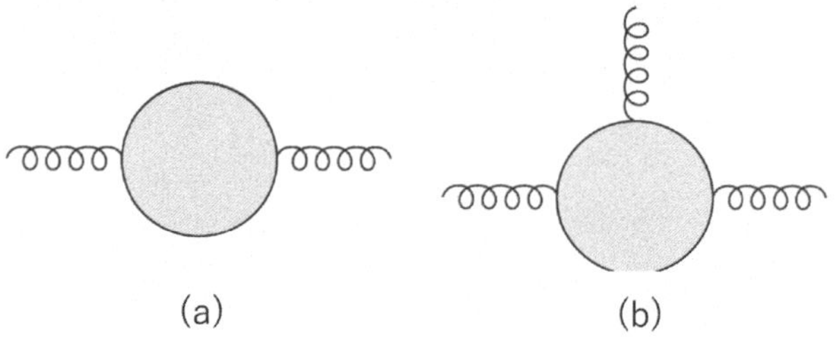

Fig. 5.6 Loop corrections for 2- and 3-point functions of the traceless tensor field

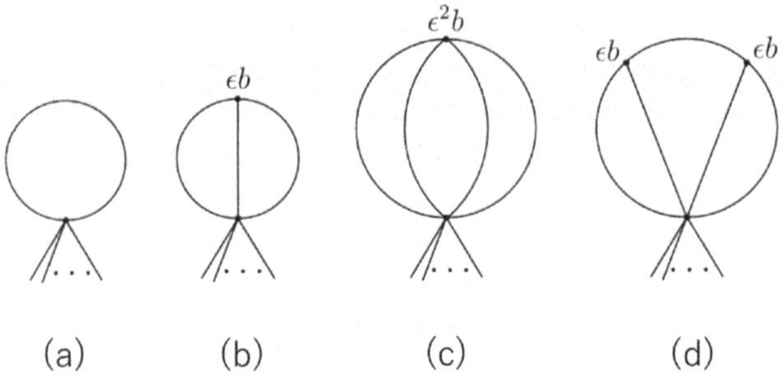

Fig. 5.7 Loop corrections up to $o(1/b^3)$ for the cosmological term

$$\begin{aligned} S_{\rm EH} &= -\frac{M_0^2}{2} \int d^D x \sqrt{g} R \\ &= -\frac{M_0^2}{2} \int d^D x e^{(D-2)\phi} \left\{ \bar{R} - (D-1)\bar{\nabla}^2 \phi \right\}. \end{aligned}$$

Since the conformal-factor field does not undergo renormalization, namely, $Z_\phi = 1$ (5.8), the renormalization calculation can be performed simply by replacing the bare cosmological constant and Planck mass squared as in

$$\begin{aligned} \Lambda_0 &= \mu^{D-4} Z_\Lambda \left(\Lambda + L_M M^4 \right), \\ M_0^2 &= \mu^{D-4} Z_{\rm EH} M^2, \end{aligned}$$

respectively, where Λ is the renormalized cosmological constant with canonical dimension 4 and M is the renormalized Planck mass with canonical dimension 1. Z_Λ and $Z_{\rm EH}$ are their renormalization factors, and L_M is a term with only poles.

The anomalous dimension of the cosmological constant is defined by

$$\gamma_\Lambda = -\frac{\mu}{\Lambda} \frac{d\Lambda}{d\mu} = D - 4 + \bar{\gamma}_\Lambda + \frac{M^4}{\Lambda} \bar{\delta}_\Lambda,$$

where $\bar{\gamma}_\Lambda = \mu d(\log Z_\Lambda)/d\mu$ and

$$\bar{\delta}_\Lambda = \mu \frac{dL_M}{d\mu} - (D-4) L_M + (\bar{\gamma}_\Lambda - 2\bar{\gamma}_{\rm EH}) L_M.$$

We also define the anomalous dimension of the Planck mass squared as

$$\gamma_{\rm EH} = -\frac{\mu}{M^2} \frac{dM^2}{d\mu} = D - 4 + \bar{\gamma}_{\rm EH},$$

where $\bar{\gamma}_{\rm EH} = \mu d(\log Z_{\rm EH})/d\mu$.

Letting the simple pole of the renormalization factor Z_Λ be u_1, the anomalous dimension is expressed as

$$\bar{\gamma}_\Lambda = \bar{\beta}_b \frac{\partial u_1}{\partial b} + \alpha_t \frac{\partial u_1}{\partial \alpha_t},$$

from a renormalization group equation $\mu d\Lambda_0/d\mu = 0$. The contribution up to $o(1/b^3)$ to the residue u_1 is given by Feynman diagrams in Fig. 5.7, and calculating them yields [Ham14b][5]

$$\bar{\gamma}_\Lambda = -b\frac{\partial u_1}{\partial b} = \frac{4}{b} + \frac{8}{b^2} + \frac{20}{b^3} + \cdots . \tag{5.32}$$

Lastly, replacing b with b_c yields the anomalous dimension of the cosmological term.

The anomalous dimension can be determined exactly from the BRST conformal invariance, as shown in the next chapter. Compared to the ϕ-dependence of the physical state (6.22) corresponding to the cosmological term, that is, the Riegert charge γ_0 (6.21), it is given by

$$\gamma_\Lambda = \gamma_0 - 4 = 2b_c\left(1 - \sqrt{1 - \frac{4}{b_c}}\right) - 4$$

as a deviation from the canonical dimension 4. Expanding this expression with $1/b_c$ shows that the first three terms match (5.32).

Similarly, calculating the anomalous dimension of the Planck mass squared from Feynman diagrams in Fig. 5.8 yields [HM16]

$$\bar{\gamma}_{\rm EH} = \frac{1}{b} + \frac{1}{b^2} + \cdots .$$

This result is consistent with $\gamma_{\rm EH} = \gamma_2 - 2$ derived from the Riegert charge γ_2 (6.21) of the scalar-curvature state (6.23). In addition,

$$\bar{\delta}_\Lambda = -\frac{9}{8}\frac{(4\pi)^2}{b^2}$$

is obtained from the pole term L_M.

[5] It is only numerically confirmed that the third term has this value.

Fig. 5.8 Diagrams **a** and **b** are loop corrections up to $o(1/b^2)$ for the Einstein-Hilbert term. Diagram **c** is a loop correction of $o(1/b^2)$ for L_M

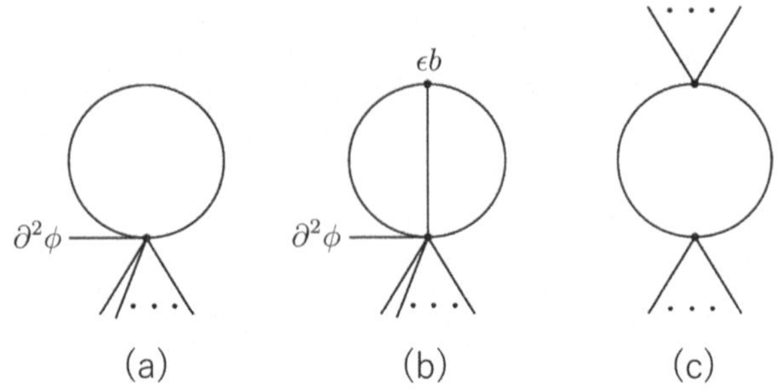

5.6 Effective Potentials and Physical Constants

Renormalizable quantum gravity has three physical mass scales, which are renormalization group invariants (2.32): one is the dynamical energy scale Λ_{QG}, and the other two are the cosmological constant $\Lambda_{\cos}$ and the Planck mass M_P.

The latter two are determined from the potential term in the effective action. Introducing the background field σ and decomposing the conformal factor field as $\phi = \sigma + \varphi$, it can be calculated as

$$\Gamma(\sigma) = \sum_n \frac{1}{n!} \int d^D x_1 \cdots d^D x_n \, \Gamma^{(n)}(x_1, \ldots, x_n)\, \sigma(x_1) \cdots \sigma(x_n)$$
$$= \sum_n \frac{1}{n!} \int \frac{d^D k_1}{(2\pi)^D} \cdots \frac{d^D k_n}{(2\pi)^D} (2\pi)^D \delta^{(D)}(k_1 + \cdots + k_n)$$
$$\times \Gamma^{(n)}(k_1, \ldots, k_n)\, \sigma(k_1) \cdots \sigma(k_n).$$

where $\Gamma^{(n)}$ is a renormalized n-point Green function obtained by summing all 1PI (1 particle irreducible) Feynman diagrams with n external legs of σ. Since the conformal-factor field does not undergo renormalization, neither does the background field σ. Therefore, $d\Gamma^{(n)}/d\mu = 0$ holds for the renormalized $\Gamma^{(n)}$.

The cosmological term is defined by the effective potential V consisting of the zero-momentum part of $\Gamma^{(n)}(k_1, \ldots, k_n)$ as

$$\Gamma(\sigma)|_V = \int d^D x \, V(\sigma) = \sum_n \frac{1}{n!} \Gamma^{(n)}(0, \ldots, 0) \int d^D x \, \sigma^n(x).$$

Diffeomorphism invariance requires $\Gamma^{(n)}(0, \ldots, 0) = \Lambda_{\cos} D^n$ so that the effective potential has the form

$$V(\sigma) = \Lambda_{\cos}\, e^{D\sigma(x)}.$$

The renormalization group equation indicates that the physical cosmological constant $\Lambda_{\cos}$ satisfies

$$\frac{d}{d\mu}\Lambda_{\cos} = 0.$$

Similarly, an effective potential containing two derivatives defines the physical Planck mass $M_{\rm P}$.

5.6.1 Derivation of Physical Cosmological Constant

The physical cosmological constant is derived by calculating the effective potential in one loop using the large-b approximation, with the background field σ as a constant [HM17]. If we decompose the conformal-factor field as $\phi = \sigma + \varphi$ and expand the action to the second order of φ, we obtain

$$S_{\rm QG}|_{\varphi^2} = \int d^D x \left\{ \frac{\mu^{D-4}}{(4\pi)^{\frac{D}{2}}} \left[2b\,\varphi\partial^4\varphi + (D-4)b\,(2\sigma+3)\varphi\partial^4\varphi \right] \right.$$
$$\left. +\mu^{D-4}\Lambda e^{D\sigma}\left(1+\frac{D^2}{2}\varphi^2\right) + (Z_\Lambda - 1)\mu^{D-4}\Lambda e^{D\sigma} \right\},$$

where the term with $D-4$ comes from the interactions in the third line of the Laurent expansion (5.19). If we redefine the field as

$$\varphi = \frac{a}{\sqrt{\mu^{D-4}}}\varphi', \quad a = \sqrt{\frac{(4\pi)^{\frac{D}{2}}}{4b[1+(D-4)(\sigma+\frac{3}{2})]}},$$

we get

$$S|_{\varphi^2} = \int d^D x \left\{ \frac{1}{2}\varphi'\partial^4\varphi' + \frac{D^2 a^2}{2}\Lambda e^{D\sigma}\varphi'^2 - \frac{1}{\bar\epsilon}\frac{2}{b}\mu^{D-4}\Lambda e^{D\sigma} \right\}.$$

The last term is the 1-loop counterterm for the cosmological constant, which eliminates the ultraviolet divergence in the $\overline{\rm MS}$ scheme with $1/\bar\epsilon = 2/(4-D) - \gamma + \log 4\pi$.

Now, let $A = D^2 a^2 \Lambda e^{D\sigma}$ and write the kinetic term operator as $\mathcal{D} = \mathcal{D}_0 + A$. Since $\mathcal{D}_0 = k^4$ in momentum space, the 1-loop effective potential is calculated in Fig. 5.9 as

$$V_{\rm loop} = -\log\left[\det\left(\mathcal{D}_0^{-1}\mathcal{D}\right)\right]^{-\frac{1}{2}} = \frac{1}{2}\int \frac{d^D k}{(2\pi)^D} \log\left(1+\frac{A}{k_z^4}\right)$$
$$= \frac{1}{2}\sum_{n=1}^{\infty}\frac{(-1)^{n-1}}{n}A^n I_n,$$

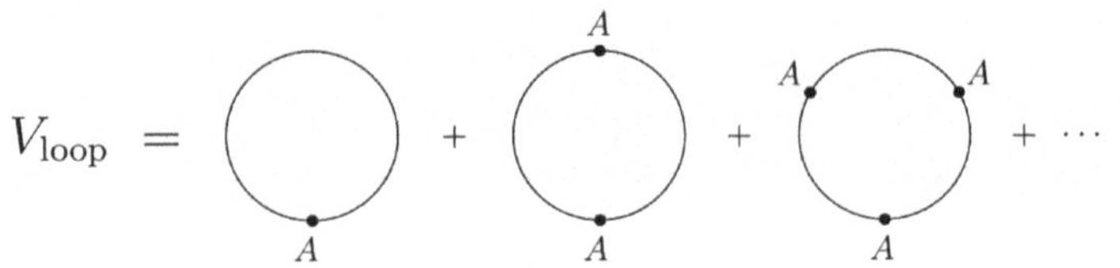

Fig. 5.9 Effective potential giving the physical cosmological constant at the one-loop level

where I_n is a tadpole-type integral defined by

$$I_n(z) = \int \frac{d^D k}{(2\pi)^D} \frac{1}{(k_z^4)^n}.$$

The integral I_1 has both ultraviolet and infrared divergences, and at $n \geq 2$ only infrared divergences appear, and they are given by

$$I_1 = \frac{1}{(4\pi)^2}\left(\frac{1}{\bar{\epsilon}} - \log z^2\right), \qquad I_{n\,(\geq 2)} = \frac{1}{(4\pi)^2}\frac{z^{2(2-2n)}}{(2n-1)(2n-2)}.$$

Substituting these integrated values yields[6]

$$\begin{aligned} V_{\rm loop} &= \frac{1}{(4\pi)^2}\left\{\frac{A}{2}\left(\frac{1}{\bar{\epsilon}} - \log z^2\right) + z^4 \sum_{n=2}^{\infty} \frac{(-1)^{n-1} A^n z^{-4n}}{2n(2n-1)(2n-2)}\right\} \\ &= \frac{1}{(4\pi)^2}\left\{\frac{A}{2}\left(\frac{1}{\bar{\epsilon}} - \log z^2\right) + \frac{1}{4}\left(z^4 - A\right)\log\left(1 + \frac{A}{z^4}\right) \right. \\ &\qquad \left. - z^2\sqrt{A}\arctan\left(\frac{\sqrt{A}}{z^2}\right) + \frac{3}{4}A\right\}. \end{aligned}$$

Paying attention to the D-dependence of A, we can see that a finite σ-dependence appears from the first term as

$$\frac{1}{(4\pi)^2}\frac{A}{2}\frac{1}{\bar{\epsilon}} = \frac{2}{b}\mu^{D-4}\Lambda e^{D\sigma}\left(\frac{1}{\bar{\epsilon}} + 2\sigma + 2 - \log 4\pi + \log \mu^2\right),$$

where in order to match the form of the counterterm, we recover the overall μ-dependence by using $\mu^{D-4}[1 - (D-4)\log\mu] = 1 + o[(D-4)^2]$. Canceling out the ultraviolet divergence with the counterterm and setting to $D = 4$, the 1-loop correction $V_{\rm loop}$ can be found as

---

[6] To calculate the sum of the series $f(x) = \sum_{n=2}^{\infty}(-1)^{n-1}x^{2n}/2n(2n-1)(2n-2)$, consider $h(x) = \partial^2\{f(x)/x\}/\partial x^2$. This can be easily found as $h(x) = [\log(1+x^2) - x^2]/2x^3$, thus the original series can be calculated as $f(x) = x\int_0^x du \int_0^u dv\, h(v)$.

$$
\frac{1}{(4\pi)^2}\left\{\frac{\bar{A}}{2}(2\sigma+2-\log 4\pi)-\frac{\bar{A}}{2}\log\frac{z^2}{\mu^2}\right.
$$

$$
\left.+\frac{1}{4}(z^4-\bar{A})\log\left(1+\frac{\bar{A}}{z^4}\right)-z^2\sqrt{\bar{A}}\arctan\left(\frac{\sqrt{\bar{A}}}{z^2}\right)+\frac{3}{4}\bar{A}\right\},
$$

where $\bar{A}=A|_{D=4}=4(4\pi)^2\Lambda e^{4\sigma}/b$.

This expression can take the limit of $z\to 0$ and obtain a finite result $V_{\text{loop}}=\bar{A}\{4\sigma+7-2\log 4\pi-\log(\bar{A}/\mu^4)\}/4(4\pi)^2$. In this way, it can be shown that the infrared divergence cancels out when we take the sum of all contributions. Adding the classical (tree) part $\Lambda e^{4\sigma}$ and putting $b=b_c$, we get the effective potential $V=\Lambda_{\cos}e^{4\sigma}$ with

$$
\Lambda_{\cos}=\Lambda\left\{1+\frac{1}{b}\left[7-2\log 4\pi-\log\left(\frac{64\pi^2}{\mu^4}\frac{\Lambda}{b}\right)\right]\right\},
$$

where note that the σ-dependence disappears except for the last $e^{4\sigma}$. The constant part $\Lambda_{\cos}$ gives the physical cosmological constant.

More generally, if we add the Einstein-Hilbert action and incorporate loop corrections by the traceless tensor field calculated in the Landau gauge, we obtain the following physical cosmological constant at the 1-loop level:

$$
\begin{aligned}
\Lambda_{\cos}=\Lambda&+(7-2\log 4\pi)\frac{\Lambda}{b_c}-\left(\frac{\Lambda}{b_c}-\frac{9\pi^2M^4}{2b_c^2}\right)\log\left(\frac{64\pi^2}{\mu^4}\frac{\Lambda}{b_c}\right)\\
&-\frac{9\pi^2}{2}\left(\frac{25}{3}-4\log 4\pi\right)\frac{M^4}{b_c^2}\\
&-6\pi\frac{M^2}{b_c}\sqrt{\frac{\Lambda}{b_c}-\frac{9\pi^2M^4}{4b_c^2}}\arccos\left(\frac{3\pi M^2}{2\sqrt{b_c\Lambda}}\right)\\
&+\frac{5}{128}\alpha_t^2M^4\left(\log\frac{\pi^2\alpha_t^2M^4}{\mu^4}-\frac{21}{5}\right),
\end{aligned}
$$

where the calculation is carried out by assuming that Λ/b_c and M^4/b_c^2 are comparable, and $\alpha_t/4\pi$ and $1/b_c$ are also so, in the large-b_c approximation.

5.7 Appendix: Hathrell's Renormalization Group Equations

In this appendix, we demonstrate how to determine the gravitational counterterms [Ham14a] using the specific renormalization group equations derived by Hathrell [Hat82a, Hat82b, Fre84]. We here describe it using QCD in curved spacetime as an example.

The action is defined by

$$S = \int d^D x \sqrt{g} \left\{ \frac{1}{g_0^2} \left[\frac{1}{4} F_{0\mu\nu}^{A} F_0^{A\mu\nu} + \frac{1}{2\xi_0} \left(\nabla^\mu A_{0\mu}^{A} \right)^2 \right] + i \bar{\psi}_0 \gamma^\mu D_\mu \psi_0 \right.$$
$$\left. -i \partial^\mu \bar{c}_0^{A} \left(\partial_\mu c_0^{A} - f^{ABC} A_{0\mu}^{B} c_0^{C} \right) + a_0 F_D + b_0 G_4 + c_0 H^2 \right\}, \tag{5.33}$$

where $F_{0\mu\nu}^{A} = \partial_\mu A_{0\nu}^{A} - \partial_\nu A_{0\nu}^{A} - f^{ABC} A_{0\mu}^{B} A_{0\nu}^{C}$ and $D_\mu = \partial_\mu + \omega_{\mu ab} \Sigma_{ab}/2 - A_{0\mu}^{A} T^{A}$. The generators of the Lie group are normalized as $\mathrm{Tr}(T^A T^B) = -T_R \delta^{AB}$ and $f^{ACD} f^{BCD} = C_G \delta^{AB}$. For later convenience, we use the convention that the gauge coupling g is factored out, so that the field strength, the fermion and ghost actions do not manifestly depend on the coupling. The renormalization factors are then defined by $g_0 = \mu^{2-D/2} Z_g g$, $A_{0\mu}^{A} = \mu^{2-D/2} Z_A^{1/2} A_\mu^{A}$, $\psi_0 = Z_2^{1/2} \psi$, and $\xi_0 = Z_A Z_g^{-2} \xi$.

For the moment, we consider three gravitational counterterms with the bare couplings, a_0, b_0, and c_0. The end of this appendix is to see that the last two are related to each other through the Hathrell's renormalization group equations at all orders, and thus c_0 is integrated with b_0 and these two terms are combined into the form G_D (5.3).

Here note the following points. The setting in this appendix is convenient when considering quantum field theory in curved spacetime, but in the quantum gravity, the first couping constant is set as $a_0 = 1/t_0^2$, and b_0 is used only the pure-pole term L_b as (5.12), taking into account the absence of tree-level kinetic terms in the G_D term. Furthermore, the constant part existing in the residue b_1 is denoted as b as in (5.13), unlike the definition of b_0 below.

5.7.1 *Normal Products and Renormalization Group Equations*

Using $\alpha_g = g^2/4\pi$, the QCD beta function is defined by

$$\beta_g = \frac{\mu}{\alpha_g} \frac{d\alpha_g}{d\mu} = D - 4 + \bar{\beta}_g,$$

and anomalous dimensions of the fields are $\gamma_A = \mu d(\log Z_A)/d\mu$ and $\gamma_2 = \mu d(\log Z_2)/d\mu$.

The bare gravitational couplings, a_0, b_0, and c_0, are defined by

$$a_0 = \mu^{D-4}(a + L_a), \qquad L_a = \sum_{n=1}^{\infty} \frac{a_n}{(D-4)^n},$$

and similar equations for b_0 and c_0. The beta functions of the coupling constants, a, b, and c, are defined by

$$\beta_a = \mu \frac{da}{d\mu} = -(D-4)a + \bar{\beta}_a,$$

and so on. Their ordinary renormalization group equations, such as $\mu da_0/d\mu = 0$, provide

$$\bar{\beta}_a = -\frac{\partial(\alpha_g a_1)}{\partial \alpha_g}, \qquad \frac{\partial(\alpha_g a_{n+1})}{\partial \alpha_g} + \bar{\beta}_g \alpha_g \frac{\partial a_n}{\partial \alpha_g} = 0, \tag{5.34}$$

and similar equations for b_n and c_n.

In the following, we essentially use some normal products denoted by the symbol [ ]. The equation-of-motion operators defined by $E_{0A} = (A^{\mathrm{A}}_{0\mu}/\sqrt{g}) \times \delta S/\delta A^{\mathrm{A}}_{0\mu}$ and $E_{0\psi} = (\bar{\psi}_0 \delta S/\delta \bar{\psi}_0 + \psi_0 \delta S/\delta \psi_0)/\sqrt{g} \equiv \delta S/\delta \Upsilon$ are the simplest normal products. This is because $\langle E_{0A,\psi} \prod A \prod \psi \rangle$ becomes finite for any renormalized correlation function among the fundamental fields without ghost fields, denoted by $\langle \prod A \prod \psi \rangle$, which can be easily shown by carrying out the partial integral for each field variable in the path integral. Thus, we can write them as $E_{0A} = [E_A]$ and $E_{0\psi} = [E_\psi]$.

The normal product for the square of the gauge field strength generally has the following form $[F^{\mathrm{A}}_{\mu\nu} F^{\mathrm{A}\mu\nu}]/g^2 = (1 + \sum \text{poles}) F^{\mathrm{A}}_{0\mu\nu} F_0^{\mathrm{A}\mu\nu}/g_0^2 + \sum \text{poles} \times$ (other fields), which is known to be given by

$$\begin{aligned}
\frac{1}{4g^2}[F^{\mathrm{A}}_{\mu\nu} F^{\mathrm{A}\mu\nu}] &= \frac{D-4}{\beta_g} \frac{1}{4g_0^2} F^{\mathrm{A}}_{0\mu\nu} F_0^{\mathrm{A}\mu\nu} - \frac{\bar{\gamma}_A}{2\beta_g} E_{0A} - \frac{\bar{\gamma}_2}{2\beta_g} E_{0\psi} \\
&\quad + \frac{D-4}{\beta_g} \mu^{D-4} \left\{ \left(L_a + \frac{\bar{\beta}_a}{D-4} \right) F_D + \left(L_b + \frac{\bar{\beta}_b}{D-4} \right) G_4 \right. \\
&\quad \left. + \left(L_c + \frac{\bar{\beta}_c}{D-4} \right) H^2 - \frac{4(\sigma + L_\sigma)}{D-4} \nabla^2 H \right\},
\end{aligned} \tag{5.35}$$

where the anomalous dimensions with the bar are defined by $\bar{\gamma}_A = \gamma_A + \{\gamma_A - (D-4)\}\xi \partial(\log Z_A)/\partial\xi$ and $\bar{\gamma}_2 = \gamma_2 + \{\gamma_A - (D-4)\}\xi \partial(\log Z_2)/\partial\xi$. Note that $1/\beta_g$ has poles, thus

$$\frac{D-4}{\beta_g} = 1 + \sum_{n=1}^{\infty} \frac{(-\bar{\beta}_g)^n}{(D-4)^n}. \tag{5.36}$$

In order to see that (5.35) is a normal product, we consider a finite correlation function given by $(\alpha_{\mathrm{g}}\partial/\partial\alpha_{\mathrm{g}} - \xi\partial/\partial\xi)\langle\prod A \prod \psi\rangle$, according to [Hat82b, Fre84].[7] Indeed, this correlation function can be expressed using (5.35) in the form $\langle\int [F^{\mathrm{A}}_{\mu\nu} F^{\mathrm{A}\mu\nu}] \prod A \prod \psi\rangle / 4\mathrm{g}^2$, up to the terms of gauge-fixing origin that becomes BRST trivial. In this way, we can determine the form of the normal product (5.35), apart from the last divergence term.

The energy-momentum tensor is defined by $T^{\mu\nu} = (2/\sqrt{g})\delta S/\delta g_{\mu\nu}$ and its trace is denoted by $\Theta = T^{\mu}_{\ \mu} = \delta S/\delta\Omega$. This is also the normal product because $\langle T^{\mu\nu} \prod A \prod \psi\rangle$ obtained by the variation of a renormalized correlation function is trivially finite, as shown in Sect. 5.3. As a convention, we do not use the symbol [ ] for the energy-momentum tensor.

First, consider the two-point function of Θ. Since the partition function is finite, its gravitational variations are also finite. Therefore, carrying out the variation two times, we obtain the following condition:

$$\langle\Theta(x)\Theta(y)\rangle - \left\langle \frac{\delta\Theta(x)}{\delta\Omega(y)} \right\rangle = \text{finite}.$$

Taking the flat-space limit in momentum space, this yields $\langle\Theta(p)\Theta(-p)\rangle|_{\text{flat}} - 8c_0 p^4 = \text{finite}$. For later convenience, we introduce a combination $\bar{\Theta} = \Theta - (D-1)[E_\psi]/2$. Since the two-point function with $[E_\psi]$ vanishes,[8] we obtain

$$\langle\bar{\Theta}(p)\bar{\Theta}(-p)\rangle|_{\text{flat}} - 8p^4\mu^{D-4}L_c = \text{finite}.$$

Here, introduce a composite operator defined in flat space as

$$\begin{aligned}\{A^2\} &\equiv \frac{D-4}{\beta_{\mathrm{g}}} \frac{1}{4\mathrm{g}_0^2} F^{\mathrm{A}}_{0\mu\nu} F^{\mathrm{A}\mu\nu}_0 \\ &= \frac{1}{4\mathrm{g}^2}\left[F^{\mathrm{A}}_{\mu\nu} F^{\mathrm{A}\mu\nu}\right] + \frac{1}{2\beta_{\mathrm{g}}}\left(\bar{\gamma}_A[E_A] + \bar{\gamma}_2[E_\psi]\right),\end{aligned}$$

and consider its two-point function defined by $\Gamma_{AA}(p^2) = \langle\{A^2(p)\}\{A^2(-p)\}\rangle|_{\text{flat}}$ in momentum space. Although $\{A^2\}$ is not a normal product because the terms with $1/\beta_{\mathrm{g}}$ yield poles, contributions from them vanish due to the property of the equation-of-motion operators. Therefore, Γ_{AA} is given by the two-point function of the normal product $[F^{\mathrm{A}}_{\mu\nu} F^{\mathrm{A}\mu\nu}]$. Since correlation functions among normal products do not yield nonlocal poles in general, it can be expressed in the following form:

---

[7] Note here that the differential operators $\alpha_{\mathrm{g}}\partial/\partial\alpha_{\mathrm{g}}$ and $\xi\partial/\partial\xi$ do not act on the bare fields, $A_{0\mu}$ and ψ_0, that are integration variables.

[8] Since one-point functions are dimensionally regularized to zero for first- and second-order massless theories in flat space, $\langle[E_\psi(x)]P(y)\rangle|_{\text{flat}} = \langle\delta P(y)/\delta\Upsilon(x)\rangle|_{\text{flat}} = 0$ is satisfied for a polynomial composite $P(y)$ in the fields $\bar{\psi}(y)$ and $\psi(y)$, where $\delta/\delta\Upsilon = (\bar{\psi}_0\delta/\delta\bar{\psi}_0 + \psi_0\delta/\delta\psi_0)/\sqrt{g}$.

$$\Gamma_{AA}(p^2) - p^4\mu^{D-4}\left(\frac{D-4}{\beta_{\mathrm{g}}}\right)^2 L_x = \text{finite}, \tag{5.37}$$

where $L_x = \sum_{n=1}^{\infty} x_n/(D-4)^n$ is a new pure-pole term defined by this equation. The factor $(D-4)^2/\beta_{\mathrm{g}}^2$ before L_x is introduced for later convenience.

Since $\bar{\Theta}|_{\mathrm{flat}} = \beta_{\mathrm{g}}\{A^2\}$ up to the term of gauge-fixing origin, $\beta_{\mathrm{g}}^2\Gamma_{AA} = \langle\bar{\Theta}\bar{\Theta}\rangle|_{\mathrm{flat}}$ is satisfied. Thus, from (5.37), we obtain the relation

$$(D-4)^2 L_x - 8L_c = \text{finite}. \tag{5.38}$$

This implies that the residues of L_c can be determined from the residues of L_x as $c_n = x_{n+2}/8$. This indicates that if x_3 is obtained, c_1 can be determined. Since the renormalization group equation for c_0 has already been given as in (5.34), this means that if x_3 is obtained, all c_n will be determined when $\bar{\beta}_{\mathrm{g}}$ is given. At the same time, x_n $(n \geq 4)$ are also determined.

To determine c_n, we consider a renormalization group equation that relates x_3 with x_1. This can be derived using the following fact: if F is a finite quantity, $\beta_{\mathrm{g}}^{-n}\mu d(\beta_{\mathrm{g}}^n F)/d\mu$ is also finite in spite of the presence of the pole factor β_{g}^{-n} because of $\beta_{\mathrm{g}}^{-n}\mu d\beta_{\mathrm{g}}^n/d\mu = n\alpha_{\mathrm{g}}\partial\bar{\beta}_{\mathrm{g}}/\partial\alpha_{\mathrm{g}}$. Applying this fact for $n = 2$ to the finite equation (5.37) as F, we obtain

$$\frac{1}{\beta_{\mathrm{g}}^2}\mu\frac{d}{d\mu}\left\{\mu^{D-4}(D-4)^2 L_x\right\} = \text{finite}, \tag{5.39}$$

where $\mu d(\beta_{\mathrm{g}}\{A^2\})/d\mu = 0$ is used, which follows from the fact that $\beta_{\mathrm{g}}\{A^2\}$ is defined in terms of bare quantities. Expanding this equation in Laurent series using (5.36) and extracting finiteness conditions, we can derive recursion relations among the residues x_n. Solving them, we obtain

$$x_3(\alpha_{\mathrm{g}}) = -\frac{\bar{\beta}_{\mathrm{g}}(\alpha_{\mathrm{g}})}{\alpha_{\mathrm{g}}}\int_0^{\alpha_{\mathrm{g}}} d\alpha\left\{\alpha^2 x_1(\alpha)\frac{\partial}{\partial\alpha}\left(\frac{\bar{\beta}_{\mathrm{g}}(\alpha)}{\alpha}\right)\right\}. \tag{5.40}$$

As shown later, the lowest term of x_1 is $o(1)$, and thus x_3 and c_1 start from $o(\alpha_{\mathrm{g}}^3)$. Note here that as can be seen from (5.38), the recursion relation for x_n with $n \geq 3$ reduces to the same equation as (5.34) for c_n.

Next, we consider the three-point function of Θ given as follows:

$$\langle\Theta(x)\Theta(y)\Theta(z)\rangle - \left\langle\frac{\delta\Theta(x)}{\delta\Omega(y)}\Theta(z)\right\rangle - \left\langle\frac{\delta\Theta(y)}{\delta\Omega(z)}\Theta(x)\right\rangle$$
$$-\left\langle\frac{\delta\Theta(z)}{\delta\Omega(x)}\Theta(y)\right\rangle + \left\langle\frac{\delta\Theta(x)}{\delta\Omega(y)\delta\Omega(z)}\right\rangle = \text{finite}.$$

In terms of $\bar{\Theta}$, this is expressed in the flat-space limit as

$$\langle\bar{\Theta}(x)\bar{\Theta}(y)\bar{\Theta}(z)\rangle|_{\text{flat}} - \langle\bar{\Theta}(x)\bar{\Theta}_2(y,z)\rangle|_{\text{flat}} - \langle\bar{\Theta}(y)\bar{\Theta}_2(z,x)\rangle|_{\text{flat}}$$
$$-\langle\bar{\Theta}(z)\bar{\Theta}_2(x,y)\rangle|_{\text{flat}} + \left\langle \frac{\delta^3 S}{\delta\Omega(x)\delta\Omega(y)\delta\Omega(z)} \right\rangle\Bigg|_{\text{flat}} = \text{finite},$$

where $\bar{\Theta}_2(x,y) = \delta\bar{\Theta}(x)/\delta\Omega(y) - (D-1)\delta\bar{\Theta}(x)/2\delta\Upsilon(y)$.

The three-point function of $\{A^2\}$ in flat space is denoted by Γ_{AAA} as before. Since $\bar{\Theta}|_{\text{flat}} = \beta_g\{A^2\}$ and $\bar{\Theta}_2(x,y)|_{\text{flat}} = -4\beta_g\{A^2\}\delta^D(x-y) + 8c_0\partial^4\delta^D(x-y)$, the finiteness condition above can be written in momentum space as

$$\beta_g^3\Gamma_{AAA}(p_x^2,p_y^2,p_z^2) + 4\beta_g^2\left\{\Gamma_{AA}(p_x^2) + \Gamma_{AA}(p_y^2) + \Gamma_{AA}(p_z^2)\right\}$$
$$+b_0 B(p_x^2,p_y^2,p_z^2) + c_0 C(p_x^2,p_y^2,p_z^2) = \text{finite},$$

where the last two functions are given by $B(p_x^2,p_y^2,p_z^2) = -2(D-2)(D-3)(D-4)\{p_x^4+p_y^4+p_z^4 - 2(p_x^2p_y^2+p_y^2p_z^2+p_z^2p_x^2)\}$ and $C(p_x^2,p_y^2,p_z^2) = -4\{(D+2)(p_x^4+p_y^4+p_z^4) + 4(p_x^2p_y^2+p_y^2p_z^2+p_z^2p_x^2)\}$.

Consider the special cases that some momenta are taken to be on shell. Combining (5.37) and (5.38), we obtain

$$\beta_g^3\Gamma_{AAA}(p^2,0,0) - p^4\mu^{D-4}\{2(D-2)(D-3)(D-4)L_b + 4(D-6)L_c\}$$
$$= \text{finite} \tag{5.41}$$

and $\beta_g^3\Gamma_{AAA}(p^2,p^2,0) - 8(D-4)p^4\mu^{D-4}L_c = \text{finite}$.

In general, Γ_{AAA} has the following form:

$$\Gamma_{AAA}(p_x^2,p_y^2,p_z^2) - \sum \text{poles} \times \left\{\Gamma_{AA}(p_x^2) + \Gamma_{AA}(p_y^2) + \Gamma_{AA}(p_z^2)\right\}$$
$$-\mu^{D-4}\sum \text{poles} \times \left\{\text{terms in } p_i^2p_j^2\right\} = \text{finite}. \tag{5.42}$$

This expression cannot be obtained by dividing (5.41) by $1/\beta_g^3$ because $1/\beta_g$ has poles.[9] In order to determine the pure-pole factor in front of Γ_{AA}, we consider an equation obtained by applying $\alpha_g\partial/\partial\alpha_g$ to (5.37), which yields an equation for $\Gamma_{AAA}(p^2,p^2,0)$ because of $\alpha_g\partial S/\partial\alpha_g|_{\text{flat}} = -\int d^Dx\{A^2\}$ up to the gauge-fixing term origin and

$$\alpha_g\frac{\partial\{A^2\}}{\partial\alpha_g} = -\left[1 + \frac{\alpha_g^2}{\beta_g}\frac{\partial}{\partial\alpha_g}\left(\frac{\bar{\beta}_g}{\alpha_g}\right)\right]\{A^2\}.$$

The pole factor before Γ_{AA} in (5.42) can be extracted from this equation and we find that $\Gamma_{AAA}(p^2,0,0)$ has the following form:

---

9 Since three-point functions with the equation-of-motion operators do not vanish, the operators with $1/\beta_g$ in $\{A^2\}$ produce nonlocal poles. Thus, Γ_{AAA} has nonlocal poles, unlike Γ_{AA}. The second term in (5.42) plays a role in canceling out such nonlocal poles.

$$\Gamma_{AAA}(p^2, 0, 0) - \frac{\alpha_g^2}{\beta_g}\frac{\partial}{\partial \alpha_g}\left(\frac{\bar{\beta}_g}{\alpha_g}\right)\Gamma_{AA}(p^2)$$
$$-p^4\mu^{D-4}\left(\frac{D-4}{\beta_g}\right)^3 L_y = \text{finite}. \tag{5.43}$$

Here, the last pure-pole term $L_y = \sum_{n=1}^{\infty} y_n/(D-4)^n$ cannot be deduced from the argument above, which is defined through this equation.

Multiplying (5.43) by β_g^3, we obtain another equation of $\beta_g^3\Gamma_{AAA}$, independent of (5.41). By eliminating $\beta_g^3\Gamma_{AAA}$ from these equations, we obtain the following pole relation:

$$2(D-2)(D-3)(D-4)L_b + 4\left[D - 6 - 2\alpha_g^2\frac{\partial}{\partial \alpha_g}\left(\frac{\bar{\beta}_g}{\alpha_g}\right)\right]L_c - (D-4)^3 L_y$$
$$= \text{finite}. \tag{5.44}$$

where (5.37) and (5.38) are used.

The renormalization group equation for L_y can be derived as similar to the derivation of (5.39). Consider an equation obtained by applying $\beta_g^{-3}\mu d/d\mu$ to (5.43) multiplied by β_g^3. Since that $\mu d(\beta_g^3\Gamma_{AAA})/d\mu = \mu d(\beta_g^2\Gamma_{AA})/d\mu = 0$, we obtain the following:

$$\left(\frac{D-4}{\beta_g}\right)^3\left[(D-4)L_y + \beta_g\alpha_g\frac{\partial}{\partial \alpha_g}L_y\right] + \alpha_g^2\frac{\partial^2 \bar{\beta}_g}{\partial \alpha_g^2}\left(\frac{D-4}{\beta_g}\right)^2 L_x$$
$$= \text{finite}, \tag{5.45}$$

where (5.37) is used. This provides a recursion relation between x_n and y_n.

5.7.2 *Determination of Gravitational Counterterms*

The four renormalization group equations, (5.38), (5.39), (5.44), and (5.45), indicate that the pole terms, L_b and L_c, are related to each other at all orders.[10] Thus, we can combine the last two terms in (5.33) into $b_0 G_D$ with $G_D = G_4 + (D-4)\chi(D)H^2$ (5.3), such that

$$L_c - (D-4)\chi(D)L_b = \text{finite} \tag{5.46}$$

is satisfied, where the coupling constant c in c_0 is removed. By finding the relations between the residues, b_n and c_n, from the renormalization group equations and sub-

[10] The derived renormalization group equations have the same forms as those obtained for the QED in curved space.

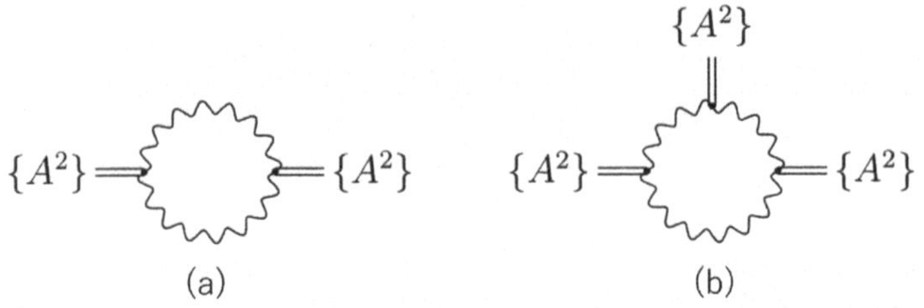

Fig. 5.10 Loop corrections for Γ_{AA} and Γ_{AAA}, which give the residues X_1 and Y_1, respectively

stituting them into (5.46), we can determine the coefficient $\chi(D) = \sum_{n=1}^{\infty} \chi_n (D-4)^{n-1}$ [Ham14a].

Letting the QCD beta function $\bar{\beta}_g$ be given, information needed to solve the renormalization group equations is the simple poles of L_x and L_y, expanded as $x_1 = X_1 + o(\alpha_g)$ and $y_1 = Y_1 + o(\alpha_g)$. Using these values, we can obtain the residue x_n from (5.39), thus c_n from (5.38), and the residue y_n from (5.45). In addition, given c_n and y_n, we can determine the residue b_n from (5.44), which provides the following recursion relation:

$$4b_{n+1} + 6b_{n+2} + 2b_{n+3} - 8\left[1 + \alpha_g^2 \frac{\partial}{\partial \alpha_g}\left(\frac{\bar{\beta}_g}{\alpha_g}\right)\right] c_n + 4c_{n+1} - y_{n+3} = 0.$$

In this way, we can determine the coefficient χ through (5.46) order by order. The solution for the first two terms of χ is then given by

$$\chi_1 = \frac{1}{2}, \qquad \chi_2 = 1 - \frac{Y_1}{4X_1}.$$

The explicit values of X_1 and Y_1 are obtained by calculating 1-loop corrections for Γ_{AA} and Γ_{AAA} (see Fig. 5.10), respectively, which are given by

$$\Gamma_{AA}(p^2) = -\frac{r}{2}\frac{\mu^{D-4}}{(4\pi)^2} p^4 \frac{1}{D-4}, \qquad \Gamma_{AAA}(p^2, 0, 0) = -\frac{r}{2}\frac{\mu^{D-4}}{(4\pi)^2} p^4 \frac{1}{D-4},$$

where r is the dimension of the Lie group. From these, we obtain $X_1 = Y_1 = -r/2(4\pi)^2$. Thus, χ_2 is determined to be 3/4, as shown in (5.4).

Thus, we can see that at least χ_1 and χ_2 are the universal coefficients independent of gauge groups and fermion representations. At present, it is not confirmed that the coefficient χ_n ($n \geq 3$) has a universal value independent of gauge theories. In any case, χ_n can in principle be determined at all orders.

The explicit value of b_1', which is the coupling-dependent part of b_1, can be calculated as follows.[11] From $\chi_1 = 1/2$, we obtain the relation $b_2 = 2c_1 + o(\alpha_g^4)$. The residue $c_1 = x_3/8$ is obtained from x_1 through (5.40). Since $x_1 = X_1 + o(\alpha_g)$,

[11] Note that the α_g-independent part cannot be determined from the renormalization group equation.

we obtain $c_1 = -\beta_1\beta_2 X_1 \alpha_g^3/96 + o(\alpha_g^4)$ and thus b_2, where $\bar{\beta}_g = \beta_1\alpha_g + \beta_2\alpha_g^2 + o(\alpha_g^3)$. Further, using (5.34) for b_n, we obtain

$$b_1' = \frac{\beta_2 X_1}{24}\alpha_g^2 + o(\alpha_g^3). \tag{5.47}$$

Thus, the α_g-dependent part of the residue b_1 starts from $o(\alpha_g^2)$.

5.7.3 Forms of Conformal Anomalies

If we use the gravitational counterterms, $a_0 F_D$ and $b_0 G_D$, then the trace of the energy-momentum tensor, or conformal anomaly, is given by

$$\begin{aligned}\Theta &= \frac{D-4}{4g_0^2} F_{0\mu\nu}^{A} F_0^{A\mu\nu} + \frac{1}{2}(D-1)E_{0\psi} + (D-4)(a_0 F_D + b_0 E_D) \\ &= \frac{\beta_g}{4g^2}\left[F_{\mu\nu}^{A} F^{A\mu\nu}\right] + \frac{1}{2}(D-1+\bar{\gamma}_2)[E_\psi] + \frac{1}{2}\bar{\gamma}_A[E_A] \\ &\quad -\mu^{D-4}(\beta_a F_D + \beta_b E_D),\end{aligned}$$

apart from the terms associated with gauge-fixing, where

$$E_D = G_D - 4\chi(D)\nabla^2 H. \tag{5.48}$$

In order to determine this anomaly form, we use finiteness conditions such as $\beta_g^{-1}\mu d(\beta_g[F_{\mu\nu}^{A} F^{A\mu\nu}]/g^2)d\mu =$ finite, so that the normal product (5.35) reduces to the form

$$\begin{aligned}\frac{1}{4g^2}\left[F_{\mu\nu}^{A} F^{A\mu\nu}\right] &= \frac{D-4}{\beta_g}\frac{1}{4g_0^2} F_{0\mu\nu}^{A} F^{A\mu\nu} - \frac{\bar{\gamma}_A}{2\beta_g}E_{0A} - \frac{\bar{\gamma}_2}{2\beta_g}E_{0\psi} \\ &\quad + \frac{D-4}{\beta_g}\mu^{D-4}\left\{\left(L_a + \frac{\bar{\beta}_a}{D-4}\right)F_D + \left(L_b + \frac{\bar{\beta}_b}{D-4}\right)E_D\right\}.\end{aligned}$$

The combination E_D (5.48) becomes E_4 (2.13) in four dimensions due to $\chi_1 = 1/2$. In this way, we can derive the conformal-anomaly form proposed by Riegert [Rie84].

Chapter 6
BRST Conformal Algebra and Physical States

This chapter is devoted to describing physical states of quantum gravity that appear at the high-energy limit beyond the Plancks scale. They are defined as BRST conformally invariant states, and we construct and classify them concretely by employing the cylindrical background $\mathbb{R} \times S^3$. We find that there are an infinite number of physical states, and only states called primary scalars are allowed.

The purpose of this chapter is to briefly show how what we have described so far are expressed. As with the previous chapter, you can skip this chapter without significantly affecting your understanding of the rest of this book. For mathematical details on conformal algebra, see the references cited or the book [Ham18].

6.1 Canonical Quantization on $\mathbb{R} \times S^3$

The quantum gravity world far beyond the Planck scale is described by the Riegert and Weyl actions. In this section, we quantize the gravitational field by using these two actions.

Since the Minkowski background M^4 and $\mathbb{R} \times S^3$ can be changed by conformal transformations, the result will be the same no matter which background is used,[1] but the latter allows for more detailed discussions without worrying about infrared divergence because of the compact S^3 space. In addition, there is a similarity with two-dimensional quantum gravity on $\mathbb{R} \times S^1$, as in Appendix Sect. 6.4, which is helpful for understanding.

The metric of the background $\mathbb{R} \times S^3$ is described by the coordinate $x^\mu = (\eta, x^i)$ using Euler angles $x^i = (\alpha, \beta, \gamma)$ as

[1] For quantization on the Minkowski background, see [Ham12a, Ham12b].

K. Hamada, *Trans-Planckian Physics and Inflation*, Fundamental Theories of Physics 223, https://doi.org/10.1007/978-981-96-3476-7_6

$$
\begin{aligned}
d\hat{s}^2_{\mathbb{R}\times S^3} &= \hat{g}_{\mu\nu}dx^\mu dx^\nu = -d\eta^2 + \hat{\gamma}_{ij}dx^i dx^j \\
&= -d\eta^2 + \frac{1}{4}\left(d\alpha^2 + d\beta^2 + d\gamma^2 + 2\cos\beta d\alpha d\gamma\right),
\end{aligned}
$$

where the ranges of α, β, and γ are $[0, 2\pi]$, $[0, \pi]$, and $[0, 4\pi]$, respectively. The radius is normalized to unity, such that $\hat{R}_{ijkl} = \hat{\gamma}_{ik}\hat{\gamma}_{jl} - \hat{\gamma}_{il}\hat{\gamma}_{jk}$ and $\hat{R} = 6$. The volume element of the unit S^3 is $d\Omega_3 = \sqrt{\hat{\gamma}}\, d^3x = \sin\beta d\alpha d\beta d\gamma/8$, and the volume is $\mathrm{V}_3 = \int d\Omega_3 = 2\pi^2$.

First, decompose the traceless tensor field into components, h_{00}, h_{0i}, and $h_{ij} = h^{\mathbf{tr}}_{ij} + \hat{\gamma}_{ij}h_{00}/3$, where $h^{\mathbf{tr}}_{ij}$ is a component that satisfies the spatial traceless condition ${h^{\mathbf{tr}}}^i{}_i = \hat{\gamma}^{ij}h^{\mathbf{tr}}_{ij} = 0$. As a gauge-fixing condition to fix (3.4), we here adopt the radiation gauge

$$
h_{00} = 0, \quad \hat{\nabla}^i h_{0i} = 0, \quad \hat{\nabla}^i h^{\mathbf{tr}}_{ij} = 0. \tag{6.1}
$$

If the transverse-traceless tensor and the transverse vector components satisfying this condition are written in the Gothic style as h_{ij} and h_i, respectively, then the gauge-fixed action is given by

$$
\begin{aligned}
S_{\mathrm{QG}} = \int d\eta \int_{S^3} d\Omega_3 \biggl\{ &-\frac{2b_c}{(4\pi)^2}\phi\left(\partial_\eta^4 - 2\Box_3\partial_\eta^2 + \Box_3^2 + 4\partial_\eta^2\right)\phi \\
&-\frac{1}{2}\mathsf{h}_{ij}\left(\partial_\eta^4 - 2\Box_3\partial_\eta^2 + \Box_3^2 + 8\partial_\eta^2 - 4\Box_3 + 4\right)\mathsf{h}^{ij} \\
&+\mathsf{h}_i\left(\Box_3 + 2\right)\left(-\partial_\eta^2 + \Box_3 - 2\right)\mathsf{h}^i \biggr\},
\end{aligned}
$$

where $\Box_3 = \hat{\gamma}^{ij}\hat{\nabla}_i\hat{\nabla}_j$ is the Laplacian on S^3. The first line is derived from the Riegert action, and the others are derived from the kinetic term of the Weyl action. At that time, the t-dependence is removed by performing a rewrite of $\mathsf{h} \to t\mathsf{h}$.

Furthermore, since the mode satisfying $(\Box_3 + 2)\mathsf{h}_i = 0$ in the transverse vector field has a vanishing kinetic term, it should be removed as a non-dynamical mode. Since this mode can be written with the $J = 1/2$ component of the vector harmonics introduced below, we add a condition

$$
\mathsf{h}_i|_{J=\frac{1}{2}} = 0. \tag{6.2}
$$

The radiation gauge (6.1) with this additional condition is called the radiation$^+$ gauge [HH03].

Each gravitational field is expanded in modes using spherical harmonics on S^3 [RO84]. The symmetric-transverse-traceless (ST^2) n-th rank tensor harmonics are classified by using the $(J + \varepsilon_n, J - \varepsilon_n)$ representation of the rotation (isometry) group $SU(2) \times SU(2)$ on S^3, and are denoted as $Y^{i_1\cdots i_n}_{J(M\varepsilon_n)}$ [AMM97a, HH03], where $J \geq n/2$ is given as an integer and a half-integer, and $\varepsilon_n = \pm n/2$ is a polarization index. $M = (m, m')$ is an index representing degeneracy of the representation for each polarization, and takes the following values:

$$m = -J - \varepsilon_n, -J - \varepsilon_n + 1, \ldots, J + \varepsilon_n - 1, J + \varepsilon_n,$$
$$m' = -J + \varepsilon_n, -J + \varepsilon_n + 1, \ldots, J - \varepsilon_n - 1, J - \varepsilon_n.$$

From this, the degeneracy of the harmonic function is $(2J + 1)^2$ for $n = 0$ and $2(2J + n + 1)(2J - n + 1)$ for $n > 0$.

The spherical tensor harmonics are eigenfunctions of the Laplacian on S^3, satisfying

$$\Box_3 Y^{i_1 \cdots i_n}_{J(M\varepsilon_n)} = [-2J(2J + 2) + n]\, Y^{i_1 \cdots i_n}_{J(M\varepsilon_n)}.$$

The complex conjugates are defined by $Y^{i_1 \cdots i_n *}_{J(M\varepsilon_n)} = (-1)^n \epsilon_M Y^{i_1 \cdots i_n}_{J(-M\varepsilon_n)}$ and are used to normalize the harmonics to

$$\int_{S^3} d\Omega_3 Y^{i_1 \cdots i_n *}_{J_1(M_1 \varepsilon^1_n)} Y_{i_1 \cdots i_n\, J_2(M_2 \varepsilon^2_n)} = \delta_{J_1 J_2} \delta_{M_1 M_2} \delta_{\varepsilon^1_n \varepsilon^2_n},$$

where $\epsilon_M = (-1)^{m - m'}$ is a sign factor that satisfies $\epsilon_M^2 = 1$, and $\delta_{M_1 M_2} = \delta_{m_1 m_2} \delta_{m'_1 m'_2}$. In the following, we particularly use the symbols $y = \varepsilon_1 = \pm 1/2$ and $x = \varepsilon_2 = \pm 1$ for the polarization indices of the vector and second-rank tensor, respectively.

The specific expressions of spherical harmonics on S^3 can be expressed using Wigner's D-function [VMK88]. For example, scalar harmonics are expressed as

$$Y_{JM} = \sqrt{\frac{2J + 1}{V_3}} D^J_{mm'}.$$

The spherical tensor harmonics can be expressed by combining the D-function with Clebsch-Gordan coefficients.

From the equation of motion, it can be seen that the conformal-factor field ϕ can be mode-expanded using the scalar harmonics as [AMM97b]

$$\begin{aligned}\phi = \frac{\pi}{2\sqrt{b_c}} \Bigg\{ & 2(\hat{q} + \hat{p}\eta) Y_{00} \\ & + \sum_{J \geq \frac{1}{2}} \sum_M \frac{1}{\sqrt{J(2J + 1)}} \left(a_{JM} e^{-i2J\eta} Y_{JM} + a^\dagger_{JM} e^{i2J\eta} Y^*_{JM} \right) \\ & + \sum_{J \geq 0} \sum_M \frac{1}{\sqrt{(J + 1)(2J + 1)}} \left(b_{JM} e^{-i(2J+2)\eta} Y_{JM} + b^\dagger_{JM} e^{i(2J+2)\eta} Y^*_{JM} \right) \Bigg\},\end{aligned}$$

where $Y_{00} = 1/\sqrt{V_3} = 1/\pi\sqrt{2}$. If canonical quantization is performed,[2] then the commutation relation of each mode becomes

$$\left[\hat{q}, \hat{p}\right] = i, \quad \left[a_{J_1 M_1}, a^{\dagger}_{J_2 M_2}\right] = \delta_{J_1 J_2}\delta_{M_1 M_2}, \quad \left[b_{J_1 M_1}, b^{\dagger}_{J_2 M_2}\right] = -\delta_{J_1 J_2}\delta_{M_1 M_2},$$

and the Hamiltonian is given by

$$H = \frac{1}{2}\hat{p}^2 + b_c + \sum_{J\geq 0}\sum_{M}\{2J a^{\dagger}_{JM} a_{JM} - (2J+2) b^{\dagger}_{JM} b_{JM}\}.$$

This shows that a_{JM} has a positive metric and b_{JM} has a negative metric. The energy shift b_c is a Casimir effect that depends on the coordinate system, and here its value is determined so that conformal algebra is closed.

Similarly, the transverse-traceless tensor field h_{ij} and the transverse vector field h_k are mode-expanded using the ST^2 tensor and vector harmonics as [HH03][3]

$$\begin{aligned}
\mathsf{h}^{ij} &= \frac{1}{4}\sum_{J\geq 1}\sum_{M,x}\frac{1}{\sqrt{J(2J+1)}}\left\{c_{J(Mx)} e^{-i2J\eta} Y^{ij}_{J(Mx)} + c^{\dagger}_{J(Mx)} e^{i2J\eta} Y^{ij*}_{J(Mx)}\right\} \\
&\quad + \frac{1}{4}\sum_{J\geq 1}\sum_{M,x}\frac{1}{\sqrt{(J+1)(2J+1)}}\left\{d_{J(Mx)} e^{-i(2J+2)\eta} Y^{ij}_{J(Mx)}\right. \\
&\quad \left. + d^{\dagger}_{J(Mx)} e^{i(2J+2)\eta} Y^{ij*}_{J(Mx)}\right\}, \\
\mathsf{h}^{k} &= i\frac{1}{2}\sum_{J\geq 1}\sum_{M,y}\frac{1}{\sqrt{(2J-1)(2J+1)(2J+3)}} \\
&\quad \times\left\{e_{J(My)} e^{-i(2J+1)\eta} Y^{k}_{J(My)} - e^{\dagger}_{J(My)} e^{i(2J+1)\eta} Y^{k*}_{J(My)}\right\}.
\end{aligned} \tag{6.3}$$

Here note that h_k is expanded using the vector modes of $J \geq 1$, according to the radiation$^+$ gauge condition (6.2). Executing canonical quantization method yields the commutation relations such as

$$\begin{aligned}
\left[c_{J_1(M_1 x_1)}, c^{\dagger}_{J_2(M_2 x_2)}\right] &= -\left[d_{J_1(M_1 x_1)}, d^{\dagger}_{J_2(M_2 x_2)}\right] = \delta_{J_1 J_2}\delta_{M_1 M_2}\delta_{x_1 x_2}, \\
\left[e_{J_1(M_1 y_1)}, e^{\dagger}_{J_2(M_2 y_2)}\right] &= -\delta_{J_1 J_2}\delta_{M_1 M_2}\delta_{y_1 y_2}.
\end{aligned}$$

Thus, $c_{J(Mx)}$ is a positive-metric mode, while $d_{J(Mx)}$ and $e_{J(My)}$ are negative-metric modes. The Hamiltonian is given by

---

[2] It is better to introduce a new variable $\chi = \partial_\eta \phi$ and perform Dirac quantization [Dir64] with this relation as a constraint.

[3] The reason why we use an unusual expansion formula for h^k is merely a convention to make it consistent with conformal algebra given later.

$$H = \sum_{J\ge 1}\sum_{M,x}\{2Jc^{\dagger}_{J(Mx)}c_{J(Mx)} - (2J+2)d^{\dagger}_{J(Mx)}d_{J(Mx)}\}$$
$$-\sum_{J\ge 1}\sum_{M,y}(2J+1)e^{\dagger}_{J(My)}e_{J(My)}.$$

6.2 Generators of Conformal Algebra

Let us find 15 conformal Killing vectors on $\mathbb{R}\times S^3$ that satisfy the conformal Killing equation (3.5). Writing this equation for each component yields

$$3\partial_\eta\zeta_0 + \psi = 0, \qquad \partial_\eta\zeta_i + \hat{\nabla}_i\zeta_0 = 0,$$
$$\hat{\nabla}_i\zeta_j + \hat{\nabla}_j\zeta_i - \frac{2}{3}\hat{\gamma}_{ij}\psi = 0,$$

where $\psi = \hat{\nabla}_k\zeta^k$.

First of all, the Killing vectors, which are defined by $\psi = 0$, are composed of a vector $\zeta^\mu_{\mathrm{T}} = (1,0,0,0)$ of time translation and a vector $\zeta^\mu_{\mathrm{R}} = (0,\zeta^k_{\mathrm{R}})$ representing rotation on S^3, where the latter spatial component can be written using the scalar harmonics of $J = 1/2$ as

$$\left(\zeta^k_{\mathrm{R}}\right)_{MN} = i\frac{\mathrm{V}_3}{4}\left[Y^*_{\frac{1}{2}M}\hat{\nabla}^k Y_{\frac{1}{2}N} - Y_{\frac{1}{2}N}\hat{\nabla}^k Y^*_{\frac{1}{2}M}\right].$$

The solution for $\psi \neq 0$ is expressed by $\zeta^\mu_{\mathrm{S}} = (\zeta^0_{\mathrm{S}},\zeta^k_{\mathrm{S}})$ and its complex conjugate $\zeta^{\mu*}_{\mathrm{S}}$, whose components are given by

$$(\zeta^0_{\mathrm{S}})_M = \frac{\sqrt{\mathrm{V}_3}}{2}\mathrm{e}^{i\eta}Y^*_{\frac{1}{2}M}, \quad (\zeta^k_{\mathrm{S}})_M = -i\frac{\sqrt{\mathrm{V}_3}}{2}\mathrm{e}^{i\eta}\hat{\nabla}^k Y^*_{\frac{1}{2}M}.$$

Here, the indices written only with M and N represent **4**-vectors in $SU(2)\times SU(2)$ by themselves. Even after this, for simplicity, we display only the indicies M and N with omitting $J = 1/2$ when describing the **4**-vector indices of the conformal Killing vectors and corresponding generators.

Generators of diffeomorphism that constitute conformal algebra are defined by

$$Q_\zeta = \int_{S^3} d\Omega_3\,\zeta^\mu :\hat{T}_{\mu 0}:, \tag{6.4}$$

where $\hat{T}^{\mu\nu} = (2/\sqrt{-\hat{g}})\times\delta S_{\mathrm{QG}}/\delta\hat{g}_{\mu\nu}$ and $\hat{T}_{\mu\nu} = \hat{g}_{\mu\lambda}\hat{g}_{\nu\sigma}\hat{T}^{\lambda\sigma}$ is the energy-momentum tensor that satisfies the traceless condition. When adopting canonical quantization, it is necessary to apply normal ordering : : to restore diffeomorphism invariance (see Sects. 4.2.2 and 5.3).

Denote the generators Q_ζ for the conformal Killing vectors $\zeta^\mu = \zeta^\mu_{\rm T}$, $(\zeta^\mu_{\rm R})_{MN}$, $(\zeta^\mu_{\rm S})_M$, and $(\zeta^{\mu*}_{\rm S})_M$ as H, R_{MN}, Q_M, and $Q^\dagger_M$, respectively. H is the Hamiltonian given in Sect. 6.1, which acts as a dilatation operator that counts conformal dimensions of states on the cylindrical background $\mathbb{R} \times S^3$. R_{MN} are six generators representing rotations on S^3 that satisfy the relations $R^\dagger_{MN} = R_{NM}$ and $R_{MN} = -\epsilon_M \epsilon_N R_{-N-M}$. Q_M are four generators of special conformal transformations, and their Hermitian conjugates $Q^\dagger_M$ correspond to four generators of translations, when considered in the cylindrical background. There are 15 generators in total.

The special conformal transformation provides very strong constraints on physical states. Deriving its generator in the Riegert sector according to the definition (6.4) yields [AMM97b]

$$
\begin{aligned}
Q_M = {} & \left(\sqrt{2b_c} - i\hat{p}\right) a_{\frac{1}{2}M} + \sum_{J\geq 0} \sum_{M_1} \sum_{M_2} \mathbf{C}^{\frac{1}{2}M}_{JM_1, J+\frac{1}{2}M_2} \\
& \times \Big\{ \alpha(J) \epsilon_{M_1} a^\dagger_{J-M_1} a_{J+\frac{1}{2}M_2} + \beta(J) \epsilon_{M_1} b^\dagger_{J-M_1} b_{J+\frac{1}{2}M_2} \\
& + \epsilon_{M_2} a^\dagger_{J+\frac{1}{2}-M_2} b_{JM_1} \Big\},
\end{aligned} \tag{6.5}
$$

where

$$
\alpha(J) = \sqrt{2J(2J+2)}, \qquad \beta(J) = -\sqrt{(2J+1)(2J+3)}. \tag{6.6}
$$

The $\mathbf{C}$ is a $SU(2) \times SU(2)$ Clebsch-Gordan coefficient defined by

$$
\begin{aligned}
\mathbf{C}^{JM}_{J_1M_1, J_2M_2} &= \sqrt{\mathrm{V}_3} \int_{S^3} d\Omega_3 Y^*_{JM} Y_{J_1M_1} Y_{J_2M_2} \\
&= \sqrt{\frac{(2J_1+1)(2J_2+1)}{2J+1}} C^{Jm}_{J_1m_1, J_2m_2} C^{Jm'}_{J_1m'_1, J_2m'_2},
\end{aligned}
$$

where $C^{Jm}_{J_1m_1,J_2m_2}$ is the standard Clebsch-Gordan coefficient, from which $J + J_1 + J_2$ is an integer, and a triangle inequality $|J_1 - J_2| \leq J \leq J_1 + J_2$ and $M = M_1 + M_2$ hold. The $\mathbf{C}$-coefficient of $J = 1/2$ appears in the Q_M generator.

The $\mathbf{C}$-coefficient is a real function and satisfies the relations, $\mathbf{C}^{JM}_{J_1M_1,J_2M_2} = \mathbf{C}^{JM}_{J_2M_2,J_1M_1} = \mathbf{C}^{J-M}_{J_1-M_1,J_2-M_2} = \epsilon_{M_2} \mathbf{C}^{J_1M_1}_{JM,J_2-M_2}$ and $\mathbf{C}^{00}_{JM,JN} = \epsilon_M \delta_{M-N}$. In addition, using the fact that a product of two scalar harmonics can be expanded as $Y_{J_1M_1} Y_{J_2M_2} = \sum_{J\geq 0} \sum_M \mathbf{C}^{JM}_{J_1M_1,J_2M_2} Y_{JM}/\sqrt{\mathrm{V}_3}$, we obtain the following crossing relation by evaluating the integral of a product of four scalar harmonics over S^3 in two ways:

$$
\sum_{J\geq 0} \sum_M \epsilon_M \mathbf{C}^{J_1M_1}_{J_2M_2,J-M} \mathbf{C}^{J_3M_3}_{JM,J_4M_4} = \sum_{J\geq 0} \sum_M \epsilon_M \mathbf{C}^{J_1M_1}_{J_4M_4,J-M} \mathbf{C}^{J_3M_3}_{JM,J_2M_2}. \tag{6.7}
$$

The generator of the special conformal transformation in the Weyl sector that acts on the traceless tensor field is given by [HH03]

$$
\begin{aligned}
Q_M = & \sum_{J\geq 1}\sum_{M_1,x_1}\sum_{M_2,x_2} \mathbf{E}^{\frac{1}{2}M}_{J(M_1x_1),J+\frac{1}{2}(M_2x_2)} \Big\{ \alpha(J)\epsilon_{M_1} c^{\dagger}_{J(-M_1x_1)} c_{J+\frac{1}{2}(M_2x_2)} \\
& + \beta(J)\epsilon_{M_1} d^{\dagger}_{J(-M_1x_1)} d_{J+\frac{1}{2}(M_2x_2)} + \epsilon_{M_2} c^{\dagger}_{J+\frac{1}{2}(-M_2x_2)} d_{J(M_1x_1)} \Big\} \\
& + \sum_{J\geq 1}\sum_{M_1,x_1}\sum_{M_2,y_2} \mathbf{H}^{\frac{1}{2}M}_{J(M_1x_1);J(M_2y_2)} \\
& \times \Big\{ A(J)\epsilon_{M_1} c^{\dagger}_{J(-M_1x_1)} e_{J(M_2y_2)} + B(J)\epsilon_{M_2} e^{\dagger}_{J(-M_2y_2)} d_{J(M_1x_1)} \Big\} \\
& + \sum_{J\geq 1}\sum_{M_1,y_1}\sum_{M_2,y_2} \mathbf{D}^{\frac{1}{2}M}_{J(M_1y_1),J+\frac{1}{2}(M_2y_2)} C(J)\epsilon_{M_1} e^{\dagger}_{J(-M_1y_1)} e_{J+\frac{1}{2}(M_2y_2)},
\end{aligned}
\tag{6.8}
$$

where $\alpha(J)$ and $\beta(J)$ are the same as (6.6). Other coefficients are given by

$$
A(J) = \sqrt{\frac{4J}{(2J-1)(2J+3)}}, \qquad B(J) = \sqrt{\frac{2(2J+2)}{(2J-1)(2J+3)}},
$$

$$
C(J) = \sqrt{\frac{(2J-1)(2J+1)(2J+2)(2J+4)}{2J(2J+3)}}.
$$

The various $SU(2) \times SU(2)$ Clebsch-Gordan coefficients with tensor indices are defined by

$$
\begin{aligned}
\mathbf{E}^{JM}_{J_1(M_1x_1),J_2(M_2x_2)} &= \sqrt{\mathrm{V}_3}\int_{S^3} d\Omega_3 Y^*_{JM} Y^{ij}_{J_1(M_1x_1)} Y_{ij\,J_2(M_2x_2)}, \\
\mathbf{H}^{JM}_{J_1(M_1x_1);J_2(M_2y_2)} &= \sqrt{\mathrm{V}_3}\int_{S^3} d\Omega_3 Y^*_{JM} Y^{ij}_{J_1(M_1x_1)} \hat{\nabla}_i Y_{j\,J_2(M_2y_2)}, \\
\mathbf{D}^{JM}_{J_1(M_1y_1),J_2(M_2y_2)} &= \sqrt{\mathrm{V}_3}\int_{S^3} d\Omega_3 Y^*_{JM} Y^{i}_{J_1(M_1y_1)} Y_{i\,J_2(M_2y_2)}.
\end{aligned}
$$

General formulas of these can be expressed using the Clebsch-Gordan coefficients and 6-j symbols [HH03]. Each coefficient of $J = 1/2$ appearing in the generator is simply given by

$$\mathbf{E}^{\frac{1}{2}M}_{J(M_1x_1),J+\frac{1}{2}(M_2x_2)} = \sqrt{(2J-1)(J+2)}\, C^{\frac{1}{2}m}_{J+x_1m_1,J+\frac{1}{2}+x_2m_2} C^{\frac{1}{2}m'}_{J-x_1m_1',J+\frac{1}{2}-x_2m_2'},$$
$$\mathbf{H}^{\frac{1}{2}M}_{J(M_1x_1);J(M_2y_2)} = -\sqrt{(2J-1)(2J+3)}\, C^{\frac{1}{2}m}_{J+x_1m_1,J+y_2m_2} C^{\frac{1}{2}m'}_{J-x_1m_1',J-y_2m_2'},$$
$$\mathbf{D}^{\frac{1}{2}M}_{J(M_1y_1),J+\frac{1}{2}(M_2y_2)} = \sqrt{J(2J+3)}\, C^{\frac{1}{2}m}_{J+y_1m_1,J+\frac{1}{2}+y_2m_2} C^{\frac{1}{2}m'}_{J-y_1m_1',J+\frac{1}{2}-y_2m_2'}.$$

The 15 generators of the conformal transformations form a closed $SO(4,2)$ conformal algebra as follows:

$$\begin{aligned}
\left[Q_M, Q_N^\dagger\right] &= 2\delta_{MN}H + 2R_{MN}, \\
\left[H, Q_M\right] &= -Q_M, \qquad \left[H, Q_M^\dagger\right] = Q_M^\dagger, \\
\left[H, R_{MN}\right] &= \left[Q_M, Q_N\right] = 0, \\
\left[Q_M, R_{M_1M_2}\right] &= \delta_{MM_2}Q_{M_1} - \epsilon_{M_1}\epsilon_{M_2}\delta_{M-M_1}Q_{-M_2}, \\
\left[R_{M_1M_2}, R_{M_3M_4}\right] &= \delta_{M_1M_4}R_{M_3M_2} - \epsilon_{M_1}\epsilon_{M_2}\delta_{-M_2M_4}R_{M_3-M_1} \\
&\quad -\delta_{M_2M_3}R_{M_1M_4} + \epsilon_{M_1}\epsilon_{M_2}\delta_{-M_1M_3}R_{-M_2M_4}.
\end{aligned} \tag{6.9}$$

Parametrizing a **4**-vector index $M = \{(1/2,1/2),(1/2,-1/2),(-1/2,1/2),(-1/2,-1/2)\}$ by $\{1,2,3,4\}$ and setting $A_+ = R_{31}$, $A_- = R_{31}^\dagger$, $A_3 = (R_{11}+R_{22})/2$, $B_+ = R_{21}$, $B_- = R_{21}^\dagger$, $B_3 = (R_{11}-R_{22})/2$, the last closed rotation algebra can be expressed in a familiar form of $SU(2)\times SU(2)$ algebra as $[A_+, A_-] = 2A_3$, $[A_3, A_\pm] = \pm A_\pm$, $[B_+, B_-] = 2B_3$, $[B_3, B_\pm] = \pm B_\pm$, where $A_{\pm,3}$ and $B_{\pm,3}$ are commutative. The specific expression of R_{MN} is not presented here, as it is not very important for the following discussion.

What should be noted here is that the Q_M generators of diffeomorphism given by (6.5) and (6.8) have terms where the positive- and negative-metric modes cross. Due to the presence of the cross terms, transformations by Q_M mix positive- and negative-metric modes. As a result, it can be shown that the negative-metric mode alone does not become gauge invariant, that is, it is unphysical.

6.2.1 BRST Formulation and Physical State Conditions

The BRST conformal transformation is a transformation in which the gauge degree of freedom ζ^μ in (3.6) is replaced with a gauge ghost field c^μ satisfying the conformal Killing equation $\hat\nabla_\mu c_\nu + \hat\nabla_\nu c_\mu - \hat g_{\mu\nu}\hat\nabla_\lambda c^\lambda/2 = 0$. The gauge ghost is expressed using 15 Grassmann modes $\mathrm{c}, \mathrm{c}_{MN}, \mathrm{c}_M, \mathrm{c}_M^\dagger$ as

$$c^\mu = \mathrm{c}\eta^\mu + \sum_M \left(\mathrm{c}_M^\dagger \zeta_M^\mu + \mathrm{c}_M \zeta_M^{\mu *}\right) + \sum_{M,N} \mathrm{c}_{MN}\zeta_{MN}^\mu,$$

where c is Hermite, and c_{MN} are six modes satisfying the relations $c^\dagger_{MN} = c_{NM}$ and $c_{MN} = -\epsilon_M \epsilon_N c_{-N-M}$. In addition, we introduce anti-ghost modes b, b_{MN}, b_M, $b^\dagger_M$ with the same properties. They are collectively called the (four-dimensional) *bc*-ghosts and the anti-commutative relations are set as

$$\{b, c\} = 1, \qquad \{b_{MN}, c_{LK}\} = \delta_{ML}\delta_{NK} - \epsilon_M \epsilon_N \delta_{-MK}\delta_{-NL},$$
$$\{b^\dagger_M, c_N\} = \{b_M, c^\dagger_N\} = \delta_{MN}.$$

Using these *bc*-ghost modes, we can construct 15 generators of conformal transformations as [AMM97a]

$$\begin{aligned} H^{\mathrm{gh}} &= \sum_M \left(c^\dagger_M b_M - c_M b^\dagger_M\right), \\ R^{\mathrm{gh}}_{MN} &= -c_M b^\dagger_N + c^\dagger_N b_M + \epsilon_M \epsilon_N \left(c_{-N} b^\dagger_{-M} - c^\dagger_{-M} b_{-N}\right) \\ &\quad - \sum_L \left(c_{LM} b_{LN} - c_{NL} b_{ML}\right), \\ Q^{\mathrm{gh}}_M &= -2c_M b - c b_M - \sum_L \left(2c_{LM} b_L + c_L b_{ML}\right), \\ Q^{\mathrm{gh}\dagger}_M &= 2c^\dagger_M b + c b^\dagger_M + \sum_L \left(2c_{ML} b^\dagger_L + c^\dagger_L b_{LM}\right). \end{aligned} \tag{6.10}$$

These generators satisfy the same conformal algebra as (6.9). In the following, we write the full generators including the *bc*-ghost part as

$$\mathcal{H} = H + H^{\mathrm{gh}}, \qquad \mathcal{R}_{MN} = R_{MN} + R^{\mathrm{gh}}_{MN},$$
$$\mathcal{Q}_M = Q_M + Q^{\mathrm{gh}}_M, \qquad \mathcal{Q}^\dagger_M = Q^\dagger_M + Q^{\mathrm{gh}\dagger}_M.$$

The nilpotent BRST operator that generates the conformal transformations (3.6) is given by [Ham12c]

$$\begin{aligned} Q_{\mathrm{BRST}} &= cH + \sum_M \left(c^\dagger_M Q_M + c_M Q^\dagger_M\right) + \sum_{M,N} c_{MN} R_{MN} \\ &\quad + \frac{1}{2} c H^{\mathrm{gh}} + \frac{1}{2} \sum_M \left(c^\dagger_M Q^{\mathrm{gh}}_M + c_M Q^{\mathrm{gh}\dagger}_M\right) + \frac{1}{2} \sum_{M,N} c_{MN} R^{\mathrm{gh}}_{MN}. \end{aligned}$$

This can be rewritten in the form

$$Q_{\mathrm{BRST}} = c\mathcal{H} + \sum_{M,N} c_{MN} \mathcal{R}_{MN} - bM - \sum_{M,N} b_{MN} Y_{MN} + \hat{Q},$$

where $\mathcal{H}$ and $\mathcal{R}_{MN}$ are the full generators defined above and the other operators are defined by

$$M = 2\sum_M \mathrm{c}_M^\dagger \mathrm{c}_M, \qquad Y_{MN} = \mathrm{c}_M^\dagger \mathrm{c}_N + \sum_L \mathrm{c}_{ML}\mathrm{c}_{LN},$$
$$\hat{Q} = \sum_M (\mathrm{c}_M^\dagger Q_M + \mathrm{c}_M Q_M^\dagger).$$

The nilpotency $Q_{\mathrm{BRST}}^2 = 0$ can be shown using the conformal algebra (6.9).

The BRST conformal transformations, or diffeomorphism, are expressed using commutation relations with the BRST operator as

$$i\,[Q_{\mathrm{BRST}}, \phi] = c^\mu \hat{\nabla}_\mu \phi + \frac{1}{4}\hat{\nabla}_\mu c^\mu,$$

and the same applies to the other fields.[4] For the ghost field, using an anti-commutation relation, it is given by

$$i\,\{Q_{\mathrm{BRST}}, c^\mu\} = c^\nu \hat{\nabla}_\nu c^\mu.$$

Anti-commutation relations between the BRST operator and the anti-ghost modes are calculated as

$$\left\{Q_{\mathrm{BRST}}, \mathrm{b}\right\} = \mathcal{H}, \qquad \left\{Q_{\mathrm{BRST}}, \mathrm{b}_{MN}\right\} = 2\,\mathcal{R}_{MN},$$
$$\left\{Q_{\mathrm{BRST}}, \mathrm{b}_M\right\} = \mathcal{Q}_M, \qquad \left\{Q_{\mathrm{BRST}}, \mathrm{b}_M^\dagger\right\} = \mathcal{Q}_M^\dagger. \tag{6.11}$$

This shows that the full generators are BRST trivial.

Physical states are now represented as BRST-invariant states. Hereafter, we denote it as $|\Psi\rangle$ and show that we can construct an infinite number of states satisfying

$$Q_{\mathrm{BRST}}|\Psi\rangle = 0. \tag{6.12}$$

First, let us define various vacuum states. Divide the bc-ghost and the other parts, and write an ordinary Fock vacuum that disappears for the latter annihilation operators such as a_{JM} and b_{JM} and the zero mode $\hat{p}$ as $|0\rangle$. In addition, a conformally invariant vacuum that disappears for all generators of conformal transformations excluding the bc-ghost part, H, R_{MN}, Q_M, $Q_M^\dagger$, is given by

$$|\Omega\rangle = e^{-2b_c\phi_0(0)}|0\rangle,$$

where $\phi_0(0) = \hat{q}/\sqrt{2b_c}$. From this expression, we can see that the vacuum $|\Omega\rangle$ and its Hermitian conjugate each have a background charge of $-2b_c$, giving a total of $-4b_c$.[5]

---

[4] In the transformation laws for the traceless tensor field and gauge fields, the Fradkin-Palchik term arises so that they do not conflict with the gauge-fixing condition [FP84, Ham09, Ham12a].

[5] The background charge originates from the linear term of ϕ in the Riegert action (2.21).

Next, we write a conformally invariant ghost vacuum that disappears for all generators (6.10) of the bc-ghost part as $|0\rangle_{\rm gh}$. This is a vacuum that disappears for all anti-ghosts, but not for gauge ghosts. On the other hand, a Fock ghost vacuum that disappears when the annihilation operators c_M and b_M are applied to it is expressed as $\prod \mathrm{c}_M |0\rangle_{\rm gh}$.[6]

Furthermore, we introduce a Fock vacuum with arbitrary charge γ, including the bc-ghost vacuum, as

$$|\gamma\rangle = e^{\gamma\phi_0(0)}|\Omega\rangle \otimes \prod_M \mathrm{c}_M |0\rangle_{\rm gh}. \tag{6.13}$$

This γ is called the Riegert charge. This vacuum is an eigenstate of the Hamiltonian that satisfies

$$\mathcal{H}|\gamma\rangle = (h_\gamma - 4)|\gamma\rangle, \qquad h_\gamma = \gamma - \frac{\gamma^2}{4b_c}. \tag{6.14}$$

In conformal field theory on $\mathbb{R} \times S^3$, the eigenvalue of $\mathcal{H}$ corresponds to a conformal dimension and -4 comes from the bc-ghost vacuum, where 4 represents the spacetime dimension.

Physical states are constructed by applying creation operators or zero-mode $\hat{p}$ to the Fock vacuum (6.13). Since the Fock vacuum disappears for anti-ghost modes b and b_{MN} and these modes satisfy (6.11), we can find that as physical states, it is sufficient to consider a subspace satisfying

$$\mathcal{H}|\Psi\rangle = \mathcal{R}_{MN}|\Psi\rangle = 0, \quad \mathrm{b}|\Psi\rangle = \mathrm{b}_{MN}|\Psi\rangle = 0. \tag{6.15}$$

On this subspace, the BRST-invariant state (6.12) coincides with a $\hat{Q}$-invariant state.

For the time being, we consider physical states of the following type constructed on subspace (6.15):

$$|\Psi\rangle = \mathcal{O}\big(\hat{p}, a^\dagger_{JM}, b^\dagger_{JM}, c^\dagger_{J(Mx)}, d^\dagger_{J(Mx)}, e^\dagger_{J(My)}\big)|\gamma\rangle. \tag{6.16}$$

The operator $\mathcal{O}$ and the Riegert charge γ are determined from the BRST conformal invariance condition. The state in which $\mathcal{O}$ contains the bc-ghost creation modes will be discussed later.

Since $\mathrm{c}_M|\Psi\rangle = 0$ holds for the state (6.16), the $\hat{Q}$-invariance condition is expressed as

$$\hat{Q}|\Psi\rangle = \sum_M \mathrm{c}^\dagger_M \mathcal{Q}_M |\Psi\rangle = 0.$$

---

[6] Since the Hamiltonian does not contain c, c_{MN}, b, and b_{MN}, the bc-ghost vacuum $\prod \mathrm{c}_M|0\rangle_{\rm gh}$ is degenerate. The degenerate partner is given by multiplying this vacuum by the Hermitian operator $\theta = i\mathrm{c}\prod \mathrm{c}_{MN}$. For the inner product structure, see [Ham12c].

Therefore, in this case, together with the Hamiltonian and the rotation invariance conditions in (6.15), the physical state condition (6.12) is expressed as

$$(H-4)|\Psi\rangle = R_{MN}|\Psi\rangle = Q_M|\Psi\rangle = 0, \tag{6.17}$$

where no condition is required for $Q^\dagger_M$. The condition (6.17) indicates that $|\Psi\rangle$ is given by a scalar primary state with conformal dimension 4.[7] Note here that descendant states generated by applying $Q^\dagger_M$ to the primary state become BRST trivial due to (6.11).

6.3 Construction of Physical States

In order to construct physical states, we first need to find and classify primary states that disappear when the special conformal transformation generator Q_M is applied. Among them, the states that further satisfy the Hamiltonian and rotation invariance conditions are physical states.

6.3.1 Primary States

To find primary states, we look for creation operators that commute with the generator Q_M.

First, consider the Riegert sector. Commutation relations between Q_M (6.5) and the zero modes are given by

$$\left[Q_M, \hat{q}\right] = -a_{\frac{1}{2}M}, \qquad \left[Q_M, \hat{p}\right] = 0.$$

For the positive-metric creation modes $a^\dagger_{JM}$ ($J \geq 1/2$) and the negative-metric creation modes $b^\dagger_{JM}$ ($J \geq 0$), we obtain

$$\left[Q_M, a^\dagger_{\frac{1}{2}M_1}\right] = \left(\sqrt{2b_c} - i\hat{p}\right)\delta_{MM_1},$$

$$\left[Q_M, a^\dagger_{JM_1}\right] = \alpha\left(J-\frac{1}{2}\right)\sum_{M_2} \mathbf{C}^{\frac{1}{2}M}_{JM_1, J-\frac{1}{2}M_2}\epsilon_{M_2} a^\dagger_{J-\frac{1}{2}-M_2},$$

[7] Primary states in normal conformal field theory on $\mathbb{R}\times S^3$ are defined by $H|h,\{r\}\rangle = h|h,\{r\}\rangle$, $R_{MN}|h,\{r\}\rangle = (\Sigma_{MN})_{\{r'\},\{r\}}|h,\{r'\}\rangle$, $Q_M|h,\{r\}\rangle = 0$, where h is a conformal dimension, $\{r\}$ is a representation of $SU(2)\times SU(2)$, and Σ_{MN} is its matrix representation.

and

$$[Q_M, b^\dagger_{JM_1}] = -\sum_{M_2} \mathbf{C}^{\frac{1}{2}M}_{JM_1,J+\frac{1}{2}M_2} \epsilon_{M_2} a^\dagger_{J+\frac{1}{2}-M_2}$$
$$-\beta\left(J-\frac{1}{2}\right)\sum_{M_2} \mathbf{C}^{\frac{1}{2}M}_{JM_1,J-\frac{1}{2}M_2} \epsilon_{M_2} b^\dagger_{J-\frac{1}{2}-M_2}, \quad (6.18)$$

respectively. From these formulas, it can be seen that there is no creation mode that commutes with Q_M, that is, there is no mode that is independently gauge invariant.

To construct creation operators that commute with Q_M, consider composite operators expressed as products of creation modes. The existence of such operators has been investigated. Writing only the results [AMM97b, Ham05], the Q_M-invariant creation operators with conformal dimension $2L$ are given for integer L (≥ 1) in two types as

$$S^\dagger_{LN} = \frac{\sqrt{2}(\sqrt{2b_c} - i\hat{p})}{\sqrt{(2L-1)(2L+1)}} a^\dagger_{LN}$$
$$+ \sum_{K=\frac{1}{2}}^{L-\frac{1}{2}} \sum_{M_1} \sum_{M_2} x(L,K) \mathbf{C}^{LN}_{L-KM_1,KM_2} a^\dagger_{L-KM_1} a^\dagger_{KM_2},$$
$$\mathcal{S}^\dagger_{L-1\,N} = -\sqrt{2}\left(\sqrt{2b_c} - i\hat{p}\right) b^\dagger_{L-1\,N}$$
$$+ \sum_{K=\frac{1}{2}}^{L-\frac{1}{2}} \sum_{M_1} \sum_{M_2} x(L,K) \mathbf{C}^{L-1\,N}_{L-KM_1,KM_2} a^\dagger_{L-KM_1} a^\dagger_{KM_2}$$
$$+ \sum_{K=\frac{1}{2}}^{L-1} \sum_{M_1} \sum_{M_2} y(L,K) \mathbf{C}^{L-1\,N}_{L-K-1M_1,KM_2} b^\dagger_{L-K-1M_1} a^\dagger_{KM_2},$$

where each coefficient is

$$x(L,K) = \frac{(-1)^{2K}}{\sqrt{(2L-2K+1)(2K+1)}} \sqrt{\binom{2L}{2K}\binom{2L-2}{2K-1}},$$
$$y(L,K) = -2\sqrt{(2L-2K-1)(2L-2K+1)}\, x(L,K). \quad (6.19)$$

The results are summarized in Table 6.1. In deriving and classifying these, the crossing relation (6.7) among the $SU(2) \times SU(2)$ Clebsch-Gordan coefficients plays a crucial role.

The Q_M-invariant creation operators for the Weyl sector can be constructed in the same way. Commutation relations between Q_M (6.8) and each of the tensor creation modes $c^\dagger_{J(Mx)}$, $d^\dagger_{J(Mx)}$, and $e^\dagger_{J(My)}$ with $J \geq 1$ are given by

Table 6.1 Building blocks of primary states for the Riegert sector

Rank of tensor	0
Creation operator	$S^{\dagger}_{LN}$
	$\mathcal{S}^{\dagger}_{L-1N}$
Conformal dimension	$2L$

Each exists only for an integer L (≥ 1)

Table 6.2 Building blocks of primary states for the Weyl sector

Rank of tensor	0	1	2	3	4
Creation operator	$A^{\dagger}_{LN}$	$B^{\dagger}_{L-\frac{1}{2}(Ny)}$	$c^{\dagger}_{1(Nx)}$	$D^{\dagger}_{L-\frac{1}{2}(Nz)}$	$E^{\dagger}_{L(Nw)}$
	$\mathcal{A}^{\dagger}_{L-1N}$				$\mathcal{E}^{\dagger}_{L-1(Nw)}$
Conformal dimension	$2L$	$2L$	2	$2L$	$2L$

Each exists only for an integer L (≥ 3), and here polarization indices ε_3 and ε_4 are denoted as z and w, respectively

$$
\left[Q_M, c^{\dagger}_{J(M_1x_1)}\right] = \alpha\left(J-\frac{1}{2}\right)\sum_{M_2,x_2}\epsilon_{M_2}\mathbf{E}^{\frac{1}{2}M}_{J(M_1x_1),J-\frac{1}{2}(-M_2x_2)}c^{\dagger}_{J-\frac{1}{2}(M_2x_2)},
$$

$$
\begin{aligned}
\left[Q_M, d^{\dagger}_{J(M_1x_1)}\right] = &-\sum_{M_2,x_2}\epsilon_{M_2}\mathbf{E}^{\frac{1}{2}M}_{J(M_1x_1),J+\frac{1}{2}(-M_2x_2)}c^{\dagger}_{J+\frac{1}{2}(M_2x_2)} \\
&-\beta\left(J-\frac{1}{2}\right)\sum_{M_2,x_2}\epsilon_{M_2}\mathbf{E}^{\frac{1}{2}M}_{J(M_1x_1),J-\frac{1}{2}(-M_2x_2)}d^{\dagger}_{J-\frac{1}{2}(M_2x_2)} \\
&-B(J)\sum_{M_2,y_2}\epsilon_{M_2}\mathbf{H}^{\frac{1}{2}M}_{J(M_1x_1);J(-M_2y_2)}e^{\dagger}_{J(M_2y_2)},
\end{aligned}
$$

$$
\begin{aligned}
\left[Q_M, e^{\dagger}_{J(M_1y_1)}\right] = &-A(J)\sum_{M_2,x_2}\epsilon_{M_2}\mathbf{H}^{\frac{1}{2}M}_{J(-M_2x_2);J(M_1y_1)}c^{\dagger}_{J(M_2x_2)} \\
&-C\left(J-\frac{1}{2}\right)\sum_{M_2,y_2}\epsilon_{M_2}\mathbf{D}^{\frac{1}{2}M}_{J(M_1y_1),J-\frac{1}{2}(-M_2y_2)}e^{\dagger}_{J-\frac{1}{2}(M_2y_2)}.
\end{aligned}
$$

The Weyl sector has a creation mode that commutes with Q_M, which is only the lowest positive-metric mode $c^{\dagger}_{1(Mx)}$ of the transverse-traceless tensor field. Since all other creation modes do not commutate with Q_M, we need to find the Q_M-invariant composite operators consisting of products of the creation modes. Such operators have been investigated using crossing relations for the $SU(2) \times SU(2)$ Clebsch-Gordan coefficients with tensor indices. They have fairly complicated forms with tensor indices of rank up to 4, and are finally summarized as shown in Table 6.2 [Ham05].

By using the $SU(2) \times SU(2)$ Clebsch-Gordan coefficients and combining the composite creation operators summarized in Tables 6.1 and 6.2, we can construct a basis for primary states. Moreover, it is inferred that any Q_M-invariant primary state can be reduced to such a basic form by using the crossing relations among the

Clebsch-Gordan coefficients. Therefore, the Q_M-invariant creation operators summarized in the tables are considered to form building blocks of primary states.

Let us concretely construct primary states using these building blocks. As examples, primary states with conformal dimension 2 are given by

$$\mathcal{S}^\dagger_{00}|\Omega\rangle, \qquad S^\dagger_{1N}|\Omega\rangle, \qquad c^\dagger_{1(Nx)}|\Omega\rangle,$$

where the second tensor state has nine components, and the last tensor state corresponds to the Weyl tensor $C_{\mu\nu\lambda\sigma}$ with ten components, in which the polarization $x = \pm 1$ represents the self-dual and anti-self-dual components. Primary states of conformal dimension 4 are given by three scalar states,

$$\left(\mathcal{S}^\dagger_{00}\right)^2|\Omega\rangle, \qquad \sum_N \epsilon_N S^\dagger_{1-N} S^\dagger_{1N}|\Omega\rangle, \qquad \sum_{N,x} \epsilon_N c^\dagger_{1(-Nx)} c^\dagger_{1(Nx)}|\Omega\rangle,$$

and two tensor states,

$$\mathcal{S}^\dagger_{1N}|\Omega\rangle, \qquad \sum_{N_1,x_1}\sum_{N_2,x_2} \mathbf{E}^{1N}_{1(N_1x_1),1(N_2x_2)} c^\dagger_{1(N_1x_1)} c^\dagger_{1(N_2x_2)}|\Omega\rangle.$$

Here note that some of these primary states do not satisfy the unitarity bound [FGG74, Mac77] discussed in normal conformal field theory, as a feature of higher-derivative theories. However, these states do not yet satisfy the BRST invariance conditions (6.12) or (6.17), thus they are not gauge-invariant states. The physical states are obtained by further imposing the Hamiltonian and rotation invariance conditions on these states.

6.3.2 Physical States in Quantum Gravity

The physical state $|\Psi\rangle$ (6.16) is now expressed using the building blocks in the form

$$|\Psi\rangle = \mathcal{O}(S^\dagger, \mathcal{S}^\dagger, \ldots)|\gamma\rangle, \tag{6.20}$$

instead of (6.16), where the dots represent other building blocks. The BRST invariance condition (6.17) requires that the operator $\mathcal{O}$ satisfies the following algebra:

$$[H, \mathcal{O}] = l\mathcal{O}, \quad [R_{MN}, \mathcal{O}] = 0, \quad [Q_M, \mathcal{O}] = 0,$$

where the first condition means that $\mathcal{O}$ has a conformal dimension l (≥ 0). Since the conformal dimension of the building blocks is even, l is given by an even number, which corresponds to the number of differentials acting on the gravitational field. All tensor indices are contracted using the $SU(2) \times SU(2)$ Clebsch-Gordan coefficients to be R_{MN}-invariant. Furthermore, the Riegert charge γ must satisfy

$$h_\gamma + l - 4 = 0$$

from the Hamiltonian condition (6.17). This is a quadratic equation in γ, and if we choose a solution that approaches the canonical value $4 - l$ in the classical limit of large b_c, then the Riegert charge is determined for each l to be

$$\gamma_l = 2b_c\left(1 - \sqrt{1 - \frac{4-l}{b_c}}\right). \tag{6.21}$$

Since $b_c > 4$ regardless of the contents of matter fields, γ_l is a real number.

Now, we present the physical state of quantum gravity with l up to 4. The lowest $l = 0$ state is a state in which the identity operator $\mathcal{O} = I$ is dressed by quantum gravity, and is simply expressed as

$$|\gamma_0\rangle. \tag{6.22}$$

This is a quantum state corresponding to the cosmological term which reduces to $\sqrt{-g}$ in the classical limit of $b_c \to \infty$. The physical state with $l = 2$ is the quantum scalar-curvature state, given by

$$\mathcal{S}_{00}^\dagger|\gamma_2\rangle. \tag{6.23}$$

This corresponds to $\sqrt{-g}R$ subjected to quantum corrections. There are three physical states with $l = 4$ as

$$\left(\mathcal{S}_{00}^\dagger\right)^2|\gamma_4\rangle, \qquad \sum_M \epsilon_M S_{1-M}^\dagger S_{1M}^\dagger|\gamma_4\rangle, \qquad \sum_{M,x} \epsilon_M c_{1(-Mx)}^\dagger c_{1(Mx)}^\dagger|\gamma_4\rangle.$$

Since $\gamma_4 = 0$ from (6.21), these states have no exponential factor of ϕ, and correspond to $\sqrt{-g}R^2$, $\sqrt{-g}\,(R_{\mu\nu} - g_{\mu\nu}R/4)^2$, and $\sqrt{-g}\,C_{\mu\nu\lambda\sigma}^2$, respectively.

Finally, we discuss a physical state containing the bc-ghost creation modes $\mathrm{c}_M^\dagger$ and $\mathrm{b}_M^\dagger$. For example, in the case of $l = 2$, in addition to the above, there is a BRST invariant state,

$$\left\{-\left(\sqrt{2b_c} - i\hat{p}\right)^2 \sum_M \epsilon_M \mathrm{b}_{-M}^\dagger \mathrm{c}_M^\dagger + \hat{h}\sum_M \epsilon_M a_{\frac{1}{2}-M}^\dagger a_{\frac{1}{2}M}^\dagger\right\}|\gamma_2\rangle,$$

where $\hat{h} = \hat{p}^2/2 + b_c$. However, we find that this state can be rewritten as

$$\frac{1}{2\sqrt{2}} S_{00}^{\dagger} |\gamma_2\rangle + Q_{\mathrm{BRST}} |\Upsilon\rangle$$

by introducing a new state $|\Upsilon\rangle = (\sqrt{2b_1} - i\hat{p}) \sum_M \epsilon_M \mathrm{b}_{-M}^{\dagger} a_{\frac{1}{2}M}^{\dagger} |\gamma_2\rangle$ in the subspace satisfying (6.15), and thus becomes BRST equivalent to the physical state already given. In general, the physical state containing the bc-ghost may be reduced to the standard form (6.20) up to such a BRST-trivial term.

6.4 Appendix: Two-Dimensional Quantum Gravity on $\mathbb{R} \times S^1$

In this appendix, we briefly describe two-dimensional quantum gravity, as a simpler model with the BRST conformal invariance.

6.4.1 Liouville Action, Virasoro Algebra, and Physical States

The exact solution of two-dimensional quantum gravity was found by Polyakov, Knizhnik, and Zamolodchikov [KPZ88], and then it was formulated in the conformal gauge by Distler and Kawai, David, and others [DK89, Dav88, Sei90, LZ91, BMP92].

In two dimensions, the traceless tensor field has two degrees of freedom, but it can be completely gauge-fixed using the gauge degrees of freedom ξ^μ. Thus, the conformal gauge

$$h_{\mu\nu} = 0 \tag{6.24}$$

is usually adopted as a gauge-fixing condition. The metric field is then decomposed as $g_{\mu\nu} = e^{2\phi}\hat{g}_{\mu\nu}$, where ϕ is the conformal-factor field and $\hat{g}_{\mu\nu}$ is the background metric.

In the conformal gauge, the action of two-dimensional quantum gravity is expressed as

$$S_{\mathrm{2DQG}} = S_{\mathrm{L}} + I_{\mathrm{M}} + I_{\mathrm{gh}},$$

where S_{L} is the Liouville action, which is a Wess-Zumino action for two-dimensional conformal anomaly, satisfying the same consistency condition as (2.11) (replacing $\bar{g}$ with $\hat{g}$). I_{M} is a conformally invariant matter action and I_{gh} is the gauge ghost action associated with the gauge 6.24, defined later.

Since the two-dimensional conformal anomaly is proportional to the scalar curvature, the Liouville action can be obtained by integrating it with ϕ as[8]

$$S_{\rm L} = -\frac{b_{\rm L}}{4\pi}\int d^2x \int_0^\phi d\phi \sqrt{-g}R$$
$$= -\frac{b_{\rm L}}{4\pi}\int d^2x \sqrt{-\hat{g}}\left(\phi\hat{\Delta}_2\phi + \hat{R}\phi\right) \tag{6.25}$$

where $\sqrt{-g}\Delta_2 = \sqrt{-g}(-\nabla^2)$ is a conformally invariant differential operator in two dimensions, and noting $\sqrt{-g}R = \sqrt{-\hat{g}}(2\hat{\Delta}_2\phi + \hat{R})$, we can easily perform the integration of ϕ. The coefficient $b_{\rm L}$ in front of the Liouville action is proportional to the sum of the coefficients of conformal anomalies, called central charges in two dimensions, which arise when quantizing each field. Letting c be a central charge of $I_{\rm M}$ described by conformal field theory [BPZ84, FQS84], it is given by

$$b_{\rm L} = -\frac{c-25}{6}, \tag{6.26}$$

where -25 of the numerator is the sum of 1 and -26, which are the contributions from the conformal-factor field and the gauge ghosts, respectively. Therefore, when $c < 25$, the Liouville action has a correct sign of $b_{\rm L} > 0$.

Let us quantize the gauge-fixed action $S_{\rm 2DQG}$ on the cylindrical background $\mathbb{R} \times S^1$, which is expressed using coordinates $x^\mu = (\eta, \sigma)$ with $0 < \sigma < 2\pi$ in flat background $\hat{g}_{\mu\nu} = \eta_{\mu\nu}$.

From the equation of motion $\partial^2\phi = (-\partial_\eta^2 + \partial_\sigma^2)\phi = 0$, the conformal-factor field is expanded using left- and right-moving modes and zero modes as

$$\phi(\eta,\sigma) = \frac{1}{\sqrt{2b_{\rm L}}}\left\{\hat{q} + 2\eta\hat{p} + \sum_{n\neq 0}\frac{i}{n}\left(\alpha_n^+ e^{-in(\eta+\sigma)} + \alpha_n^- e^{-in(\eta-\sigma)}\right)\right\}.$$

The conjugate momentum is given by $\Pi = (b_L/2\pi)\partial_\eta\phi$, and the equal-time commutation relation is set as $[\phi(\sigma), \Pi(\sigma')] = i\delta(\sigma-\sigma')$. The commutation relation for each mode is then given by

$$[\hat{q},\hat{p}] = i, \quad [\alpha_n^\pm, \alpha_m^\pm] = n\delta_{n+m,0}, \quad [\alpha_n^\pm, \alpha_m^\mp] = 0.$$

Since ϕ is a real field, the Hermitian conjugate of each mode is defined as $\alpha_n^{\pm\dagger} = \alpha_{-n}^\pm$.

Even if the conformal gauge (6.24) is adopted, all gauge degrees of freedom are not fixed. A gauge degree of freedom ζ^μ satisfying the conformal Killing equation

$$\partial_\mu\zeta_\nu + \partial_\nu\zeta_\mu - \eta_{\mu\nu}\partial^\lambda\zeta_\lambda = 0 \tag{6.27}$$

---

[8] Usually, the field is redefined as $\phi \to \phi/\sqrt{2b_{\rm L}}$ and the Liouville action is rewritten in the form $-(1/8\pi)\{(\partial\phi)^2 + Q\hat{R}\phi\}$ with $Q = \sqrt{2b_{\rm L}}$. However, here we quantize it without redefining the field to emphasize the resemblance to four-dimensional quantum gravity.

still remains. In fact, substituting this degrees of freedom into the gauge transformation of the traceless tensor field $\delta_\xi h_{\mu\nu} = \partial_\mu \xi_\nu + \partial_\nu \xi_\mu - \eta_{\mu\nu}\partial^\lambda \xi_\lambda$ yields $\delta_\zeta h_{\mu\nu} = 0$, which means that the gauge condition (6.24) is preserved.

The generator of conformal transformation associated with this residual gauge degree of freedom ζ^μ constitutes a conformal algebra, which is defined by

$$L_\zeta = \int_{S^1} d\sigma \zeta^\mu :\hat{T}_{\mu 0}: \qquad (6.28)$$

where $\hat{T}^{\mu\nu} = (2/\sqrt{-\hat{g}}) \times \delta S_{2\mathrm{DQG}}/\delta \hat{g}_{\mu\nu}$ is an energy-momentum tensor that satisfies the traceless condition. The symbol : : represents performing normal ordering.

The energy-momentum tensor in the Liouville sector is given by

$$\hat{T}^{\mathrm{L}}_{\mu\nu} = \frac{b_\mathrm{L}}{2\pi}\left\{\partial_\mu\phi\partial_\nu\phi - \frac{1}{2}\eta_{\mu\nu}\partial^\lambda\phi\partial_\lambda\phi + \left(\eta_{\mu\nu}\partial^\lambda\partial_\lambda - \partial_\mu\partial_\nu\right)\phi\right\},$$

where the last term is derived from the linear term $\hat{R}\phi$ of the Liouville action (6.25). Using the fact that the trace of the energy-momentum tensor vanishes in proportion to the equation of motion, we can show that the generator defined by (6.28) is conserved.

In two dimensions, there are an infinite number of conformal Killing vectors ζ^μ, given by $(e^{in(\eta+\sigma)}/2, e^{in(\eta+\sigma)}/2)$ and $(e^{in(\eta-\sigma)}/2, -e^{in(\eta-\sigma)}/2)$ for an integer n. Substituting these into (6.28), we obtain generators of two-dimensional conformal transformations, the so-called Virasoro generators, writing them as $L^\pm_n$. For the Liouville sector, we get

$$\begin{aligned} L^{\mathrm{L}\pm}_n &= e^{in\eta}\int_0^{2\pi} d\sigma e^{\pm in\sigma}\frac{1}{2} :\left(\hat{T}^{\mathrm{L}}_{00} \pm \hat{T}^{\mathrm{L}}_{01}\right): + \frac{b_\mathrm{L}}{4}\delta_{n,0} \\ &= \frac{1}{2}\sum_{m\in\mathbf{Z}} :\alpha^\pm_m \alpha^\pm_{n-m}: + i\sqrt{\frac{b_\mathrm{L}}{2}}n\alpha^\pm_n + \frac{b_\mathrm{L}}{4}\delta_{n,0}, \end{aligned}$$

where $\alpha^\pm_0 = \hat{p}$ and the Hermiticity is given by $L^{\mathrm{L}\pm\dagger}_n = L^{\mathrm{L}\pm}_{-n}$. The last term $(b_\mathrm{L}/4)\delta_{n,0}$ represents the Casimir effect on $\mathbb{R} \times S^1$, and the Hamiltonian is shifted by a constant $b_\mathrm{L}/2$ as

$$H^\mathrm{L} = L^{\mathrm{L}+}_0 + L^{\mathrm{L}-}_0 = \hat{p}^2 + \frac{b_\mathrm{L}}{2} + \sum_{n=1}^\infty \left\{\alpha^{+\dagger}_n\alpha^+_n + \alpha^{-\dagger}_n\alpha^-_n\right\}.$$

This energy shift is necessary for the Virasoro algebra to close at the quantum level.

The full Virasoro generator, including generators in the matter and gauge ghost sectors defined later, is denoted as

$$L_n^{\pm} = L_n^{\mathrm{L}\pm} + L_n^{\mathrm{M}\pm} + L_n^{\mathrm{gh}\pm},$$

which satisfies the Virasoro algebra

$$\left[L_n^{\pm}, L_m^{\pm}\right] = (n-m)L_{n+m}^{\pm} + \frac{c_{\mathrm{T}}}{12}(n^3 - n)\delta_{n+m,0}$$

and $\left[L_n^{+}, L_m^{-}\right] = 0$. The last term is a central extension of the $SO(2,2)$ conformal algebra that exists only in the two-dimensional case, where c_{T} is the central charge. It can be shown that the central charge of two-dimensional quantum gravity vanishes as

$$c_{\mathrm{T}} = 1 + 6b_{\mathrm{L}} + c - 26 = 0,$$

where c and -26 come from the matter and gauge ghost fields, respectively. The ϕ field has the central charge $1 + 6b_{\mathrm{L}}$, in which 1 comes from the fact that ϕ is a scalar-like quantum boson field, while $6b_{\mathrm{L}}$ is derived from the $\hat{R}\phi$ term of the Liouville action, which is not conformally invariant classically. In fact, the Virasoro algebra at the Poisson brackets level has the non-zero central charge $6b_{\mathrm{L}}$. Hence, conformal symmetry that is broken for quantum fields in curved space is exactly restored when gravity is quantized.

In this way, diffeomorphism invariance is realized as conformal invariance, and all theories with different background spacetimes that can be connected one another by conformal transformations become gauge equivalent. In other words, the background freedom (also called the reparameterization invariance in two dimensions) holds.

Now, describe physical states in two-dimensional quantum gravity. The ghost sector will be treated in Sect. 6.4.2, but here it is considered as it is integrated out, and its contribution is implicitly included in the central charge and a shift in energy mentioned below. First, introduce a conformally invariant vacuum defined by a state satisfying condition $L_n^{\mathrm{L}\pm}|\Omega\rangle_{\mathrm{L}} = 0$ for $n \geq -1$, which is given by

$$|\Omega\rangle_{\mathrm{L}} = e^{-b_{\mathrm{L}}\phi_0}|0\rangle_{\mathrm{L}}, \tag{6.29}$$

where $\phi_0 = \hat{q}/\sqrt{2b_{\mathrm{L}}}$ is the zero mode and $|0\rangle_{\mathrm{L}}$ is a normal Fock vacuum that vanishes for all annihilation modes of ϕ. Commonly, exponents of the zero mode are called Liouville charges. The Liouville charge $-b_{\mathrm{L}}$ possessed by the vacuum (6.29) and the charge $-b_{\mathrm{L}}$ in its conjugate vacuum, totally $-2b_{\mathrm{L}}$, is especially called the background charge, which originates from the $\hat{R}\phi$ term.

Furthermore, we introduce a Fock vacuum state $|\gamma\rangle_{\mathrm{L}} = e^{\gamma\phi_0}|\Omega\rangle_{\mathrm{L}}$ with Liouville charge γ, which satisfies an eigenequation $H^{\mathrm{L}}|\gamma\rangle_{\mathrm{L}} = h_\gamma|\gamma\rangle_{\mathrm{L}}$ with eigenvalue $h_\gamma = \gamma - \gamma^2/2b_{\mathrm{L}}$. Consider a vacuum defined by the product of this vacuum and a conformally invariant matter vacuum $|\Omega\rangle_{\mathrm{M}}$, and construct a state by applying creation modes to such a vacuum as follows:

$$|\Psi'\rangle = \mathcal{O}(\alpha_n^{\pm\dagger}, \cdots)|\gamma\rangle_{\mathrm{L}} \otimes |\Omega\rangle_{\mathrm{M}},$$

where the dots represent matter creation modes. Among these states, physical states are defined as those satisfying the Virasoro conditions

$$\left(H^{\mathrm{L}} + H^{\mathrm{M}} - 2\right)|\Psi'\rangle = 0, \quad \left(L_n^{\mathrm{L}\pm} + L_n^{\mathrm{M}\pm}\right)|\Psi'\rangle = 0 \quad (n \geq 1), \tag{6.30}$$

where $H^{\mathrm{M}} = L_0^{\mathrm{M}+} + L_0^{\mathrm{M}-}$. These are nothing but the Hamiltonian and momentum constraints (precisely combinations of the two) in two-dimensional quantum gravity. The shift -2 in the Hamiltonian condition is a contribution from the gauge ghost, and 2 reflects the spacetime dimension.

For simplicity, here we consider only the physical state in which a primary state of conformal field theory representing a matter state is dressed by gravity with Liouville charge only. The state can be expressed as $|\gamma_d\rangle_{\mathrm{L}} \otimes |d\rangle_{\mathrm{M}}$, in which $|d\rangle_{\mathrm{M}}$ denotes a real primary matter state with the same left and right conformal dimension d defined by $L_0^{\mathrm{M}\pm}|d\rangle_{\mathrm{M}} = d|d\rangle_{\mathrm{M}}$ and $L_n^{\mathrm{M}\pm}|d\rangle_{\mathrm{M}} = 0\,(n \geq 1)$, and $|\gamma_d\rangle_{\mathrm{L}}$ is a gravitational part with Liouville charge γ_d applying to it. The charge γ_d is determined by solving the equation

$$h_{\gamma_d} + 2d = 2,$$

derived from the Hamiltonian condition (6.30). Choosing a physical solution value that approaches the classical value $2 - 2d$ at $b_{\mathrm{L}} \to \infty$ gives

$$\gamma_d = b_{\mathrm{L}}\left(1 - \sqrt{1 - \frac{4 - 4d}{b_{\mathrm{L}}}}\right). \tag{6.31}$$

This state corresponds to $:e^{\gamma_d\phi}\Phi_d:$, where Φ_d represents the matter conformal field. In particular, the $d = 0$ field, $:e^{\gamma_0\phi}:$, is the cosmological term operator corresponding to $\sqrt{-g}$.

6.4.2 BRST Formulation

Lastly, we describe the BRST analysis of physical states in two-dimensional quantum gravity [LZ91, BMP92]. This type of the BRST conformal invariance has been first formulated in string theory [KO83, FMS86].

In the conformal gauge (6.24), letting c^μ be the ghost field, the anti-ghost field is represented by a symmetric traceless tensor field $b_{\mu\nu}$. These are called the bc-ghosts, whose action is given by $I_{\mathrm{gh}} = (-i/\pi)\int d^2x\sqrt{-\hat{g}}\, b_{\mu\nu}\hat{\nabla}^\mu c^\nu$.

Rewriting the ghost and anti-ghost fields as $c^\pm = c^0 \pm c^1$ and $b_{\pm\pm} = b_{00} \pm b_{01}$, the bc-ghost action can be written using $\partial_\pm = (\partial_\eta \pm \partial_\sigma)/2$ as

$$I_{\mathrm{gh}} = \frac{i}{\pi}\int d^2x\left(b_{++}\partial_- c^+ + b_{--}\partial_+ c^-\right).$$

The equations of motion are then simplified as $\partial_- c^+ = \partial_+ c^- = 0$ and $\partial_- b_{++} = \partial_+ b_{--} = 0$, where that of the ghost field is nothing but a rewrite of the conformal Killing equation (6.27) with ζ^μ replaced by c^μ. Thus, the ghost and anti-ghost fields are expanded into modes as $c^\pm = \sum_{n\in\mathbf{Z}} c_n^\pm e^{-in(\eta\pm\sigma)}$ and $b_{\pm\pm} = \sum_{n\in\mathbf{Z}} b_n^\pm e^{-in(\eta\pm\sigma)}$, respectively.

The Hermitian conjugate of each mode is defined by $c_n^{\pm\dagger} = c_{-n}^\pm$ and $b_n^{\pm\dagger} = b_{-n}^\pm$. Since the conjugate momentum of the ghost field $c^\pm$ is $ib_{\pm\pm}/2\pi$, the equal-time anti-commutation relations are defined by $\{c^\pm(\sigma), b_{\pm\pm}(\sigma')\} = 2\pi\delta(\sigma - \sigma')$ and $\{c^\pm(\sigma), b_{\mp\mp}(\sigma')\} = 0$. From these, the anti-commutation relation for each mode is given by

$$\{c_n^\pm, b_m^\pm\} = \delta_{n+m,0}, \qquad \{c_n^\pm, b_m^\mp\} = 0.$$

The Virasoro generator of the bc-ghost sector is then given by

$$L_n^{\mathrm{gh}\pm} = \sum_{m\in\mathbf{Z}} (n+m) :b_{n-m}^\pm c_m^\pm:,$$

where we choose the normal ordering defined for a conformally invariant bc-ghost vacuum $|0\rangle_{\mathrm{gh}}$ that satisfies $L_n^{\mathrm{gh}\pm}|0\rangle_{\mathrm{gh}} = 0$ $(n \geq -1)$, that is, the annihilation modes are setting as $c_n^\pm|0\rangle_{\mathrm{gh}} = 0$ $(n \geq 2)$ and $b_n^\pm|0\rangle_{\mathrm{gh}} = 0$ $(n \geq -1)$ and they are placed on the right.[9]

The BRST operator that is the generator of diffeomorphism is given by $Q_{\mathrm{BRST}} = Q^+ + Q^-$, where

$$Q^\pm = \sum_{n\in\mathbf{Z}} c_{-n}^\pm \left(L_n^{\mathrm{L}\pm} + L_n^{\mathrm{M}\pm}\right) - \frac{1}{2}\sum_{n,m\in\mathbf{Z}} (n-m) :c_{-n}^\pm c_{-m}^\pm b_{n+m}^\pm: .$$

This can be further decomposed as

$$Q^\pm = c_0^\pm L_0^\pm - b_0^\pm M^\pm + \hat{Q}^\pm, \qquad M^\pm = 2\sum_{n=1}^{\infty} n c_{-n}^\pm c_n^\pm,$$

where $L_0^\pm$ is the full Hamiltonian including the bc-ghost sector, and the last term of $Q^\pm$ that does not contain the ghost zero-modes is given by

$$\hat{Q}^\pm = \sum_{n\neq 0} c_{-n}^\pm \left(L_n^{\mathrm{L}\pm} + L_n^{\mathrm{M}\pm}\right) - \frac{1}{2}\sum_{\substack{n,m\neq 0\\ n+m\neq 0}} (n-m) :c_{-n}^\pm c_{-m}^\pm b_{n+m}^\pm:$$

[9] This is called the conformal normal-ordering. On the other hand, what is called the creation-annihilation normal-ordering is defined for the Fock ghost vacuum such that $c_n^\pm$ and $b_n^\pm$ with $n > 0$ are placed to the right. If we write it by $\ddagger\ \ddagger$, the generator is described as $L_n^{\mathrm{gh}\pm} = \sum_{m\in\mathbf{Z}}(n+m)\ddagger b_{n-m}^\pm c_m^\pm \ddagger - \delta_{n,0}$.

The nilpotency $Q_{\mathrm{BRST}}^2 = 0$ of the BRST operator is then expressed by $\hat{Q}^{\pm 2} = L_0^{\pm} M^{\pm}$ and $[\hat{Q}^{\pm}, L_0^{\pm}] = [\hat{Q}^{\pm}, M^{\pm}] = [L_0^{\pm}, M^{\pm}] = 0$. In addition, the full Virasoro generator is BRST-trivial such as $L_n^{\pm} = \{Q_{\mathrm{BRST}}, b_n^{\pm}\}$.

In order to construct the physical state, we introduce a Fock ghost vacuum $c_1^{+} c_1^{-} |0\rangle_{\mathrm{gh}}$ annihilated by $c_n^{\pm}, b_n^{\pm}$ $(n > 0)$, and consider a state obtained by applying the creation modes $\alpha_n^{\pm\dagger}, c_n^{\pm\dagger}, b_n^{\pm\dagger}$ $(n > 0)$ to the full Fock vacuum with Liouville charge γ as follows:

$$|\Psi\rangle = \mathcal{O}(\alpha_n^{\pm\dagger}, c_n^{\pm\dagger}, b_n^{\pm\dagger}, \cdots)|\gamma\rangle_L \otimes |\Omega\rangle_{\mathrm{M}} \otimes c_1^{+} c_1^{-} |0\rangle_{\mathrm{gh}}.$$

The physical state is then obtained by imposing the BRST invariance condition

$$Q_{\mathrm{BRST}}|\Psi\rangle = 0.$$

Since the state $|\Psi\rangle$ annihilates when the zero-mode $b_0^{\pm}$ is applied to it, we only consider a subspace satisfying the following conditions:

$$b_0^{\pm}|\Psi\rangle = 0, \qquad L_0^{\pm}|\Psi\rangle = 0, \tag{6.32}$$

where the second condition comes from the fact that the full Hamiltonian is BRST-trivial such as $L_0^{\pm} = \{Q_{\mathrm{BRST}}, b_0^{\pm}\}$. Therefore, the BRST invariance condition on the subspace (6.32) reduces to

$$\hat{Q}^{\pm}|\Psi\rangle = 0. \tag{6.33}$$

As discussed before, if we consider the cases that $\mathcal{O}$ does not include the ghost modes, the condition (6.33) reduces to the Virasoro condition (6.30), where the energy shift -2 arises from the algebra $L_0^{\mathrm{gh}\pm} c_1^{\pm}|0\rangle_{\mathrm{gh}} = -c_1^{\pm}|0\rangle_{\mathrm{gh}}$.

In two-dimensional quantum gravity, there are various physical states not discussed here, called discrete states and grand ring states [Wit92]. Furthermore, combining these states we can construct the W_∞ current. By applying this current to correlation functions, non-linear structures called the W-algebra constraints [FKN91] can be derived as Ward identities of W_∞ symmetry [Ham94].

Chapter 7
Quantum Gravity Inflation

The idea of inflation had been developed to solve the "horizon problem" of why there were larger correlations than the horizon size in the early universe and the "flatness problem" of why curvature of the universe remains almost zero even after more than 10 billion years [Gut81, Sat81, Sta80, Lin82, AS82]. Taking those thoughts for granted, this book focuses on another important role of inflation: the generation of primordial fluctuations, which give the initial conditions of the current universe. It is known that their spectra must be scalar-like and almost scale-invariant with very small amplitudes, at least for fluctuations in size involved in the subsequent structure formation of the universe.

In this chapter, we first show that the quantum gravity theory has a stable inflationary solution. We then present a scenario for the evolution of the early universe from the world far beyond the Planck scale until the spacetime phase transition occurs. Furthermore, we show that the entropy of the current universe is generated from fluctuations in quantum gravity.

7.1 Stable Inflationary Solution

The equation of motion that describes the evolution of spacetime is derived from a variation of the quantum gravity effective action with respect to the gravitational field as

$$\begin{aligned}\delta\Gamma &= \frac{1}{2}\int d^4x\sqrt{-g}\,T^{\mu\nu}\delta g_{\mu\nu} \\ &= \int d^4x\left(\mathbf{T}^{\mu}{}_{\mu}\delta\phi + \frac{1}{2}\mathbf{T}^{\mu}{}_{\nu}\delta h^{\nu}{}_{\mu}\right) = 0. \end{aligned} \tag{7.1}$$

K. Hamada, *Trans-Planckian Physics and Inflation*, Fundamental Theories of Physics 223, https://doi.org/10.1007/978-981-96-3476-7_7

The background metric employs the Minkowski metric $\eta_{\mu\nu}$ here, whose coordinates $x^\mu = (\eta, \mathbf{x})$ are called comoving coordinates, and η is the conformal time. Note that raising or lowering of the indices in the bold energy-momentum tensor is performed by the background metric as $\mathbf{T}_{\mu\nu} = \eta_{\mu\lambda}\mathbf{T}^\lambda{}_\nu$, in contrast to that $T^{\mu\nu}$ is done with the full metric $g_{\mu\nu}$.

The energy-momentum tensor is expressed as the sum of contributions from the Riegert, Weyl, Einstein-Hilbert, and matter sectors, and is written as

$$\mathbf{T}_{\mu\nu} = \mathbf{T}^{\mathrm{R}}_{\mu\nu} + \mathbf{T}^{\mathrm{W}}_{\mu\nu} + \mathbf{T}^{\mathrm{EH}}_{\mu\nu} + \mathbf{T}^{\mathrm{M}}_{\mu\nu} = 0.$$

It is known that an inflationary solution exists if we set $M_{\mathrm{P}} > \Lambda_{\mathrm{QG}}$, that is, if we assume that the spacetime phase transition occurs below the Planck scale [HY05, HHY06a, HHY10].

Let us decompose the conformal-factor field ϕ into a spatially homogeneous component $\hat{\phi}$ and fluctuations (perturbations) around it. Here consider only $\hat{\phi}$, which is an average value of spacetime fluctuations. The homogeneous nonlinear equation is given from the trace equation $\mathbf{T}^\mu{}_\mu = 0$ as

$$-\frac{b_c}{4\pi^2} B \partial_\eta^4 \hat{\phi} + 6M_{\mathrm{P}}^2 e^{2\hat{\phi}} \left(\partial_\eta^2 \hat{\phi} + \partial_\eta \hat{\phi} \partial_\eta \hat{\phi}\right) = 0. \tag{7.2}$$

The first term is from the Riegert action (2.21) and the second term is from the Einstein-Hilbert action. On the other hand, contributions from the Weyl and matter sectors occur near the spacetime phase transition, which will be introduced later.

The scale factor a and the proper (physical) time τ associated with it are defined as

$$a = e^{\hat{\phi}}, \qquad d\tau = a d\eta. \tag{7.3}$$

Furthermore, introducing the Hubble variable H and its derivatives as

$$H = \frac{\partial_\tau a}{a}, \qquad Y = \partial_\tau H, \qquad Z = \partial_\tau^2 H, \tag{7.4}$$

and rewriting (7.2) using these variables yields

$$B\left(\partial_\tau Z + 7\, HZ + 4Y^2 + 18\, H^2 Y + 6\, H^4\right) - 3H_{\mathrm{D}}^2\left(Y + 2\, H^2\right) = 0. \tag{7.5}$$

At the initial stage where the coupling constant t vanishes and thus B (2.22) is unity, this equation has an inflationary solution

$$H = H_{\mathrm{D}}, \quad H_{\mathrm{D}} = \sqrt{\frac{8\pi^2}{b_c}} M_{\mathrm{P}}, \tag{7.6}$$

where the value of the coefficient b_c (2.23) depends on the matter contents, but it is around 10, thus it can be seen that the value of $H_{\rm D}$ is between the Planck mass $m_{\rm pl}$ and the reduced Planck mass $M_{\rm P}$. Therefore, this is also regarded as one of the Planck mass scales. This expresses that the scale factor of the universe expands exponentially as

$$a(\tau) \propto e^{H_{\rm D}\tau}, \tag{7.7}$$

and thus starts to increase rapidly from the Planck time $\tau_{\rm P} = 1/H_{\rm D}$.

The inflationary solution (7.6) is stable, and we can see that the solution converges to (7.6) even if the initial conditions are changed significantly. In fact, if we substitute $H = H_{\rm D}(1+\delta)$ into (7.5) of $B = 1$ with δ as a small deviation from the solution and ignore $o(\delta^2)$, we get

$$(\partial_\tau^3 + 7H_{\rm D}\partial_\tau^2 + 15H_{\rm D}^2\partial_\tau + 12H_{\rm D}^3)\delta = 0.$$

If we solve it by setting as $\delta = e^{v\tau}$, we find three solutions for v, $-4H_{\rm D}$ and $(-3/2 \pm i\sqrt{3}/2)H_{\rm D}$. The real parts of them are all negative and thus the deviation reduces immediately, so that the initial value problem is significantly improved. Furthermore, it can be shown later that this inflationary solution is stable against spacetime fluctuations around it, that is, the fluctuations reduce in amplitude over time.

If we employ the perfect-fluid type for the traceless energy-momentum tensor $\mathbf{T}^{\rm M}_{\mu\nu}$ and let ρ be an energy density of matter fields, then it is expressed under the inflationary solution as

$$\mathbf{T}^{\rm M}_{00} = e^{4\hat{\phi}}\rho, \qquad \mathbf{T}^{\rm M}_{ij} = \frac{1}{3}\delta_{ij}e^{4\hat{\phi}}\rho.$$

Using this expression, the energy conservation equation is derived from $\mathbf{T}_{00} = 0$ as

$$\frac{b_c}{8\pi^2}B\big(2\partial_\eta^3\hat{\phi}\partial_\eta\hat{\phi} - \partial_\eta^2\hat{\phi}\partial_\eta^2\hat{\phi}\big) - 3M_{\rm P}^2 e^{2\hat{\phi}}\partial_\eta\hat{\phi}\partial_\eta\hat{\phi} + e^{4\hat{\phi}}\rho = 0.$$

Rewriting this equation in terms of the Hubble variables yields

$$B\big(2HZ - Y^2 + 6H^2Y + 3H^4\big) - 3H_{\rm D}^2H^2 + \frac{8\pi^2}{b_c}\rho = 0. \tag{7.8}$$

Substituting (7.6) into this equation reveals that $\rho = 0$ at the initial stage when the coupling constant t disappears. Thus, matter is generated when t increases so that B becomes less than unity.

7.2 Big Bang as Spacetime Phase Transition

As the universe begins to expand, the running coupling constant $\bar{t}$ (2.28) increases accordingly. As a result, it gradually deviates from the conformally-flat spacetime, and inflation eventually terminates at the dynamical energy scale Λ_{QG} where $\bar{t}$ diverges. At this time, the conformal gravity dynamics disappears completely and the classical spacetime phase emerges.

In order to express its dyanmics as time evolution, the running coupling constant is approximated by a time-dependent mean field [HHY06a]. That is defined by replacing the physical momentum Q in (2.28) with the reciprocal of the proper time τ as

$$\bar{t}^2(\tau) = \frac{1}{\beta_0 \log(\tau_\Lambda^2/\tau^2)} \tag{7.9}$$

where $\tau_\Lambda = 1/\Lambda_{QG}$ is called the dynamical time. This running coupling constant increases logarithmically as inflation progresses, and diverges rapidly at the dynamical time τ_Λ. The Weyl part (2.26) of the effective action then disappears in proportional to the reciprocal of $\bar{t}^2$.

The effective action of the Riegert sector, involving the strong coupling dynamics, is also described by replacing t^2 in B (2.22) with $\bar{t}^2(\tau)$. Here, denote it as $\bar{B}(\tau)$ and consider equations in which B in the equations of motion is simply replaced with $\bar{B}(\tau)$. Furthermore, in order to express that the dynamics of the Riegert sector also disappears when $\bar{t}^2$ diverges, let us consider boldly summing up it in the following form called dynamical factor:

$$\bar{B}(\tau) = \left[1 + \frac{\gamma_1}{\kappa} \frac{\bar{t}^2(\tau)}{4\pi} \right]^{-\kappa}, \tag{7.10}$$

where κ is a phenomenological parameter, assuming that it takes a value within $0 < \kappa \leq 1$. In addition, along with the modeling of the strong coupling dynamics, β_0 and γ_1 are also treated as positive phenomenological parameters below. These three are called dynamical-factor parameters.

In this way, the spacetime phase transition can be expressed as a process in which all of the fourth-derivative gravitational terms that are responsible for conformal gravity dynamics disappear at the dynamical time τ_Λ. At that time, energy of conformal gravity disappeared is transferred to energy of matter fields and the Big Bang occurs. The energy conservation equation (7.8), in which B is replaced with $\bar{B}$, describes such a process. The matter energy, initially $\rho = 0$, is generated in the process of $\bar{B}$ disappearing as the running coupling constant increases. The change occurs rapidly near the phase transition, and $\rho(\tau_\Lambda) = 3M_P^2 H^2(\tau_\Lambda)$ is obtained. One of the characteristic of the spacetime phase transition is that no matter existed before the Big Bang.

Since inflation starts at the Planck time τ_{P} and ends at the dynamical time τ_Λ, the number of e-foldings is defined by

$$\mathcal{N}_e = \log \frac{a(\tau_\Lambda)}{a(\tau_{\mathrm{P}})}. \tag{7.11}$$

Most of the inflationary period is almost described by the solution (7.7) because the time-dependent running coupling constant (7.9) and the dynamical factor (7.10) are functions that change significantly near τ_Λ. Therefore, $\mathcal{N}_e$ is approximately given by the ratio of H_{D} and Λ_{QG}, which is now set as

$$N = \frac{H_{\mathrm{D}}}{\Lambda_{\mathrm{QG}}} = 60. \tag{7.12}$$

The actual number of e-foldings is slightly larger than N. Here, we fix N to this value, and the inflationary scenario is adjusted by using the dynamical-factor parameters.

In order to fix the value of H_{D}, we here adopt $b_c = 7$, which corresponds to choosing the Standard Model matter contents. The dynamical energy scale is then determined from (7.12) to be

$$\Lambda_{\mathrm{QG}} = 1.3 \times 10^{17}\,\mathrm{GeV},$$

where $M_{\mathrm{P}} = 2.4 \times 10^{18}$ GeV is used.

The number of e-foldings is not the only condition that determines the value of Λ_{QG}. The magnitude of fluctuations also limits Λ_{QG}. If we roughly estimate the magnitude of a scalar curvature fluctuation during the inflationary period from dimensional analysis, it is given by $\delta R/R \sim E^2/12H_{\mathrm{D}}^2$, where E is an energy scale and the denominator is the scalar curvature of the inflationary (de Sitter) spacetime. This suggests that the initially large fluctuation gradually reduces, and from (7.12) it becomes

$$\left.\frac{\delta R}{R}\right|_{\tau_\Lambda} \sim \frac{\Lambda_{\mathrm{QG}}^2}{12H_{\mathrm{D}}^2} \sim 10^{-4} - 10^{-5} \tag{7.13}$$

at the energy scale Λ_{QG} where the phase transition occurs. The square of this value is consistent with the amplitude of the primordial scalar spectrum revealed by CMB observations. How spacetime fluctuations around the inflationary solution actually reduces in amplitude will be demonstrated in Chap. 10.

Figures 7.1 and 7.2 shows a numerical solution of the homogeneous equations. The dynamical-factor parameters are set as $\beta_0 = 0.4$, $\gamma_1 = 0.1$, and $\kappa = 1$.The horizontal axis is a dimensionless time $t = H_{\mathrm{D}}\tau$ (not to be confused with the coupling constant). The Hubble variable H is obtained by solving (7.5) with $\bar{B}$ instead of B.[1] The matter

[1] At the phase transition point τ_Λ, the third derivative of H diverges, but the physical quantity ρ remains finite. Numerical calculations are stopped at $t_\Lambda - \epsilon$ just before the running coupling

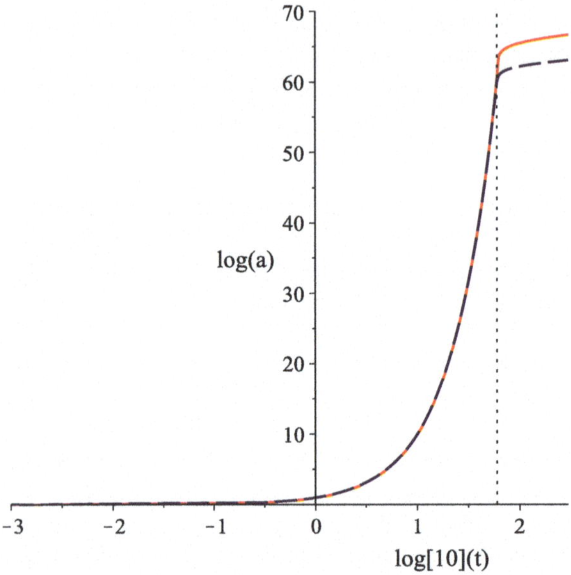

Fig. 7.1 Time evolution of the scale factor a in quantum gravity inflation (solid red line), where H_D is normalized to unity, and then $\tau_\Lambda = 60$. The vertical dotted line represents the phase transition time. The subsequent behavior is derived from the low-energy effective theory of gravity, discussed later. For comparison, the dashed violet line obtained by smoothly connecting $a \propto e^{H_D \tau}$ and the ordinary Friedmann solution at the transition point is displayed

energy density ρ (not depicted in figures) can be obtained by substituting the result of H into (7.8) with $\bar{B}$.

To better understand the physical meaning of the energy scales, we introduce the following new scales, comoving Planck mass and comoving dynamical energy:

$$m = \frac{a(\tau_i)}{a_0} H_D, \qquad \lambda = \frac{a(\tau_i)}{a_0} \Lambda_{QG}, \tag{7.14}$$

where $a(\tau_i)$ is the scale factor at a time $\tau_i = 1/E_i$ ($E_i > H_D$) before the Planck time when the scale factor starts to increase, and a_0 is the current scale factor.

Considering that this λ is the scale that explains the sharp falloff at the $l = 2$ multipole component of the CMB anisotropy spectrum (1.4), it is close to the Hubble constant H_0, thus $a(\tau_i)/a_0 \sim 10^{-59}$ is derived. Furthermore, if the current Friedmann universe was created after the Big Bang occurred around 10^{17} GeV, then the universe has expanded approximately 10^{29} times since the Big Bang. Therefore, to be a total of 10^{59}, it is required to expand by 10^{30} during the inflationary period, which corresponds to $\mathcal{N}_e \simeq 70$ in terms of the number of e-foldings.

---

constant diverges, and ϵ is reduced until the result does not change. Figures 7.1 and 7.2 shows the results calculated in this way.

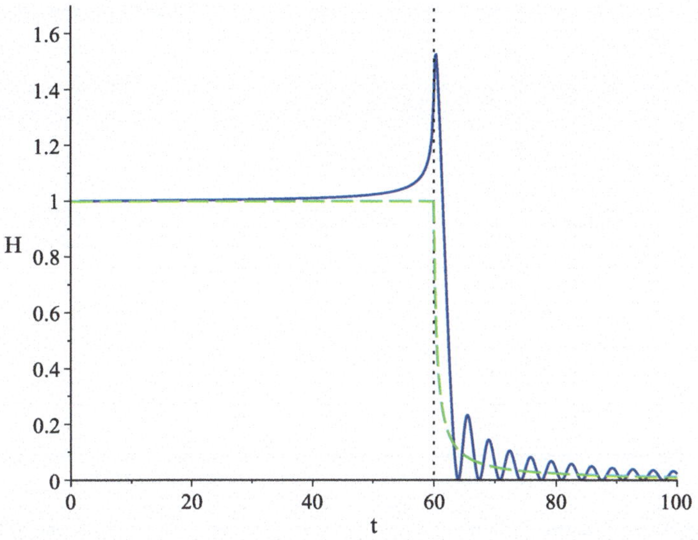

Fig. 7.2 Time evolution of the Hubble variable H (solid blue line). The parameter settings are the same as in Fig. 7.1. After the phase transition, H converges to the Friedmann solution while oscillating. The dashed green line displays $H = H_D$ and the ordinary Friedmann solution

The value $\mathcal{N}_e \simeq 70$ includes not only the e-foldings number (7.11) up to the phase transition but also a contribution from the accelerated expansion immediately after the phase transition (Figs. 7.1 and 7.2). Since the value of $\mathcal{N}_e$ itself depends on the dynamical-factor parameters and becomes larger than the scale ratio N (7.12), we find that N cannot be made very large. Under this consideration, we have constructed the scenario with $N = 60$ fixed here. Hence, the novel physical constant Λ_{QG} is a relatively tight parameter when considering the scenario of the evolution of the universe.

Figure 7.3 gives an outline of the quantum gravity inflation scenario. It shows that most of the universe we see today, measured by the Hubble distance, was originally the size of a quantum gravity excitation given by the correlation length ξ_Λ $(= 1/\Lambda_{QG})$, i.e., $1/H_0 \simeq 10^{59}\xi_\Lambda$ or $H_0 \simeq \lambda$. This suggests that the universe began as an entangled quantum system of gravity, and that the origin of the scale-invariant spectrum lies in the conformal invariance that existed there.

A consideration of spacetime fluctuations around the inflationary solution is given in Chap. 10. There, if we take, for example, $m = 0.02\,\mathrm{Mpc}^{-1}$ as a reference value for the size of spacetime fluctuations involved in the subsequent large-scale structure formation of the universe, we obtain λ $(= m/N) = 0.00033\,\mathrm{Mpc}^{-1}$ close to the Hubble constant.[2] Using these values, we solve evolution equations of the fluctuations and demonstrate that the fluctuations initially large actually reduce over time. The spectrum of the reduced fluctuation evaluated at the phase transition point is what is

[2] Using $1\mathrm{Mpc} = 3.1 \times 10^{24}\,\mathrm{cm}$ and $1\mathrm{cm} = 5.1 \times 10^{13}\,\mathrm{GeV}^{-1}$, we get $a(\tau_i)/a_0 = 1.6 \times 10^{-59}$.

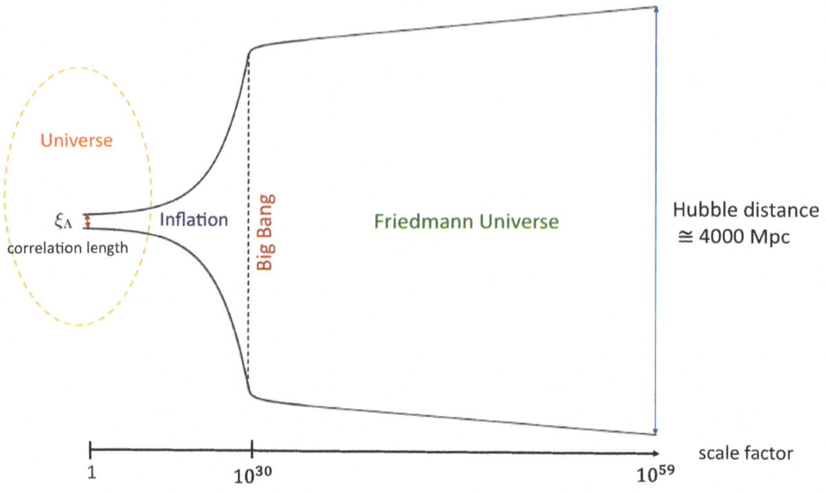

Fig. 7.3 Evolution scenario of quantum gravity inflation. The universe began as one of bubbles of the correlation length size excited by quantum gravity. Reprinted from [Ham24b] under the Creative Commons Attribution (CC-BY) license

called the primordial spectrum that provides the initial condition of the Friedmann universe. At this time, the overall amplitude is affected by the dynamical-factor parameters, but its spectral pattern is almost determined when the coupling constant is small, so that it is not affected by these parameters. This shows that the modeling made here is also quantitatively valid.

During the inflationary period, quantum fluctuations of gravity reduce their amplitude while being stretched. At the same time, fluctuations of various sizes are also generated at any time to fill the expanded space. Matters that fill spacetime after the phase transition are generated from local fluctuations of the conformal-factor field. Interactions that cause this change are served by the Wess-Zumino actions such as

$$\phi\sqrt{-g}\,\mathrm{Tr}(F_{\mu\nu}^2),$$

which opens rapidly as the running coupling constant increases. There are no gravitational waves originating from the trans-Planckian physics, which are thought to extend currently to the size of the Hubble distance, but localized ones will be generated at the Big Bang through the interaction $\phi\sqrt{-g}\,C_{\mu\nu\lambda\sigma}^2$. Furthermore, many excited states of quantum gravity, which will be discussed in the next chapter, may be generated.

Finally, we note that the condition $M_\mathrm{P} > \Lambda_\mathrm{QG}$ to exist the inflationary solution also serves as a unitarity condition to ensure that ghosts are not generated as particles in the present world during the spacetime phase transition. In quantum spacetime, all ghost modes are constrained by the BRST conformal invariance, but this is not applied to the world after the phase transition. If a spacetime in which conventional perturbation theory can be applied were realized in a region higher than the Planck energy, massive gravitons with the Planck mass, which are the root cause of the ghost problem in higher-derivative gravity, would be generated. In order to prevent their generation energetically, the phase transition needs to occur below the Planck scale.

7.2.1 After Phase Transition

Reducing the amplitude of gravitational fluctuations means that time and distance can be measured more accurately. That is to say that the reduction in the large-sized fluctuations involved in the structure formation indicates that classical spacetime emerges as an entity. On the other hand, small fluctuations in the correlation length size locally generated during the spacetime phase transition can be regarded as particles that propagate in the formed classical spacetime.

Now, let us consider how the low-energy spacetime after the phase transition can be described. In QCD, when the gluon-field action composed of second derivatives disappears as its running coupling constant diverges, the gluon dynamics completely disappears. Therefore, its low-energy effective theory is described with new dynamical variables, mesons and baryons, which are composite fields of fermions tightly bound by gluons. On the other hand, in quantum gravity, even if the fourth-derivative gravitational actions that are responsible for the conformal gravity dynamics disappears, the second-derivative Einstein-Hilbert action remains, and provides the kinetic term of the gravitational field at low energy. Thus, Einstein's theory of gravity appears as a low-energy effective theory of gravity. Since the conformal-factor field and the traceless tensor field are tightly coupled, the full gravitational field that combines them becomes a new field variable.

A quantity that changes significantly before and after the phase transition is the scalar curvature. The inflationary spacetime has $R \neq 0$, while the Friedmann universe has $R = 0$. Hence, this is an order parameter.

The actual low-energy effective theory of gravity is given by a derivative expansion of the gravitational field with the Einstein-Hilbert action as the lowest order like

$$I_{\rm Low} = \int d^4x \sqrt{-g} \left\{ \mathcal{L}_2 + \mathcal{L}_4 + \cdots \right\},$$

where the subscript number denotes the number of derivatives. The cosmological term without derivatives can be ignored in the early universe. The second-derivative term consists of the Einstein-Hilbert and matter field actions as

$$\mathcal{L}_2 = \frac{M_{\rm P}^2}{2} R + \mathcal{L}_{\rm M},$$

and the kinetic terms of fermions are also included here. Higher-derivative terms are expanded in the reciprocal of the reduced Planck mass, and the correctness of this expansion method is guaranteed by $M_\mathrm{P} > \Lambda_\mathrm{QG}$.

If we here consider up to the fourth-derivative term $\mathcal{L}_4$, there are five possible candidates: R^2, $R^2_{\mu\nu}$, $R^2_{\mu\nu\lambda\sigma}$, $R^{\mu\nu}T^\mathrm{M}_{\mu\nu}/M_\mathrm{P}^2$, and $T^{\mu\nu}_\mathrm{M}T^\mathrm{M}_{\mu\nu}/M_\mathrm{P}^4$, where $T^\mathrm{M}_{\mu\nu}$ is the traceless energy-momentum tensor of generated matter. Since the low-energy effective theory adopts Einstein's theory of gravity as a first approximation, if we consider that the higher-derivative terms are not independent as they are connected by the Einstein equation $M_\mathrm{P}^2 R_{\mu\nu} = T^\mathrm{M}_{\mu\nu}$, we can reduce the number of $\mathcal{L}_4$. The Einstein equation represents $R = 0$ as well, and furthermore, by eliminating the square of the Riemann tensor using the Euler relation,[3] the only independent action becomes

$$\mathcal{L}_4 = \frac{\alpha}{(4\pi)^2} R^{\mu\nu} R_{\mu\nu},$$

where α is a phenomenological parameter and is taken to be positive.

Assuming that the space is homogeneous and isotropic, we obtain the equation of motion,

$$M_\mathrm{P}^2\left(\dot{H} + 2H^2\right) + \frac{\alpha}{4\pi^2}\left(\dddot{H} + 7H\ddot{H} + 4\dot{H}^2 + 12H^2\dot{H}\right) = 0, \tag{7.15}$$

and the energy conservation equation,

$$-3M_\mathrm{P}^2 H^2 + \rho + \frac{\alpha}{4\pi^2}\left(-6H\ddot{H} + 3\dot{H}^2 - 18H^2\dot{H}\right) = 0. \tag{7.16}$$

Now, let us simply connect the inflationary solution and a solution of the equations of motion obtained from the low-energy effective theory at the phase transition time $\tau = \tau_\Lambda$. The initial values of a, H, $\dot{H}$, and ρ on the low-energy effective theory side are chosen so that they are smoothly connected to the inflation side. The initial value of $\ddot{H}$ required to solve (7.15) is given by substituting the other initial values into the conservation equation (7.16). Figures 7.1 and 7.2 shows the results of a numerical calculation where the parameter is set to $\alpha = 1$. According to this, the Hubble variable increases slightly after the phase transition, the number of e-folding increases further, and then rapidly approaches the Friedmann solution while oscillating.

7.3 Entropy of The Universe

The origin of the enormous amount of entropy that exists in the universe is one of the great mysteries. Most of that is currently carried out by the cosmic microwave background radiation (CMB). Within the framework of Einstein's theory of gravity,

[3] At this time, pay attention to Footnote 2 in Chap. 2.

it cannot be satisfactorily explained without introducing an unknown scalar field as a source of all matter. Moreover, the standard cosmology is based on an ansatz that the whole entropy of the universe is conserved during evolution [KT90]. This is usually explained thermodynamically by assuming that the early universe was in thermal equilibrium. However, this explanation can only be applied after matter is created, and thus does not link to the origin of entropy.

The observed spectra of CMB temperature fluctuations have revealed that the whole universe is entangled in every corner. This shows that their original fluctuations, now far apart, were correlated long ago when they were still close to each other. Inflation theory suggests that such an event occurred in the trans-Planckian world. Here, we will elucidate the origin of the entropy and its conservation law from the perspective of quantum gravity [Ham24b].

First of all, recall that in renormalizable quantum field theory, the effective action is finite and renormalization group invariant, and if it is diffeomorphism invariant, then the energy-momentum tensor is also finite and renormalization group invariant. Furthermore, as shown in Chap. 4, if gravity is quantized, then the energy-momentum tensor vanishes exactly.

The partition function of quantum gravity represents sum over states of quantum spacetime. Since the effective action is defined by the logarithm of the partition function, it is an entropy itself if the total Hamiltonian of the system vanishes. Thus, the quantum gravity effective action provides a statistical entropy S_{Univ} for the state of the universe, that is,

$$\Gamma_{\mathrm{QG}} = -S_{\mathrm{Univ}}. \tag{7.17}$$

Hence, we can show that the entropy is conserved as a renormalization group invariant. This is a general conclusion derived from diffeomorphism invariance and renormalizability.

It may be helpful to recall that the effective action corresponds to free energy in ordinary quantum field theory. If we impose a condition for energy to vanish for that, we can formally obtain the relation like above. However, note that there are no ordinary particle states in quantum gravity considered here, thus no concept of temperature or thermal equilibrium. The formula (7.17) is thus defined as purely counting the number of states, where the inability to define temperature and the vanishing of the Hamiltonian are complementary. If you want to get a statistical mechanical image of quantum gravity, refer to lattice quantum gravity using the dynamical triangulation method, discussed in Chap. 12.

Here, we demonstrate that the current entropy of the universe can be actually derived by evaluating (7.17). The asymptotic background freedom shows that most of the entropy of quantum gravity is carried by the conformal-factor field. Inflation is also caused by its dynamics. The effective action for the homogeneous component $\hat{\phi}$ of the conformal-factor field, which plays the main part of the inflation dynamics, is given by

$$\Gamma_{\rm QG} = V_3 \int d\eta \left\{ -\frac{b_c}{8\pi^2} B\hat{\phi}\partial_\eta^4\hat{\phi} - 3M_{\rm P}^2 e^{2\hat{\phi}} \left(\partial_\eta^2\hat{\phi} + \partial_\eta\hat{\phi}\partial_\eta\hat{\phi} \right) \right\}, \tag{7.18}$$

where $V_3 = \int d^3\mathbf{x}$, and the first and second terms are the contributions from the Riegert action (2.21) and the Einstein-Hilbert action, respectively.

The homogeneous equation of motion is given by (7.2), which has a stable inflationary solution $H = H_{\rm D}$ (7.6) at the vanishing coupling limit of $B = 1$. The scale factor expands exponentially as $a\,(= e^{\hat{\phi}}) = e^{H_{\rm D}\tau}$, where the scale factor at $\tau = 0$, which corresponds to infinite energy, is set to unity so that V_3 is the spatial volume before inflation starts.

Since $C_{\mu\nu\lambda\sigma} \simeq 0$ during most of the inflationary period, contributions from the Weyl effective action (2.26) to entropy are sufficiently small and can be ignored. The main role of the Weyl sector is to transfer the entropy carried by the conformal-factor field to matter fields during the spacetime phase transition, and the running coupling constant $\bar{t}$ plays this role. In order to incorporate the dynamics, consider replacing the coefficient B in the effective action (7.18) and the equation of motion (7.2) with $\bar{B}$ (7.10) containing the running effect, as in Sect. 7.2.

In order that the entropy can be calculated analytically, we here simplify $\bar{t}^2$ to a function that remains mostly zero and suddenly diverges at τ_Λ, as a first approximation. That is, $\bar{B}$ is approximated by a step function that is mostly unity and suddenly becomes zero at τ_Λ. In this case, the number of e-foldings $\mathcal{N}_e$ (7.11) is given by the ratio of the two scales, N (7.12).

The dynamical energy scale $\Lambda_{\rm QG}$ is determined by the value of N, which is determined from the number of e-foldings and the magnitude of scalar fluctuations (7.13) at the time of spacetime phase transition. Based on these conditions, N is determined to be 60, and $\Lambda_{\rm QG}$ is on the order of 10^{17}GeV. Furthermore, explaining the sharp falloff in the low multipole component of the CMB spectrum by this scale leads to $a_0 \simeq 10^{59}$.

Transforming the effective action (7.18) using the equation of motion (7.2) yields

$$\begin{aligned} \Gamma_{\rm QG} &= -3M_{\rm P}^2 V_3 \int d\eta (\hat{\phi} - 1) e^{2\hat{\phi}} \left(\partial_\eta^2\hat{\phi} + \partial_\eta\hat{\phi}\partial_\eta\hat{\phi} \right) \\ &= -3M_{\rm P}^2 V_3 \int d\tau a^3 (\log a - 1)(\partial_\tau H + 2H^2). \end{aligned} \tag{7.19}$$

If the inflationary state of (7.6) continues until the spacetime phase transition and suddenly ends at $\tau_\Lambda = 1/\Lambda_{\rm QG}$, then from (7.17), the entropy of the universe due to quantum gravity is given by

$$\begin{aligned} S_{\rm Univ} &= 6M_{\rm P}^2 H_{\rm D}^2 V_3 \int_0^{\tau_\Lambda} d\tau e^{3H_{\rm D}\tau} (H_{\rm D}\tau - 1) \\ &= 2M_{\rm P}^2 H_{\rm D} V_3 \left[e^{3H_{\rm D}\tau_\Lambda} \left(H_{\rm D}\tau_\Lambda - \frac{4}{3} \right) + \frac{4}{3} \right]. \end{aligned}$$

Rewriting with the scale ratio N and omitting the last negligible term, we finally get

$$S_{\text{Univ}} = 2\sqrt{\frac{8\pi^2}{b_c}M_{\text{P}}^3 e^{3N}\left(N-\frac{4}{3}\right)}V_3. \tag{7.20}$$

The scalar curvature changes to zero during the spacetime phase transition, and matter is produced with energy density $\rho(\tau_\Lambda) = 3M_{\text{P}}^2 H^2(\tau_\Lambda)$. After that, the universe evolves according to the Friedmann solution. Since it a solution in which the Einstein-Hilbert action vanishes, spacetime itself has no entropy, thus various matters propagating through the spacetime, including gravitons, hold most of the entropy. Only after changing to this phase, we can define the normal thermal state.

Matter is created at the phase transition, at which the entropy of quantum gravity is released into matter. Since S_{Univ} is conserved as a renormalization group invariant, it should match the current entropy of the universe, whose density is given by [KT90]

$$s = \frac{S_{\text{Univ}}}{V_3^{\text{today}}} = \frac{2\pi^2}{45} g_{*S} T_{\text{cmb}}^3 \simeq 2.91 \times 10^3\,\text{cm}^{-3}, \tag{7.21}$$

where $g_{*S} = 3.91$ and $T_{\text{cmb}} = 2.73\,\text{K} = 11.9\,\text{cm}^{-1}$.

When comparing (7.20) and (7.21), note that V_3 is the three-dimensional volume before inflation begins. The universe has gone the inflationary period and the subsequent Friedmann era, and the scale factor has increased to a_0 today, so that $V_3^{\text{today}} = a_0^3 V_3$. Thus, the current entropy density is expressed as

$$s = 2\sqrt{\frac{8\pi^2}{b_c}M_{\text{P}}^3 e^{3N}\left(N-\frac{4}{3}\right)}a_0^{-3}.$$

Now, let us find the value of N while fixing the parameters to $b_c = 7$ and $a_0 = 10^{59}$ so that this matches (7.21). Then we get $N = 62.2$, where $M_{\text{P}} = 1.235 \times 10^{32}\,\text{cm}^{-1}$ is used. In this way, we can show that the current entropy of the universe is consistent with the scenario of quantum gravity inflation.

The ratio between the current CMB temperature 2.73K and the energy scale Λ_{QG} at the phase transition suggests that the universe will expand by a factor of about 10^{29} after settling into the Friedmann spacetime. Therefore, in order to reach a total of 10^{59}, the universe needs to expand by about 10^{30} times during the inflationary period, as described in Sect. 7.2. This corresponds to 70 in terms of the e-folding number, which is larger than 60. This discrepancy can be improved by solving the equation of motion without simplifying $\bar{B}$ with a step function. In fact, if we find a solution of $N = 60$ by choosing the phenomenological parameters as $\beta_0 = 0.171$, $\gamma_1 = 0.1$, and $\kappa = 1$ and numerically evaluate the effective action (7.19) using it, we can derive a result consistent with (7.21). In any case, quantum gravity can generate sufficient entropy.

Chapter 8
Localized Massive Excitations of Quantum Gravity

A point-like particle with mass greater than the Planck mass is nothing but a black hole in Einstein's theory of gravity. This is because information of such a particle is hidden inside the horizon and is lost, as mentioned in Chap. 1. In contrast, the asymptotically background-free quantum gravity suggests that there exists a localized excited state with mass exceeding the Planck mass, which has the size of the correlation length, has no singularity, and looks like a particle from the outside.

8.1 Equations Describing Inside of Excitations

We have already mentioned that there exists a novel dynamical energy scale Λ_{QG} of about 10^{17} GeV in the theory. This value is two orders of magnitude smaller than the Planck energy scale, and has been chosen so that the scenario of inflation caused by quantum gravity effects only is consistent with the CMB observations. We proceed with the discussion adopting this value here as well.

Spacetime is then substantially quantized by the correlation length ξ_Λ $(= 1/\Lambda_{QG})$. Its length is about 100 times the Planck length. At longer distances than this length, it is possible to measure physical distances based on Einstein's theory of gravity, but at shorter than this, quantum fluctuations in the gravitational field become so large that a background-free world appears where the measurement of distance itself no longer has any physical meaning. This suggests that if excited states of quantum gravity exist, they appear as localized states whose diameter is the correlation length ξ_Λ, which remind us of glueballs in QCD. Although it is difficult to imagine what they would look like because what is fluctuating inside is spacetime itself, they can be thought to look just like particles moving in classical spacetime when viewed from the outside.

Now, we consider a spherical object whose mass is about the Planck mass as an excited state [Ham20a]. The classical horizon size of such an object is about the

K. Hamada, *Trans-Planckian Physics and Inflation*, Fundamental Theories of Physics 223, https://doi.org/10.1007/978-981-96-3476-7_8

Planck length. This means that when approaching the center of the object from a distance, quantum gravity effects begin to be active at a point 100 times away from the horizon. As a result, the horizon is buried in quantum fluctuations and disappears.

For excitations on the order of the Planck mass, the gravitational field is still small enough at the edge and is expected not to become extremely large even toward the center, in contrast to classical theory. Therefore, we apply a linear approximation to examine it. A spherical excitation can be described by scalar fluctuations in the gravitational field, the scalar components of which are denoted as

$$ds^2 = -(1 + 2\Psi)d\eta^2 + (1 + 2\Phi)d\mathbf{x}^2,$$

where Φ and Ψ are variables called the gravitational potentials or Bardeen potentials. A fluctuation of the conformal-factor field ϕ is here denoted as φ, and further a scalar component h of the traceless tensor field $h_{\mu\nu}$ is introduced. If we define them as

$$\phi = \varphi, \qquad h_{00} = h, \qquad h_{ij} = \frac{1}{3}\delta_{ij}h,$$

then the gravitational potentials are expressed as

$$\Phi = \varphi + \frac{h}{6}, \qquad \Psi = \varphi - \frac{h}{2}.$$

The gravitational potential around a point mass can be approximated by $\Phi = r_g/2r$ with $\Psi = -\Phi$ when its mass M is small, where r is the radial coordinate and $r_g = 2GM$ is the Schwarzschild radius. This is a solution to the linearized Einstein equation, $-6M_{\mathrm{P}}^2\partial^2\Phi = 0$, where ∂^2 is the spatial Laplacian such that $\partial^2\Phi = \partial_r^2\Phi + (2/r)\partial_r\Phi$ when assuming spherical symmetry and writing the radial differential as $\partial_r = \partial/\partial r$.

Here, consider an excited state whose outside smoothly connects to this Schwarzschild solution.[1] Assuming that the diameter of the excited state is the correlation length and writting the radius as $R_h = \xi_\Lambda/2$, the magnitude of the gravitational potential at the edge $r = R_h$ is r_g/ξ_Λ. If this value is sufficiently smaller than unity, the linear approximation will be valid. This condition is expressed as $M \ll m_{\mathrm{pl}}^2/2\Lambda_{\mathrm{QG}}$, and the right-hand side is approximately 50 times m_{pl}. The mass M for which the approximation actually holds to the inside is, however, about several times m_{pl}.

The equation of motion is expressed as $\mathbf{T}_{\mu\nu} = \mathbf{T}^{\mathrm{R}}_{\mu\nu} + \mathbf{T}^{\mathrm{W}}_{\mu\nu} + \mathbf{T}^{\mathrm{EH}}_{\mu\nu} + \mathbf{T}^{\mathrm{M}}_{\mu\nu} = 0$ from (7.1). The first term in the energy-momentum tensor is derived from the Riegert action, the second from the Weyl action, and the third from the Einstein-Hilbert action. See Appendix C for the specific expression of the energy-momentum tensor for each gravitational sector. By expanding each of them up to linear terms, we obtain

[1] The outside should be described by a low-energy effective gravity theory defined by derivative expansion around the Einstein-Hilbert action, but since its equation of motion also includes the Schwarzschild solution, this solution is adopted as it is.

$$\mathbf{T}^{\mathrm{R}}_{\mu\nu} = \frac{b_c}{8\pi^2} B \left\{ \frac{2}{3}\partial_\mu \partial_\nu \Box\varphi - \frac{2}{3}\eta_{\mu\nu}\Box^2\varphi - \frac{1}{9}\partial_\mu\partial_\nu\partial_\lambda\chi^\lambda + \frac{1}{9}\eta_{\mu\nu}\Box\partial_\lambda\chi^\lambda \right\},$$

$$\mathbf{T}^{\mathrm{W}}_{\mu\nu} = -\frac{2}{t^2}\left\{ \Box^2 h_{\mu\nu} - \Box\partial_\mu\chi_\nu - \Box\partial_\nu\chi_\mu + \frac{2}{3}\partial_\mu\partial_\nu\partial_\lambda\chi^\lambda + \frac{1}{3}\eta_{\mu\nu}\Box\partial_\lambda\chi^\lambda \right\},$$

$$\mathbf{T}^{\mathrm{EH}}_{\mu\nu} = M_{\mathrm{P}}^2 \left\{ 2\partial_\mu\partial_\nu\phi - 2\eta_{\mu\nu}\Box\phi - \partial_{(\mu}\chi_{\nu)} + \frac{1}{2}\Box h_{\mu\nu} + \frac{1}{2}\eta_{\mu\nu}\partial_\lambda\chi^\lambda \right\},$$

where $\chi_\mu = \partial_\lambda h^\lambda{}_\mu$ and $\Box = \partial^\lambda\partial_\lambda = -\partial_\eta^2 + \partial\!\!\!{}^{-}{}^2$ is the d'Alembertian in flat spacetime. As the matter energy-momentum tensor, we employ the traceless perfect-fluid type,

$$\mathbf{T}^{\mathrm{M}}_{00} = \rho, \qquad \mathbf{T}^{\mathrm{M}}_{ij} = \frac{1}{3}\delta_{ij}\rho,$$

where ρ is matter energy density.

First, let us derive two equations of motion consisting only of the gravitational potentials by choosing appropriate combinations of $\mathbf{T}_{\mu\nu}$ in which ρ from the matter sector disappears. Such combinations are given by

$$\mathbf{T}^\mu{}_\mu = 0,$$

$$\mathbf{T}^i{}_i - 3\frac{\partial^i\partial^j}{\partial^2}\mathbf{T}_{ij} = 0.$$

In the first trace equation, the contribution from the Weyl sector also disappears, yielding

$$\frac{b_c}{8\pi^2}B\left(-2\partial_\eta^4\Phi + \frac{10}{3}\partial_\eta^2\partial^2\Phi - \frac{4}{3}\partial^4\Phi + \frac{2}{3}\partial_\eta^2\partial^2\Psi - \frac{2}{3}\partial^4\Psi \right)$$
$$+ M_{\mathrm{P}}^2\left(6\partial_\eta^2\Phi - 4\partial^2\Phi - 2\partial^2\Psi \right) = 0, \tag{8.1}$$

where the term with the coefficient b_c is from the Riegert action, and the term with M_{P}^2 is from the Einstein-Hilbert action. From the second formula combining the spatial components of the energy-momentum tensor, we obtain

$$\frac{2}{t^2}\left(4\partial_\eta^2\partial^2\Phi - \frac{4}{3}\partial^4\Phi - 4\partial_\eta^2\partial^2\Psi + \frac{4}{3}\partial^4\Psi \right) + \frac{b_c}{8\pi^2}B\left(\frac{4}{3}\partial_\eta^2\partial^2\Phi \right.$$
$$\left. - \frac{8}{9}\partial^4\Phi - \frac{4}{9}\partial^4\Psi \right) - 2M_{\mathrm{P}}^2\left(\partial^2\Phi + \partial^2\Psi \right) = 0, \tag{8.2}$$

where the term with $1/t^2$ is from the Weyl action.

In addition, the energy conservation equation involving ρ is obtained from $\mathbf{T}_{00} = 0$ as

$$\frac{2}{t^2}\left(-\frac{4}{3}\partial^4\Phi+\frac{4}{3}\partial^4\Psi\right)+\frac{b_c}{8\pi^2}B\left(-\frac{2}{3}\partial_\eta^2\partial^2\Phi+\frac{4}{9}\partial^4\Phi+\frac{2}{9}\partial^4\Psi\right)$$
$$+2M_P^2\partial^2\Phi+\rho=0. \tag{8.3}$$

These equations do not yet include dynamics involving the correlation length ξ_Λ. The running coupling constant (2.28) is a collection of nonlinear and nonperturbative quantum gravity effects, and is necessary to connect between the inside of the excited state and the outside described by the Schwarzschild solution. Here, by analogy with the mean field approximation adopted when finding the inflationary solution, we replace the physical momentum Q with the reciprocal of $2r$ and express the running coupling constant as a function of r as

$$\bar{t}^2(r)=\frac{1}{\beta_0\log(R_h^2/r^2)}, \tag{8.4}$$

which diverges at the radius of excitation, $r=R_h\,(=\xi_\Lambda/2)$. Then, we replace the coupling constant t^2 in the equations of motion (8.1), (8.2), and (8.3) with $\bar{t}^2(r)$, and replace B with the dynamical factor

$$\bar{B}(r)=\left[1+\gamma_1\frac{\bar{t}^2(r)}{4\pi}\right]^{-1}. \tag{8.5}$$

The equations of motion obtained in this way maintain linearity while incorporating the effects of strong coupling dynamics.

The running coupling constant (8.4) is a function that vanishes at the origin, gradually increases as moving away from it, and rapidly diverges at the edge of $r=R_h$. This expresses that the excited state is connected to the Schwarzschild solution at the edge due to the disappearance of conformal gravity dynamics involving $1/\bar{t}^2(r)$ and $\bar{B}(r)$. The second equation (8.2) serves as a constraint that expresses exactly how this is achieved. That is, when $\bar{t}\to 0$, since the part inside the first parentheses disappears, the configuration in which the Weyl action vanishes becomes dominant, whereas $\bar{t}\to\infty$ indicates that the fourth-derivative term disappears and the Einstein-Hilbert term becomes dominant.

8.2 Static Spherical Excitations

From now on, we consider a static solution. In this case, if we introduce new variables

$$X=2\Phi+\Psi,\quad Y=\Phi-\Psi,$$

then the coupled equations of motion, with the above modifications to (8.1) and (8.2), are separated into two parts, given by

$$\bar{B}(r)\,\bar{\partial}^4 X(r) + 3H_{\rm D}^2\bar{\partial}^2 X(r) = 0,$$
$$\frac{4}{\bar{t}^2(r)}\bar{\partial}^4 Y(r) - M_{\rm P}^2\,\bar{\partial}^2 Y(r) = 0, \tag{8.6}$$

where $H_{\rm D} = M_{\rm P}\sqrt{8\pi^2/b_c}$. In addition, such a static solution satisfies the energy conservation equation (8.3) with $\rho = 0$, i.e., it represents a purely gravitational excited state.

Further rewrite the variables as

$$X(r) = f(r)\frac{r_g}{2r}, \quad Y(r) = g(r)\frac{r_g}{r}, \tag{8.7}$$

where r_g is an arbitrary constant at this point, but later it becomes the Schwarzschild radius for excitation mass. Applying the spatial Laplacian to these variables gives $\bar{\partial}^2 X = (r_g/2r)\,\partial_r^2 f$ and $\bar{\partial}^4 X = (r_g/2r)\,\partial_r^4 f$. The same applies to the variable Y. Expressing (8.6) with the f and g variables gets the differential equations,

$$\partial_r^4 f(r) + 3H_{\rm D}^2\left[1 + \gamma_1\frac{\bar{t}^2(r)}{4\pi}\right]\partial_r^2 f(r) = 0,$$
$$\partial_r^4 g(r) - \frac{1}{4}M_{\rm P}^2\bar{t}^2(r)\partial_r^2 g(r) = 0. \tag{8.8}$$

Next, we give boundary conditions for solving these differential equations. As conditions for variables X and Y and $\bar{\partial}^2 X$ and $\bar{\partial}^2 Y$ to not diverge at the origin, we obtain

$$f(0) = 0, \quad g(0) = 0, \quad \partial_r^2 f(0) = 0, \quad \partial_r^2 g(0) = 0. \tag{8.9}$$

As emphasized in Chap. 2, solutions with singularities are unphysical, thus those where f and g are finite at the origin are excluded as fake ones caused by employing linear approximation. The third and fourth conditions given by the second derivative (not first derivative) of r are requirements for each term of the energy equation (8.3) to be finite. Furthermore, as conditions for smoothly connecting to the Schwarzschild solution at the edge, we set

$$f(R_h) = 1, \quad g(R_h) = 1, \quad \partial_r f(R_h) = 0, \quad \partial_r g(R_h) = 0. \tag{8.10}$$

The first two indicate that $\Phi = -\Psi$ holds and the Einstein equation is satisfied at the edge. The third and fourth conditions are for smooth connection.

The excited state mass is defined by

$$M = \int_{|\mathbf{x}|\le R_h} d^3\mathbf{x}\,\mathbf{T}_{00}^{(4)}(\mathbf{x}), \tag{8.11}$$

where $\mathbf{T}^{(4)}_{00}$ is the fourth derivative part of (8.3). Since the static solution satisfies the conservation equation (8.3) of $\rho = 0$, this expression can be rewritten as

$$M = -2M_{\mathrm{P}}^2 \int\limits_{|\mathbf{x}| \leq R_h} d^3\mathbf{x}\, \partial\!\!\!{}^{-}{}^2 \Phi$$

Evaluating this using $\Phi = (X + Y)/3 = r_g(f + 2g)/6r$ and boundary conditions (8.9) and (8.10) yields

$$\begin{aligned} M &= -\frac{4\pi}{3} M_{\mathrm{P}}^2 r_g \int_0^{R_h} dr \frac{\partial}{\partial r}\left[r \frac{\partial(f + 2g)}{\partial r} - (f + 2g) \right] \\ &= \frac{4\pi}{3} M_{\mathrm{P}}^2 r_g \big[f(R_h) + 2g(R_h) \big] = 4\pi M_{\mathrm{P}}^2 r_g . \end{aligned}$$

In this way, the formula for Schwarzschild radius $r_g = M/4\pi M_{\mathrm{P}}^2 = 2GM$ is derived.

First, we examine the behavior of the solution near the origin. For that, introduce variables $\zeta = \partial_r^2 f$ and $\theta = \partial_r^2 g$ and look for a solution that satisfies $\zeta(0) = \theta(0) = 0$. In addition, since the running coupling constant is small near the origin, it is approximated with a small constant such as $\bar{t}(r) = t\,(\ll 1)$. The equations then reduce to $\partial_r^2 \zeta + K^2 \zeta = 0$ and $\partial_r^2 \theta - L^2 \theta = 0$, where $K = \sqrt{3(1 + \gamma_1 t^2/4\pi)} H_{\mathrm{D}} \simeq \sqrt{3} H_{\mathrm{D}}$ and $L = M_{\mathrm{P}} t/2\,(\ll 1)$. From these, $\zeta \simeq \sin(Kr)$ and $\theta \simeq \sinh(Lr)$ are obtained. Integrating these twice with r and finding f and g satisfying the conditions (8.9) reveals that they behave like

$$f(r) \simeq c \sin(\sqrt{3} H_{\mathrm{D}} r) + dr \tag{8.12}$$

and $g \simeq c' \sinh(Lr) + d'r$ near the origin. Since $L \ll 1$,

$$g(r) \simeq d'r \tag{8.13}$$

is obtained by considering only the part proportional to r. Here, the coefficients c, d, d' cannot be determined unless the equations are completely solved by imposing the boundary conditions (8.10) at the edge. This behavior of g near the origin shows that $\partial\!\!\!{}^{-}{}^2 \Phi = \partial\!\!\!{}^{-}{}^2 \Psi$ holds when expressed in terms of the gravitational potentials. In other words, this shows that a field configuration where the Weyl action vanishes arises.

Now, let us specifically solve the equations (8.8) for variables f and g including the running effect of the coupling constant. Here, we take the dynamical-factor parameters to be $\beta_0 = 0.4$ and $\gamma_1 = 0.1$ as before. When these equations are numerically solved as a boundary value problem under the boundary conditions (8.9) and (8.10), we obtain the results shown in Fig. 8.1, where the ratio of the two mass scales is (7.12) and H_{D} is normalized to unity. Substituting the obtained solutions into (8.7) to find the gravitational potentials, Φ and $-\Psi$, yields Fig. 8.2 for the case of $M = 4m_{\mathrm{pl}}$,

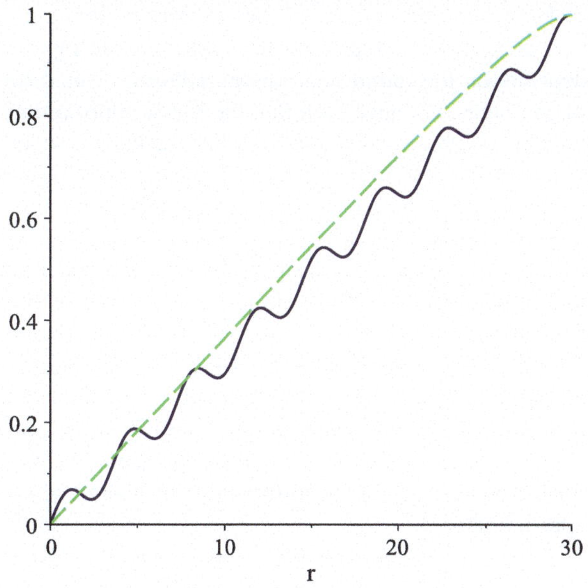

Fig. 8.1 Results of numerical calculations of functions f (solid violet line) and g (dashed green line), where $H_{\rm D}$ is normaized to unity. The radius of the excitation is set to R_h $(= \xi_\Lambda/2) = 30$

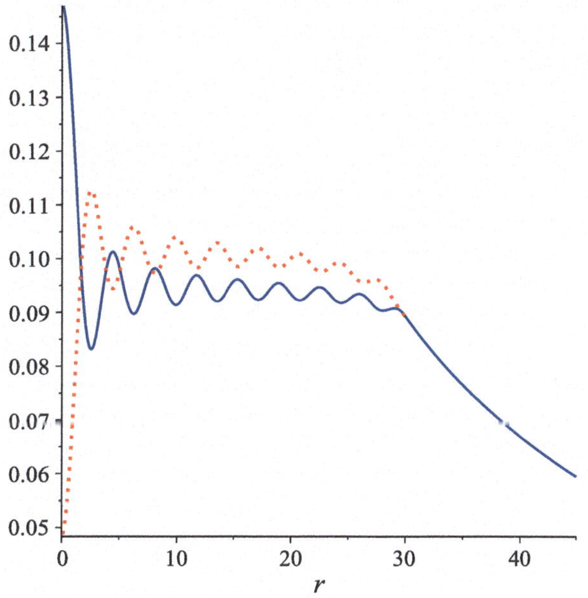

Fig. 8.2 Localized excitation of quantum gravity with mass $M = 4m_{\rm pl}$. The gravitational potential Φ is shown as a solid blue line and $-\Psi$ as a dotted red line. The outside of the excitation $r \geq R_h$ $(= 30)$ is the Schwarzschild solution $\Psi = -r_g/2r$ $(\Phi = -\Psi)$

namely, $r_g = 2GM = 8/m_{\rm pl}$, as an example. It can be seen that the behavior of $(r \leq 5)$ near the origin of this numerical solution is $c = 0.0342$, $d = 0.0328$, and $d' = 0.0365$.

Here, we have addressed the inherent nonlinearity and nonlocality of the equation of motion by replacing the running coupling constant with a mean field that depends on the radial coordinate. Although it is difficult to manage them over all mass ranges, we have seen that the linear approximation to the gravitational potentials can be successfully applied when the mass is several times the Planck mass. This mass is sufficiently larger than Λ_{QG} so that it can be considered that quantum gravity dynamics are activated inside, even though the magnitude of the gravitational potentials are still small. When the mass is larger than what examined here, it is necessary to take into account nonlinear effects other than the running coupling constant, which makes calculation difficult. In any case, although it is still incomplete, we believe that this solution captures the characteristics of excitations successfully.

This object is more appropriately called a particle rather than a black hole. Such gravitational particles with mass only are called "dark particles". If the mass is lower than Λ_{QG}, the quantum gravity effect is not activated, thus such an excitation does not occur. Moreover, the mass spectrum of quantum gravity is expected to be quantized by Λ_{QG}, but unfortunately this cannot be confirmed within the mean field approximation.

The excited state probably decays soon with a lifetime of about the reciprocal of Λ_{QG}. The fact that the running coupling constant increases at the edge means that the interaction with matter fields is open there. Therefore, it is thought that matter is generated at the edge, so that the gravitational energy gradually transfers to matter energy, causing the state to decay.

On the other hand, for objects with masses more than 100 times larger than the Planck mass, the classical horizon size $2r_g$ becomes longer than the correlation length ξ_Λ. When viewed from the outside, such an object is nothing other than a black hole, but we can believe that the gravitational field near the center fluctuates greatly in quantum mechanical way, as shown here.

In such a large black hole, gravity can be treated classically except at the center. In the 1970s, Hawking showed that under such a semiclassical approximation, the so-called evaporation phenomenon occurs, in which a black hole gradually shrinks while emitting radiation [Haw74, Haw75]. In the final stage of the evaporation, the horizon disappears and the dark particle state described above will appear, which eventually collapse. Thus, the black hole is expected to disappear completely.

We believe that dark particles were produced in abundance and at high density in the early universe immediately after the spacetime phase transition. In order for these to survive into the present day as so-called "dark matter", they must be long-lived. Since dark particles are thought to initially have no angular momentum, they may coalesce before decaying even if they individually have short lifetimes. If they become large enough to form a horizon, their lifetime would be extended as matter produced during the decay process would be trapped and returned back inside. They could be then candidates for dark matter, and would seed the formation of a large number of primordial black holes and galaxies.

Chapter 9
What the Cosmological Constant Problem Is

Some physicists say that the cosmological constant problem is a big problem, while others say that the cosmological constant is just a physical parameter, like Newton's constant. The root cause of the problem is that the existence of the world beyond the Planck scale is ignored. Here, we present an answer to this problem from the perspective of quantum gravity.

9.1 Discovery of Cosmological Constant

First, the history of the cosmological constant, which is the coefficient of the cosmological term, is mentioned briefly. The year after the general theory of relativity was published, an extra term was introduced by Einstein himself in 1917 [Ein1917], in order to realize a static universe. Although it was an impossible endeavor by nature, this is the beginning of what is called the cosmological term. This attempt itself is nothing surprising, since many researchers in those days, including Einstein, believed that the universe did not evolve over time.

The view that the universe is stationary was shattered by observations by astronomer Hubble published in 1929 [Hub1929]. He measured velocities of many galaxies and found that all galaxies were moving away from the earth. It has been shown that the farther the galaxy is, the faster its recession velocity is, that is, the space is expanding everywhere and distance between all objects is increasing. A few years after this discovery, Einstein allegedly described the introduction of the cosmological constant as "the biggest blunder in his life".

It was not until the late 1990s that the cosmological constant came into limelight again. Observations of brightness of Type Ia supernovae, which are used as standard candles, have shown that the expansion rate of the universe is increasing [Rie+98, Per+99], where a standard candle is a celestial body whose absolute luminosity, corresponding to wattage of a light bulb, is known, and difference in brightness can be

K. Hamada, *Trans-Planckian Physics and Inflation*, Fundamental Theories of Physics 223, https://doi.org/10.1007/978-981-96-3476-7_9

regarded as difference in distance from the earth. This fact indicating that the universe is currently accelerated expanding can be explained using the positive cosmological constant. The existence of the cosmological constant was further confirmed by analysis of the CMB temperature fluctuations observed by the WMAP and Planck satellites [WMAP, PLANCK]. In those days, many researchers still believed that the cosmological constant did not exist, and field theories with supersymmetry that makes the cosmological constant exactly zero were being actively studied, so this discovery came as a major shock.

The observed cosmological constant has a dimension of the fourth power of energy, which is about 120 orders of magnitude smaller than the value of the fourth power of the Planck mass. It is extremely small and negligible in the early universe, and does not contribute to the inflation scenario described in Chap. 7. Nevertheless, the reason why the small cosmological constant contributes greatly to the current accelerated expansion is as follows. In the Einstein equation that determines the evolution of the universe after the Big Bang, energy density of matters is overwhelmingly large at the beginning, but its contribution becomes smaller and smaller as the universe expands. On the other hand, since the contribution of the cosmological constant remains constant, their contributions are significantly reversed near the present (see Fig. 9.1).

In recent years, as an alternative to the cosmological constant, the term "dark energy" is often used. The cosmological constant is literally a physical constant, but this term is used more general cases, including a possibility that it is not a constant.

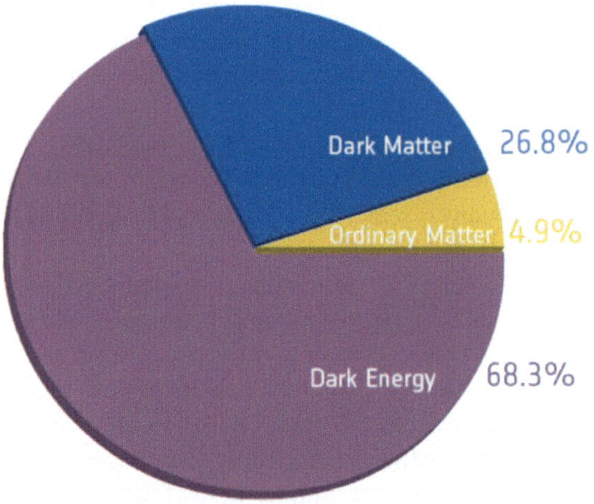

Fig. 9.1 Cosmic recipe pie-chart by Planck. Observations have revealed that the cosmological constant, or dark energy, which is responsible for accelerating the expansion of the universe, makes up just under three-quarters of the mass/energy density of the present universe. Dark matter, which is detected indirectly by its gravitational influence on nearby matter, accounts for about one-quarter, with normal matter being a comparatively small amount. Reprint from https://www.sci.esa.int/web/planck/ with permission ©ESA and the Planck Collaboration

It is impressive when used in pair with the word dark matter. In addition, inflation theory discussed in the framework of Einstein's theory of gravity generally uses the cosmological term, and in this case the cosmological constant has to change significantly. The introduction of the alternative name may have a theoretical background that wants to distinguish between the inflationary period and the current cosmological constant.

9.2 The Cosmological Constant Problem and Its Answer

Until the cosmological constant was discovered, whether it had a finite value or not was a major issue. However, it is a subject that has nothing to do with quantum theory and can be discussed within the framework of general relativity. The cosmological constant problem, which is the main subject of this chapter, is what will be described below.

The first to point out the cosmological constant problem is said to be Pauli.[1] That was in the 1920s when quantum mechanics was born. When he calculated a repulsive force due to zero-point energy, he got a very large value. He recognized it as a serious problem because it implies that the universe would expand rapidly and celestial bodies would soon be far apart and could not recognize each other. In those days, it seemed that nobody really took the story severely because it was too eccentric.

The first concrete calculation of zero-point energy was done by Zel'dovich in the 1960s when quantum field theory became widely accepted [Zel67]. As a result, a very large cosmological constant that did not match the observation was obtained, and then the cosmological constant has come to be discussed as a problem of quantum field theory.

The description so far is just a summary of what is widely known (see Weinberg's review in 1989 [Wei89]). The question here is not the magnitude of the value, but how the calculation has been carried out. The essence of the cosmological constant problem is that when an ultraviolet cutoff is introduced to the Planck scale and zero-point energy is calculated, a large cosmological constant proportional to the fourth power of the Planck mass arises.

In this calculation method, the cutoff is a finite value and has a physical meaning. In other words, it is presumed that there is no field in a world beyond the Planck scale. Needless to say, it comes from the prejudice that gravity is not renormalizable. Considering spacetime discretized by the Planck length is often called quantization of gravity, but denying field continuity like this breaks diffeomorphism invariance. Thus, it should be noted that the problem with diffeomorphism invariance is hidden behind the cosmological constant problem.

---

[1] See, for example, "Cosmic Conundrum" Scientific American, February 2021 and N. Straumann, *The History of the Cosmological Constant Problem*, arXiv:gr-qc/0208027.

In fact, when the cutoff method is applied to gauge field theory, a mass term of gauge field proportional to the cutoff squared appears and gauge invariance is broken. Mass terms also appear in other matter fields. At that time, a choice is often made to manually remove the terms that obviously break gauge invariance and leave the terms that do not. The cosmological constant in question is of that kind, but since diffeomorphism invariance is broken in the first place, there is no guarantee that the cutoff dependence will take the form of the cosmological term.

Despite the fact that the existence of the cosmological term is predicted by diffeomorphism invariance, if the quantization method violates this invariance, it is a wrong way for the end. Actually, if we use dimensional regularization, which can make the theory finite while preserving gauge invariance and diffeomorphism invariance, no ultraviolet divergence appears corresponding to a power of the cutoff that breaks the invariance.

To solve the cosmological constant problem, we have to step into the trans-Planckian regions. In general, quantum theory of gravity that holds even in such regions shows that zero-point energy vanishes as a result of diffeomorphism invariance, as was shown in Chap. 4 using the Schwinger-Dyson equation. It means that the cause of the problem itself disappears.

Well then, what is the cosmological constant? In the renormalizable and background-free quantum gravity, it is one of physical constants, $\Lambda_{\cos}$ [HM17, Ham22]. To be precise, it is a renormalization group invariant, i.e., a true constant whose value does not change in the evolution of the universe. It is a constant that must be determined experimentally, like Newton's constant.

Chapter 10
Amplitude Reduction of Spacetime Fluctuations and Primordial Spectra

This chapter reveals how quantum fluctuations of gravity evolve around the inflationary solution obtained in Chap. 7. The goal is to investigate the evolution of the fluctuations from long before the Planck time until just before the spacetime phase transition, and to derive the primordial spectrum which gives the initial condition of the current universe. The universe before inflation has large quantum fluctuations of gravity, and thus spacetime is not substantially fixed. Here, we demonstrate that as inflation begins, the initially large fluctuations with scale invariance gradually reduce in amplitude, eventually reaching the magnitude suggested by tiny CMB anisotropies. Since fluctuations in gravity represent fluctuations in time and distance, the reduction of them implies that the universe becomes a tangible world with definite spacetime.

10.1 Nonlinear Evolution Equations for Scalar Fluctuations

Let us derive a set of equations of motion for fluctuations (perturbations) around the inflationary solution from the effective action of quantum gravity. Actually, we should discuss time evolution of physical scalars given by composite fields of fluctuations, such as the scalar curvature. However, since fluctuations reduce during inflation, as mentioned earlier, and information we ultimately want to know is their reduced values at the phase transition point, it turns out that it is sufficient to consider equations of motion for fluctuations [HHY06a, HHY10, Ham24a].

A fluctuation φ of the conformal-factor field around the inflationary solution $\hat{\phi}$ is defined as

$$\phi(\eta, \mathbf{x}) = \hat{\phi}(\eta) + \varphi(\eta, \mathbf{x}). \tag{10.1}$$

The asymptotic background freedom indicates that the scalar fluctuation φ becomes dominant in the early universe beyond the Planck scale. The initial spectrum is thus

K. Hamada, *Trans-Planckian Physics and Inflation*, Fundamental Theories of Physics 223, https://doi.org/10.1007/978-981-96-3476-7_10

given by the two-point correlation function (2.24) of φ and is expressed at the equal time as

$$\langle \varphi(\eta, \mathbf{x}) \varphi(\eta, \mathbf{y}) \rangle = -\frac{1}{4b_c} \log(\mathbf{x} - \mathbf{y})^2, \tag{10.2}$$

where note that the logarithmic function reflects that φ is a dimensionless field.

The spectrum is expressed using Fourier transform in the three-dimensional comoving space. Letting $f(\mathbf{x})$ be a dimensionless field, here we define it so that the transformed quantity $f(\mathbf{k})$ is also dimensionless as

$$f(\mathbf{x}) = \int \frac{d^3\mathbf{k}}{(2\pi)^3} \frac{1}{k^3} f(\mathbf{k})\, e^{i\mathbf{k}\cdot\mathbf{x}},$$

where $k = |\mathbf{k}|$. Then its dimensionless mean square is defined through $\langle f(\mathbf{k}) f(\mathbf{k}') \rangle = \langle |f(\mathbf{k})|^2 \rangle k^3 (2\pi)^3 \delta^3(\mathbf{k} + \mathbf{k}')$. From this, the initial spectrum of φ is given by a scale-invariant

$$\langle |\varphi(\mathbf{k})|^2 \rangle = \frac{\pi^2}{b_c}, \tag{10.3}$$

called the Harrison-Zel'dovich type [Har70, Zel72, Pee93]. The dimensionless spectrum is often normalized as $\langle |\varphi(\mathbf{k})|^2 \rangle / 2\pi^2$, and is $1/2b_c$ in this case. Due to the positive definiteness of the Riegert action $b_c > 0$, the amplitude becomes a positive value.

In order to describe the process by which this scalar fluctuation is gradually deformed with time, it is also necessary to consider the scalar component h of the traceless tensor field, which is defined by

$$h_{00} = h, \qquad h_{ij} = \frac{1}{3}\delta_{ij} h, \tag{10.4}$$

as introduced in Chap. 8. This component is initially zero and acquires value over time.

In the following, we derive simultaneous differential equations that describe time evolution of the fluctuations φ and h. The size considered is initially of about the Planck length and expands significantly during the inflationary period. For these fluctuations, we demonstrate that the amplitude eventually reduces.

Therefore, linear approximation becomes better with time, but at the initial stage, the fluctuation φ is still large, thus it is difficult to estimate accurate spectra unless nonlinear terms expressed by its multipoint functions are incorporated. The exponential factor of φ present in the Einstein-Hilbert action is one such term. On the other hand, we assume that h is not very large and treat it by linear approximation throughout. This is because h is initially small, then grows, but reduces again before the coupling constant t becomes large enough.

In general, scalar fluctuations contain nonlinear terms that combine other tensor modes into scalars, and these modes may also contribute to the equation of motion as nonlinear terms. However, since the initial smallness of these tensor modes is maintained until almost the end [HHY06a], they will contribute little to the evolution of the scalar fluctuations.

The equation of motion is expressed as $\mathbf{T}_{\mu\nu} = \mathbf{T}^{\mathrm{R}}_{\mu\nu} + \mathbf{T}^{\mathrm{W}}_{\mu\nu} + \mathbf{T}^{\mathrm{EH}}_{\mu\nu} + \mathbf{T}^{\mathrm{M}}_{\mu\nu} = 0$, and the energy-momentum tensors of each sector required here will be presented below from time to time (see also Appendix C). The matter energy-momentum tensor adopts the traceless perfect-fluid type.

First, consider the trace part of the equation of motion. The Riegert sector is given by

$$\mathbf{T}^{\mathrm{R}\mu}{}_{\mu} = -\frac{b_c}{16\pi^2} B(4\bar{\Delta}_4\phi + \bar{E}_4),$$

and expanding to the first order of the traceless tensor field yields

$$\begin{aligned}
\mathbf{T}^{\mathrm{R}\mu}{}_{\mu} &= \frac{b_c}{8\pi^2} B\Big\{ -2\Box^2\phi + 4\, h^{\mu\nu}\Box\partial_\mu\partial_\nu\phi + 4\partial^\lambda h^{\mu\nu}\partial_\lambda\partial_\mu\partial_\nu\phi \\
&\quad +4\chi^\mu\partial_\mu\Box\phi + 4\Box h^{\mu\nu}\partial_\mu\partial_\nu\phi + 2\Box\chi^\mu\partial_\mu\phi + \frac{4}{3}\partial_\mu\chi^\mu\Box\phi \\
&\quad -\frac{2}{3}\partial^\mu\partial_\nu\chi^\nu\partial_\mu\phi + \frac{1}{3}\Box\partial_\mu\chi^\mu \Big\},
\end{aligned}$$

where $\chi_\mu = \partial_\nu h^\nu{}_\mu$ and $\Box = \partial^\mu\partial_\mu = -\partial_\eta^2 + \eth^2$ is the d'Alembertian in flat background.

Further expand this equation according to (10.1) and (10.4). Dividing it into the linear part of φ and h and the nonlinear part of $o(\varphi h)$, and denoting it as $\mathbf{T}^{\mathrm{R}\mu}{}_{\mu} = \mathbf{T}^{\mathrm{R}\mu}{}_{\mu}|_L + \mathbf{T}^{\mathrm{R}\mu}{}_{\mu}|_{NL}$, we obtain

$$\begin{aligned}
\mathbf{T}^{\mathrm{R}\mu}{}_{\mu}|_L &= \frac{b_c}{8\pi^2} B\Big\{ -2\partial_\eta^4\varphi + 4\partial_\eta^2\eth^2\varphi - 2\eth^4\varphi - 4\partial_\eta^4\hat{\phi}h - 8\partial_\eta^3\hat{\phi}\partial_\eta h \\
&\quad -\frac{16}{3}\partial_\eta^2\hat{\phi}\partial_\eta^2 h + \frac{32}{9}\partial_\eta^2\hat{\phi}\eth^2 h - \frac{4}{3}\partial_\eta\hat{\phi}\partial_\eta^3 h + \frac{20}{9}\partial_\eta\hat{\phi}\partial_\eta\eth^2 h \\
&\quad -\frac{1}{3}\partial_\eta^4 h + \frac{2}{9}\partial_\eta^2\eth^2 h + \frac{1}{9}\eth^4 h \Big\}
\end{aligned}$$

and

$$
\begin{aligned}
\mathbf{T}^{\mathrm{R}\mu}{}_{\mu}|_{NL} &= \frac{b_c}{8\pi^2}B\Big\{-4\,h\partial_\eta^4\varphi + \frac{8}{3}h\partial_\eta^2\reflectbox{∂}^2\varphi + \frac{4}{3}h\reflectbox{∂}^4\varphi - 8\partial_\eta h\partial_\eta^3\varphi \\
&\quad + \frac{8}{3}\partial^i h\partial_\eta^2\partial_i\varphi + \frac{8}{3}\partial_\eta h\partial_\eta\reflectbox{∂}^2\varphi + \frac{8}{3}\partial^i h\partial_i\reflectbox{∂}^2\varphi \\
&\quad - \frac{16}{3}\partial_\eta^2\,h\partial_\eta^2\varphi + \frac{32}{9}\reflectbox{∂}^2 h\partial_\eta^2\varphi + \frac{16}{9}\reflectbox{∂}^2 h\reflectbox{∂}^2\varphi - \frac{4}{3}\partial_\eta^3\,h\partial_\eta\varphi \\
&\quad + \frac{20}{9}\partial_\eta\reflectbox{∂}^2 h\partial_\eta\varphi - \frac{4}{3}\partial_\eta^2\partial^i h\partial_i\varphi + \frac{4}{9}\partial^i\reflectbox{∂}^2 h\partial_i\varphi\Big\}.
\end{aligned}
$$

On the other hand, the trace part of the Weyl sector disappears in the first order of h, thus

$$
\mathbf{T}^{\mathrm{W}\mu}{}_{\mu}|_L = 0.
$$

The trace part of the Einstein-Hilbert sector is given by $\mathbf{T}^{\mathrm{EH}\mu}{}_{\mu} = M_\mathrm{P}^2\sqrt{-g}R = M_\mathrm{P}^2 e^{2\phi}(\bar{R} - 6\bar{\nabla}^2\phi - 6\bar{\nabla}^\mu\phi\bar{\nabla}_\mu\phi)$. Expanding it to the first order of the traceless tensor field, we get

$$
\begin{aligned}
\mathbf{T}^{\mathrm{EH}\mu}{}_{\mu} &= M_\mathrm{P}^2\,e^{2\phi}\Big\{-6\Box\phi - 6\partial^\mu\phi\partial_\mu\phi + \partial_\mu\chi^\mu + 6\,h^{\mu\nu}\partial_\mu\partial_\nu\phi + 6\chi^\mu\partial_\mu\phi \\
&\quad + 6\,h^{\mu\nu}\partial_\mu\phi\partial_\nu\phi\Big\}.
\end{aligned}
$$

Expanding this further yields terms up to infinite order of φ due to the presence of the exponential factor $e^{2\phi}$, which is expressed as

$$
\mathbf{T}^{\mathrm{EH}\mu}{}_{\mu} = \sum_{n=1}^{\infty}\mathbf{T}^{\mathrm{EH}\mu}_{(n)\ \ \mu},
$$

where $n = 0$ is removed because it is the homogeneous component that gives the second term of (7.2). $\mathbf{T}^{\mathrm{EH}\mu}_{(n)\ \ \mu}$ consists of the $o(\varphi^n)$ and $o(\varphi^{n-1}h)$ terms, and the linear part is given by

$$
\begin{aligned}
\mathbf{T}^{\mathrm{EH}\mu}_{(1)\ \ \mu} &= M_\mathrm{P}^2\,e^{2\hat{\phi}}\Big\{6\partial_\eta^2\varphi - 6\reflectbox{∂}^2\varphi + 12\partial_\eta\hat{\phi}\partial_\eta\varphi + 12(\partial_\eta^2\hat{\phi} + \partial_\eta\hat{\phi}\partial_\eta\hat{\phi})\varphi \\
&\quad + \partial_\eta^2\,h + \frac{1}{3}\reflectbox{∂}^2 h + 6\partial_\eta\hat{\phi}\partial_\eta h + 6(\partial_\eta^2\hat{\phi} + \partial_\eta\hat{\phi}\partial_\eta\hat{\phi})h\Big\},
\end{aligned}
$$

the $n = 2$ nonlinear part is

$$\begin{aligned}
\mathbf{T}^{\mathrm{EH}\mu}_{(2)\ \mu} &= M_{\mathrm{P}}^2 e^{2\hat{\phi}} \Big\{ 12\varphi\partial_\eta^2\varphi - 12\varphi\bar{\partial}^2\varphi + 6\partial_\eta\varphi\partial_\eta\varphi - 6\partial^i\varphi\partial_i\varphi + 24\partial_\eta\hat{\phi}\,\varphi\partial_\eta\varphi \\
&\quad +12\big(\partial_\eta^2\hat{\phi} + \partial_\eta\hat{\phi}\partial_\eta\hat{\phi}\big)\varphi^2 + 2\varphi\partial_\eta^2 h + \frac{2}{3}\varphi\bar{\partial}^2 h + 6\partial_\eta^2\varphi h + 2\bar{\partial}^2\varphi h \\
&\quad +6\partial_\eta\varphi\partial_\eta h + 2\partial^i\varphi\partial_i h + 12\partial_\eta\hat{\phi}\,\varphi\partial_\eta h + 12\partial_\eta\hat{\phi}\,\partial_\eta\varphi h \\
&\quad +12\big(\partial_\eta^2\hat{\phi} + \partial_\eta\hat{\phi}\partial_\eta\hat{\phi}\big)\varphi h \Big\},
\end{aligned}$$

and the $n = 3$ nonlinear part is

$$\begin{aligned}
\mathbf{T}^{\mathrm{EH}\mu}_{(3)\ \mu} &= M_{\mathrm{P}}^2 e^{2\hat{\phi}} \Big\{ 12\varphi^2\partial_\eta^2\varphi - 12\varphi^2\bar{\partial}^2\varphi + 12\varphi\partial_\eta\varphi\partial_\eta\varphi - 12\varphi\partial^i\varphi\partial_i\varphi \\
&\quad +24\partial_\eta\hat{\phi}\,\varphi^2\partial_\eta\varphi + 8\big(\partial_\eta^2\hat{\phi} + \partial_\eta\hat{\phi}\partial_\eta\hat{\phi}\big)\varphi^3 + 2\varphi^2\partial_\eta^2 h + \frac{2}{3}\varphi^2\bar{\partial}^2 h \\
&\quad +12\varphi\partial_\eta^2\varphi h + 4\varphi\bar{\partial}^2\varphi h + 12\varphi\partial_\eta\varphi\partial_\eta h + 4\varphi\partial^i\varphi\partial_i h \\
&\quad +12\partial_\eta\hat{\phi}\,\varphi^2\partial_\eta h + 24\partial_\eta\hat{\phi}\,\varphi\partial_\eta\varphi h + 12\big(\partial_\eta^2\hat{\phi} + \partial_\eta\hat{\phi}\partial_\eta\hat{\phi}\big)\varphi^2 h \\
&\quad +6\partial_\eta\varphi\partial_\eta\varphi h + 2\partial^i\varphi\partial_i\varphi h \Big\}.
\end{aligned}$$

The case of $n \geq 4$ can be easily derived as well.

Since there are two field variables of φ and h, the trace equation alone is incomplete. Therefore, as in Chap. 8, in order to derive another equation consisting of only their two fields, we consider an equation combining the energy-momentum tensor as

$$\frac{1}{\bar{\partial}^2}\left(\mathbf{T}^i{}_i - 3\frac{\partial^i\partial^j}{\bar{\partial}^2}\mathbf{T}_{ij} \right) = 0, \tag{10.5}$$

where it is reduced to a second-order differential equation by further multiplying by the inverse of the spatial Laplacian.

The equation resulting from this combination has a contribution from the Weyl sector. To begin with, consider this equation at the initial stage when the coupling constant t is still small, then the necessary part of the Weyl energy-momentum tensor is given by

$$\mathbf{T}^{\mathrm{W}}_{\mu\nu}|_L = -\frac{2}{t^2}\left\{ \Box^2 h_{\mu\nu} - \Box\partial_\mu\chi_\nu - \Box\partial_\nu\chi_\mu + \frac{2}{3}\partial_\mu\partial_\nu\partial_\lambda\chi^\lambda + \frac{1}{3}\eta_{\mu\nu}\Box\partial_\lambda\chi^\lambda \right\}.$$

Since there is $1/t^2$ here, as we can see by multiplying the whole energy-momentum tensor by t^2, the other sectors become $o(t^2)$ and thus can be treated in linear. Therefore, the contributions from the Riegert and Einstein-Hilbert sectors to (10.5) can be obtained by substituting (10.1) and (10.4) into their energy-momentum tensors and extracting the linear parts of φ and h.

The expression of the energy-momentum tensor of the Riegert sector required here is a little complicated, thus it is only shown in (C.3) in Appendix C and see that. The energy-momentum tensor of the Einstein-Hilbert sector is given by

$$\begin{aligned}\mathbf{T}^{\mathrm{EH}}_{\mu\nu} &= M_{\mathrm{P}}^2 e^{2\phi}\Big\{2\partial_\mu\partial_\nu\phi - 2\partial_\mu\phi\partial_\nu\phi + \eta_{\mu\nu}(-2\Box\phi - \partial^\lambda\phi\partial_\lambda\phi) - \partial_{(\mu}\chi_{\nu)} \\ &+\frac{1}{2}\Box h_{\mu\nu} - 2h^\lambda_{(\mu}\partial_{\nu)}\partial_\lambda\phi + 2h^\lambda_{(\mu}\partial_{\nu)}\phi\partial_\lambda\phi - 2\partial_{(\mu}h^\lambda_{\nu)}\partial_\lambda\phi + \partial^\lambda h_{\mu\nu}\partial_\lambda\phi \\ &+\eta_{\mu\nu}\left(\frac{1}{2}\partial_\lambda\chi^\lambda + 2h^{\lambda\sigma}\partial_\lambda\partial_\sigma\phi + h^{\lambda\sigma}\partial_\lambda\phi\partial_\sigma\phi + 2\chi^\lambda\partial_\lambda\phi\right)\Big\}.\end{aligned}$$

By the time the coupling constant becomes large, the fluctuations have become sufficiently small, thus (10.5) can be treated with linear approximation from beginning to end, which is given by

$$\begin{aligned}&\frac{16}{3}\frac{1}{t^2}\left(\partial_\eta^2 h - \frac{1}{3}\partial\!\!\!^2 h\right) + \frac{b_c}{8\pi^2}B\Big\{\frac{4}{3}\partial_\eta^2\varphi + \frac{8}{3}\partial_\eta\hat{\phi}\,\partial_\eta\varphi + \left(8\partial_\eta^2\hat{\phi} - \frac{4}{3}\partial\!\!\!^2\right)\varphi \\ &+\frac{2}{9}\partial_\eta^2 h + \frac{4}{3}\partial_\eta\hat{\phi}\,\partial_\eta h + \left(\frac{20}{9}\partial_\eta^2\hat{\phi} - \frac{16}{9}\partial_\eta\hat{\phi}\partial_\eta\hat{\phi} + \frac{2}{27}\partial\!\!\!^2\right)h\Big\} \\ &+M_{\mathrm{P}}^2 e^{2\hat{\phi}}\left(-4\varphi + \frac{2}{3}h\right) = 0. \end{aligned} \tag{10.6}$$

Therefore, the equation of motion up to the first order of h can be expressed as a simultaneous differential equation with the nonlinear

$$\mathbf{T}^{\mathrm{R}\mu}{}_\mu|_L + \mathbf{T}^{\mathrm{R}\mu}{}_\mu|_{NL} + \sum_{n=1}^{\infty}\mathbf{T}^{\mathrm{EH}\mu}_{(n)\ \mu} = 0 \tag{10.7}$$

and the linear (10.6).

However, it is not yet in a form that can be solved numerically. It is necessary to rewrite the $\partial_\eta^4\hat{\phi}\,h$ term in the $\mathbf{T}^{\mathrm{R}\mu}{}_\mu|_L$ part to terms with lower time-derivatives of $\hat{\phi}$ using the homogeneous equation (7.2). In addition, it is necessary to rewrite the $h\partial_\eta^4\varphi$ term in the $\mathbf{T}^{\mathrm{R}\mu}{}_\mu|_{NL}$ part to terms with lower time-derivatives of φ. This can be done by noting the fact that the trace equation has a structure

$$\begin{aligned}&-\frac{b_c}{4\pi^2}B(\partial_\eta^4\varphi - 2\partial_\eta^2\partial\!\!\!^2\varphi + \partial\!\!\!^4\varphi) + M_{\mathrm{P}}^2 e^{2\hat{\phi}}\{6\partial_\eta^2\varphi - 6\partial\!\!\!^2\varphi + 12\partial_\eta\hat{\phi}\,\partial_\eta\varphi \\ &+12(\partial_\eta^2\hat{\phi} + \partial_\eta\hat{\phi}\partial_\eta\hat{\phi})\varphi + 12\varphi\partial_\eta^2\varphi - 12\varphi\partial\!\!\!^2\varphi + 6\partial_\eta\varphi\partial_\eta\varphi - 6\partial^i\varphi\partial_i\varphi \\ &+24\partial_\eta\hat{\phi}\,\varphi\partial_\eta\varphi + 12(\partial_\eta^2\hat{\phi} + \partial_\eta\hat{\phi}\partial_\eta\hat{\phi})\varphi^2 + o(\varphi^3)\} = o(h)\end{aligned}$$

and ignoring the $o(h^2)$ term. Furthermore, we introduce a field variable

$$\Phi = \varphi + \frac{1}{6} h$$

and treat Φ and h as independent variables. Using these variables, the fourth-derivative term of h in the linear part is absorbed into a fourth-derivative term of Φ, so that the trace equation can be reduced to at most the third time-derivative of h. Here, Φ is the gravitational potential when the line element is expressed as $ds^2 = a^2[-(1+2\Psi)d\eta^2 + (1+2\Phi)d\mathbf{x}^2]$. The other one is given by $\Psi = \varphi - h/2$.

The nonlinear terms contributing at the initial stage are incorporated as in (10.7), while the nonlinear effects controlling the conformal gravity dynamics are described by the running coupling constant. As done in Chap. 7, it is expressed by adopting a kind of mean field approximation that rewrites t^2 and B to the time-dependent average $\bar{t^2}(\tau)$ (7.9) and the dynamical factor $\bar{B}(\tau)$ (7.10), respectively.

This is a fairly simplified model, but since the size of the fluctuations considered here becomes much larger than the correlation length ξ_Λ at the time of spacetime phase transition, the spectral pattern of such fluctuations is not affected by the strong-coupling dynamics during the phase transition. Therefore, it seems that no further information is needed beyond the disappearance of the conformal gravity dynamics.

In what follows, we consider only the $n = 2$ nonlinear term. The system of the evolution equation obtained in the way above [Ham24a] is then given by a nonlinear equation,

$$\begin{aligned}
&\frac{b_c}{8\pi^2}\bar{B}(\tau)\Big\{-2\partial_\eta^4\Phi + 4\partial_\eta^2\not\partial^2\Phi - 2\not\partial^4\Phi - \frac{4}{9}\partial_\eta^2\not\partial^2 h + \frac{4}{9}\not\partial^4 h - 8\partial_\eta^3\hat{\phi}\,\partial_\eta h \\
&-\frac{16}{3}\partial_\eta^2\hat{\phi}\,\partial_\eta^2 h - \frac{4}{3}\partial_\eta\hat{\phi}\,\partial_\eta^3 h + \frac{32}{9}\partial_\eta^2\hat{\phi}\not\partial^2 h + \frac{20}{9}\partial_\eta\hat{\phi}\,\partial_\eta\not\partial^2 h \\
&-\frac{16}{3}h\partial_\eta^2\not\partial^2\Phi + \frac{16}{3}h\not\partial^4\Phi - 8\partial_\eta h\partial_\eta^3\Phi + \frac{8}{3}\partial^i h\partial_\eta^2\partial_i\Phi + \frac{8}{3}\partial_\eta h\partial_\eta\not\partial^2\Phi \\
&+\frac{8}{3}\partial^i h\partial_i\not\partial^2\Phi - \frac{16}{3}\partial_\eta^2 h\partial_\eta^2\Phi + \frac{32}{9}\not\partial^2 h\partial_\eta^2\Phi + \frac{16}{9}\not\partial^2 h\not\partial^2\Phi - \frac{4}{3}\partial_\eta^3 h\partial_\eta\Phi \\
&+\frac{20}{9}\partial_\eta\not\partial^2 h\partial_\eta\Phi - \frac{4}{3}\partial_\eta^2\partial^i h\partial_i\Phi + \frac{4}{9}\partial^i\not\partial^2 h\partial_i\Phi\Big\} \\
&+M_{\rm P}^2 e^{2\hat{\phi}}\Big\{6\partial_\eta^2\Phi + 12\partial_\eta\hat{\phi}\,\partial_\eta\Phi - 6\not\partial^2\Phi + 12\big(\partial_\eta^2\hat{\phi} + \partial_\eta\hat{\phi}\partial_\eta\hat{\phi}\big)\Phi \\
&+4\partial_\eta\hat{\phi}\,\partial_\eta h + \frac{4}{3}\not\partial^2 h - 8\big(\partial_\eta^2\hat{\phi} + \partial_\eta\hat{\phi}\partial_\eta\hat{\phi}\big)h \\
&+12\Phi\partial_\eta^2\Phi + 6\partial_\eta\Phi\partial_\eta\Phi + 24\partial_\eta\hat{\phi}\,\Phi\partial_\eta\Phi + 12\big(\partial_\eta^2\hat{\phi} + \partial_\eta\hat{\phi}\partial_\eta\hat{\phi}\big)\Phi^2 \\
&-12\Phi\not\partial^2\Phi - 6\partial^i\Phi\partial_i\Phi - 8h\partial_\eta^2\Phi + 4\partial_\eta h\partial_\eta\Phi - 16\partial_\eta\hat{\phi}\,h\partial_\eta\Phi + 8\partial_\eta\hat{\phi}\,\partial_\eta h\Phi \\
&-16\big(\partial_\eta^2\hat{\phi} + \partial_\eta\hat{\phi}\partial_\eta\hat{\phi}\big)h\Phi + 16h\not\partial^2\Phi + 4\partial^i h\partial_i\Phi + \frac{8}{3}\not\partial^2 h\Phi\Big\} = 0, \qquad (10.8)
\end{aligned}$$

and a linear equation containing at most second derivative of the field,

$$
\begin{aligned}
&\frac{16}{3}\frac{1}{\bar{t}^2(\tau)}\left(\partial_\eta^2 h - \frac{1}{3}\partial^2 h\right) + \frac{b_c}{8\pi^2}\bar{B}(\tau)\left\{\frac{4}{3}\partial_\eta^2\Phi - \frac{4}{3}\partial^2\Phi + \frac{8}{3}\partial_\eta\hat{\phi}\,\partial_\eta\Phi + 8\partial_\eta^2\hat{\phi}\,\Phi\right. \\
&\left. + \frac{8}{27}\partial^2 h + \frac{8}{9}\partial_\eta\hat{\phi}\,\partial_\eta h + \left(\frac{8}{9}\partial_\eta^2\hat{\phi} - \frac{16}{9}\partial_\eta\hat{\phi}\partial_\eta\hat{\phi}\right)h\right\} + M_{\mathrm{P}}\,e^{2\hat{\phi}}\left(-4\Phi + \frac{4}{3}h\right) \\
&= 0. \qquad (10.9)
\end{aligned}
$$

We solve these equations simultaneously, but in doing that, it is further necessary to rewrite $\partial_\eta^3 h$ in (10.8) to the third derivative of Φ and other lower time-derivative terms using an equation obtained by differentiating (10.9) with time.

The second equation (10.9) plays an important role as a constraint for connecting between the inflation and Friedmann phases. When the running coupling constant $\bar{t}$ is small, h is small, indicating that φ is dominant. Conversely, when $\bar{t}$ diverges, $1/\bar{t}^2$ and the dynamical factor $\bar{B}$ vanish, and the last Einstein term becomes dominant, this results in a transition to the Friedmann universe satisfying $\Phi = h/3$. In terms of the gravitational potentials, it expresses that the initial fluctuation of $\Phi = \Psi$ changes to $\Phi = -\Psi$.

10.2 How to Handle Nonlinear Terms

In order to obtain momentum representations of the nonlinear terms, consider Fourier transform of a dimensionless field product $f(\mathbf{x})g(\mathbf{x})$ defined by

$$
\begin{aligned}
(fg)(\mathbf{k}) &= k^3 \int d^3\mathbf{x} f(\mathbf{x})g(\mathbf{x})e^{-i\mathbf{k}\cdot\mathbf{x}} \\
&= k^3 \int d^3\mathbf{x} \int \frac{d^3\mathbf{p}}{(2\pi)^3}\frac{d^3\mathbf{q}}{(2\pi)^3}\frac{1}{p^3 q^3} f(\mathbf{p})g(\mathbf{q})e^{i(\mathbf{p}+\mathbf{q}-\mathbf{k})\cdot\mathbf{x}}.
\end{aligned}
$$

Furthermore, the exponential part is rewritten using the expansion formula based on the spherical Bessel function,

$$
e^{i\mathbf{k}\cdot\mathbf{x}} = 4\pi \sum_{l=0}^{\infty}\sum_{m=-l}^{l} i^l j_l(kx) Y_{lm}^*(\Omega_k) Y_{lm}(\Omega_x),
$$

where Ω denotes angular components such as $d\mathbf{x}^3 = x^2 dx d\Omega_x$ and the spherical harmonics are normalized as $\int d\Omega\, Y_{lm}^*(\Omega) Y_{l'm'}(\Omega) = \delta_{ll'}\delta_{mm'}$.

Henceforth, we consider the case of

$$
f(\mathbf{p}) = f(p), \qquad g(\mathbf{q}) = g(q), \qquad (10.10)
$$

assuming that isotropic components of each function in the linear and nonlinear terms predominantly contribute to the result. In this case, the angular integration can be easily performed and we obtain

$$(fg)(k) = \frac{8}{(2\pi)^3}k^3 \int \frac{dp}{p}\frac{dq}{q} f(p)g(q) \int dx x^2 j_0(kx) j_0(px) j_0(qx),$$

where $j_0(kx) = \sin(kx)/kx$. The x-integral yields

$$\int_0^\infty dx x^2 j_0(kx) j_0(px) j_0(qx) = \begin{cases} \frac{\pi}{4}\frac{1}{kpq} & \text{for } |p-q| < k < p+q \\ 0 & \text{otherwise} \end{cases}$$

and thus we get

$$(fg)(k) = \frac{k^2}{(2\pi)^2} \int \frac{dp}{p^2}\frac{dq}{q^2} \theta(k - |p-q|)\theta(p+q-k) f(p)g(q). \quad (10.11)$$

Fourier transform of a product of functions with spatial derivatives can be calculated similarly. Here it is sufficient to consider the type of

$$\begin{aligned}(\partial_i f \partial^i g)(k) &= k^3 \int d^3\mathbf{x} \partial_i f(\mathbf{x}) \partial^i g(\mathbf{x}) e^{-i\mathbf{k}\cdot\mathbf{x}} \\ &= k^3 \int d^3\mathbf{x} \int \frac{d^3\mathbf{p}}{(2\pi)^3}\frac{d^3\mathbf{q}}{(2\pi)^3}\frac{-\mathbf{p}\cdot\mathbf{q}}{p^3q^3} f(\mathbf{p})g(\mathbf{q}) e^{i(\mathbf{p}+\mathbf{q}-\mathbf{k})\cdot\mathbf{x}}.\end{aligned}$$

Under the isotropic condition (10.10), the integration of the angular components of the momenta $\mathbf{p}$ and $\mathbf{q}$ can be easily performed by expressing the vector $\mathbf{p} = (p_x, p_y, p_z)$ in terms of spherical harmonics as

$$\begin{aligned} p_x &= \sqrt{\frac{2\pi}{3}} p \left\{Y_{1-1}(\Omega_p) - Y_{11}(\Omega_p)\right\}, \\ p_y &= i\sqrt{\frac{2\pi}{3}} p \left\{Y_{1-1}(\Omega_p) + Y_{11}(\Omega_p)\right\}, \\ p_z &= \sqrt{\frac{4\pi}{3}} p Y_{10}(\Omega_p). \end{aligned}$$

Performing the integration of the remaining angular component Ω_x using

$$\int d\Omega_x Y^*_{l_1m_1}(\Omega_x) Y_{l_2m_2}(\Omega_x) Y_{l_3m_3}(\Omega_x) = \sqrt{\frac{(2l_2+1)(2l_3+1)}{4\pi(2l_1+1)}} C^{l_10}_{l_20l_30} C^{l_1m_1}_{l_2m_2l_3m_3}$$

yields

$$
\begin{aligned}
(\partial_i f \partial^i g)(k) &= \frac{8}{(2\pi)^3} k^3 \int dp dq f(p) g(q) \int dx x^2 j_0(kx) j_1(px) j_1(qx) \\
&= \frac{k^2}{(2\pi)^2} \int \frac{dp}{p^2} \frac{dq}{q^2} \theta(k - |p-q|) \theta(p+q-k) \\
&\quad \times \frac{1}{2}(p^2 + q^2 - k^2) f(p) g(q).
\end{aligned} \tag{10.12}
$$

The lower bound of the comoving momentum integral is given by the comoving scale λ (7.14) of the physical scale Λ_{QG}, while there is no upper bound on the integration. Using the formulas (10.11) and (10.12) for the field products, Fourier transforms of the second-order nonlinear terms in the trace Equation (10.8) are given by

$$
\begin{aligned}
&\frac{b_c}{8\pi^2} \bar{B} \frac{k^2}{(2\pi)^2} \int_\lambda^\infty \frac{dp}{p^2} \frac{dq}{q^2} \theta(k - |p-q|) \theta(p+q-k) \Big\{ -8 \partial_\eta h(p) \partial_\eta^3 \Phi(q) \\
&- \frac{16}{3} \partial_\eta^2 h(p) \partial_\eta^2 \Phi(q) - \frac{4}{3} \partial_\eta^3 h(p) \partial_\eta \Phi(q) + \left(-\frac{20}{9} p^2 + \frac{20}{3} q^2 - \frac{4}{3} k^2 \right) h(p) \partial_\eta^2 \Phi(q) \\
&- \left(\frac{20}{9} p^2 + \frac{8}{3} q^2 \right) \partial_\eta h(p) \partial_\eta \Phi(q) + \left(-\frac{2}{3} p^2 - \frac{2}{3} q^2 + \frac{2}{3} k^2 \right) \partial_\eta^2 h(p) \Phi(q) \\
&+ \left[-\frac{2}{9} p^4 + 4 q^4 + \frac{2}{9} p^2 q^2 + \left(\frac{2}{9} p^2 + \frac{4}{3} q^2 \right) k^2 \right] h(p) \Phi(q) \Big\}
\end{aligned} \tag{10.13}
$$

and

$$
\begin{aligned}
&M_P^2 e^{2\hat{\phi}} \frac{k^2}{(2\pi)^2} \int_\lambda^\infty \frac{dp}{p^2} \frac{dq}{q^2} \theta(k - |p-q|) \theta(p+q-k) \Big\{ 12 \Phi(p) \partial_\eta^2 \Phi(q) \\
&+ 6 \partial_\eta \Phi(p) \partial_\eta \Phi(q) + 24 \partial_\eta \hat{\phi}\, \Phi(p) \partial_\eta \Phi(q) + 12 \big(\partial_\eta^2 \hat{\phi} + \partial_\eta \hat{\phi} \partial_\eta \hat{\phi} \big) \Phi(p) \Phi(q) \\
&+ 3 \big(p^2 + q^2 + k^2 \big) \Phi(p) \Phi(q) - 8\, h(p) \partial_\eta^2 \Phi(q) + 4 \partial_\eta h(p) \partial_\eta \Phi(q) \\
&- 16 \partial_\eta \hat{\phi}\, h(p) \partial_\eta \Phi(q) + 8 \partial_\eta \hat{\phi}\, \partial_\eta h(p) \Phi(q) - 16 \big(\partial_\eta^2 \hat{\phi} + \partial_\eta \hat{\phi} \partial_\eta \hat{\phi} \big) h(p) \Phi(q) \\
&- \left(\frac{2}{3} p^2 + 14 q^2 + 2 k^2 \right) h(p) \Phi(q) \Big\}.
\end{aligned} \tag{10.14}
$$

Here note that the momentum integral in (10.14) has no problem on upper bound due to the strength of its negative power, while the absence of problems in (10.13) derived from the fourth-derivative Riegert action will be guaranteed by the property that the traceless tensor field becomes smaller in the high-momentum region.[1]

---

[1] When considering higher momentum regions, it will be necessary to introduce momentum dependence in the running coupling constant such as $\bar{t}^2(\tau, k) = \{\beta_0 \log[\tau_\Lambda^2 (1/\tau^2 + k^2/a(\tau)^2)]\}^{-1}$, even though the k-dependence disappears as soon as inflation begins.

10.3 Simplification of Nonlinear Terms and Numerical Evaluation

Let us consider numerically solving the above system of nonlinear evolution equations. To do this, it is necessary to express the momentum integrals by sums following the piecewise integral method and rewrite it to a multiple simultaneous differential equation in which many fields with different momentums are linked. However, it is not easy to actually solve such a nonlinear equation system. Here, we further simplify the equations and examine them to evaluate how the nonlinear terms affect the spectrum specifically. At this time, we focus on the spectral pattern, but do not pursue accuracy in the overall amplitude.

First of all, it is impossible to solve the evolution equation when two time variables are mixed, thus we unify them to the proper time τ (7.3). At that time, using the scale factor a, the Hubble variable H, and its time derivatives $Y = \partial_\tau H$ and $Z = \partial_\tau^2 H$ (7.4), the differential operators in conformal time η can be rewritten as

$$\begin{aligned}
\partial_\eta &= a\partial_\tau, \\
\partial_\eta^2 &= a^2 D_\tau^2, \quad D_\tau^2 = \partial_\tau^2 + H\partial_\tau, \\
\partial_\eta^3 &= a^3 D_\tau^3, \quad D_\tau^3 = \partial_\tau^3 + 3H\partial_\tau^2 + (Y + 2H^2)\partial_\tau, \\
\partial_\eta^4 &= a^4 D_\tau^4, \quad D_\tau^4 = \partial_\tau^4 + 6H\partial_\tau^3 + (4Y + 11H^2)\partial_\tau^2 + (Z + 7HY + 6H^3)\partial_\tau.
\end{aligned}$$

The variables for the inflationary background are also rewritten as

$$\begin{aligned}
\partial_\eta\hat{\phi} &= aH, \\
\partial_\eta^2\hat{\phi} &= a^2(Y + H^2), \\
\partial_\eta^3\hat{\phi} &= a^3(Z + 4HY + 2H^3), \\
\partial_\eta^4\hat{\phi} &= a^4(\partial_\tau Z + 7HZ + 4Y^2 + 18H^2Y + 6H^4),
\end{aligned}$$

where the last formula of $\partial_\eta^4\hat{\phi}$ was used to derive (7.5).

When solving the equation of motion, the result does not depend on the normalization of the scale factor, thus we here set the current scale factor a_0 to unity so that the comoving momentum represents the current physical wavenumber. Furthermore, decompose the scale factor as $a(t) = a(t_i) \times \bar{a}(t)$ and let the initial value of the scale factor with the bar be $\bar{a}(t_i) = 1$, where t $(= H_{\mathbf{D}}\tau)$ is a dimensionless time. Then, a dimensionless quantity obtained by dividing the physical momentum squared k^2/a^2 by the square of $H_{\mathbf{D}}$ can be expressed using the comoving Planck mass m (7.14) as

$$P_a(t, k) = \frac{k^2}{H_{\mathbf{D}}^2 a^2(\tau)} = \frac{k^2}{m^2 \bar{a}^2(t)}. \tag{10.15}$$

In addition, we define dimensionless differential operators as $D_t^n = D_\tau^n / H_{\mathbf{D}}^n = \partial_t^n + \cdots$ $(n = 2, 3, 4)$, in which the Hubble variables are eventually replaced with normalized $\bar{H} = H/H_{\mathbf{D}}$, $\bar{Y} = Y/H_{\mathbf{D}}^2$, and $\bar{Z} = Z/H_{\mathbf{D}}^3$.

Calculations of time evolution of the fluctuations are numerically performed by setting the comoving Planck mass, which provides a reference value for the calculation, to

$$m = 0.02\,\mathrm{Mpc}^{-1}. \tag{10.16}$$

The inflationary solution of $N = 60$ (7.12) shown in Chap. 7 is employed here, thus $\lambda\,(= m/N) = 0.00033\,\mathrm{Mpc}^{-1}$. Basically, m can be chosen freely, but its value cannot be changed significantly due to the consistency with the cosmic evolution scenario.

Next, the nonlinear terms are simplified as follows. Assuming that the field correlation vanishes instantaneously when distance exceeds the correlation length ξ_Λ, let the field $f(k)$ be a variable with values like a step function for $k \geq \lambda$. We then focus on the part of the integrand,

$$\frac{1}{(2\pi)^2}\frac{k^2}{p^2q^2}\theta(k - |p - q|)\theta(p + q - k), \tag{10.17}$$

which appears in the Fourier transforms of the field products. When displaying this function in the physical range $p,\ q \geq \lambda$, there are two sharp peaks at $(p, q) = (\lambda, k)$ and (k, λ) as shown in Fig. 10.1. Therefore, we extract these neighborhoods and evaluate them. Supposing that the field variable $f(k)$ is the one averaged over a

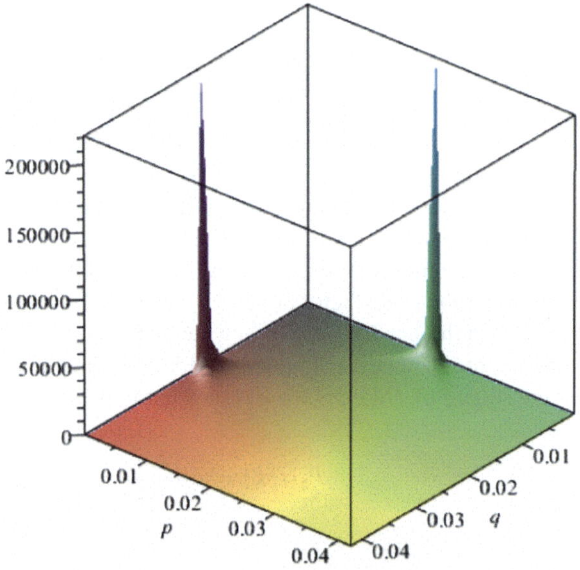

Fig. 10.1 Three-dimensional plot of (10.17) existing in the nonlinear terms when $k = m$ (10.16). The peak value is given by $1/(2\pi\lambda)^2$, and then the shape of the peak hardly depends on the value of k

narrow range Δ around the momentum k, and considering only the contribution from the narrow region near the peak where $dpdq$ is Δ^2, we simplify (10.11) as follows:

$$(fg)(k) = \frac{1}{(2\pi)^2}\frac{\Delta^2}{\lambda^2}\big[f(\lambda)g(k) + f(k)g(\lambda)\big]. \tag{10.18}$$

The formula (10.12) is also simplified in the same way. The evolution equation derived in this way is written down in Appendix Sect. 10.4, in which $TR_{[\lambda]}(k) = 0$ (10.23) is a simplified version of the nonlinear trace Equation (10.8) while $SP(k) = 0$ (10.24) is the linear constraint equation (10.9).

Thus, the nonlinear evolution equation system is simplified into the four equations

$$TR_{[\lambda]}(k) = 0, \quad SP(k) = 0, \quad TR_{[\lambda]}(\lambda) = 0, \quad SP(\lambda) = 0, \tag{10.19}$$

and we have to solve these equations simultaneously together with the homogeneous equation (7.5). Here note that the last two $TR_{[\lambda]}(\lambda) = 0$ and $SP(\lambda) = 0$ form a system of equations representing the nonlinear evolution of the field with momentum λ, and can be solved with only these two equations. In this way, although originally we have to solve a coupled multi-line equation system consisting of many fields with different momenta, it is now reduced to a two-line system consisting only of the fields with momenta k and λ, which can be solved relatively easily.

Let us solve this simplified evolution equation system in order to examine how the nonlinear terms work. Here, Δ is introduced as a phenomenological parameter representing the effective strength of the nonlinear term, and its value is assumed to be approximately $\Delta/2\pi \sim \lambda$ because the coefficient of (10.18) becomes about unity.

In solving the coupled nonlinear evolution equation (10.19), we have to pay attention to the role of $SP = 0$. As described at the end of Sect. 10.1, this equation works as a constraint that connects the initial stage, where $\bar{t}^2$ is small, to the moment of the spacetime phase transition, where it diverges. It indicates that the universe at $\bar{t}^2 \to 0$ before inflation is in a state where only the fluctuation of $\Phi = \varphi$ with $h = 0$ exists. On the other hand, at the dynamical time t_Λ when $\bar{t}^2$ diverges, the fluctuation changes to that satisfying $\Phi = h/3$.

In order to solve the evolution equation while preserving the constraint $SP = 0$, it is necessary to solve it as a boundary value problem. The initial time t_i is set to 10^{-3} sufficiently before the Planck time when inflation begins, and the magnitude of Φ at that time is given by the square root of the dimensionless amplitude (10.3). Thus, the initial value of Φ is given for $k \geq \lambda$ by

$$\Phi(t_i, k) = \frac{\pi}{\sqrt{b_c}} \tag{10.20}$$

and $\partial_t \Phi(t_i, k) = \partial_t^2 \Phi(t_i, k) = \partial_t^3 \Phi(t_i, k) = 0$. To set the boundary condition on the field h, we rewrite it as

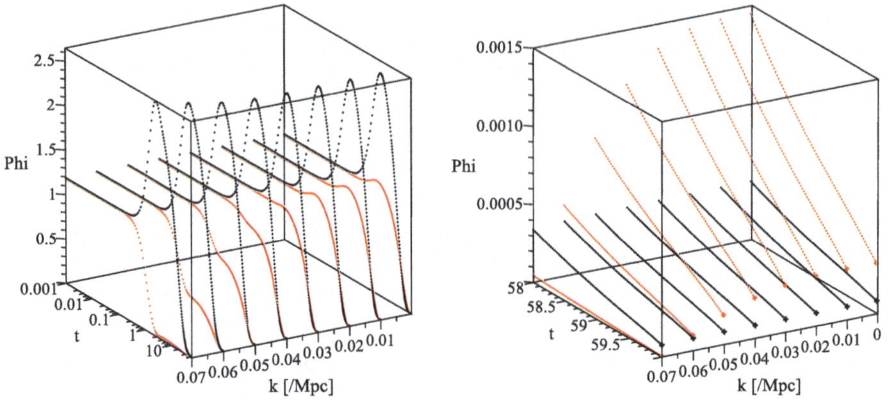

Fig. 10.2 Time evolution of the fluctuation Φ from before inflation to the spacetime phase transition. The result (black) is shown in logarithmic time when the parameters are fixed to $b_c = 7$ and $m = 0.02\,\mathrm{Mpc}^{-1}$ (see Chap. 7 for other parameters) and the strength of the nonlinear term is set to $\Delta/2\pi = 1.3\lambda$, together with the result of linear approximation ($\Delta = 0$) (red). The Planck time at which inflation begins is normalized to unity, and the phase transition occurs at 60. The horizontal axis k is the comoving momentum. The right is an enlarged view of the area around the phase transition point. Reprinted from [Ham24a] under the Creative Commons Attribution (CC-BY) license

$$h = 3(\Phi - X),$$

resulting in a system of equations for Φ and X, and then impose the boundary condition on X. Since the initial value of h is zero, $X(t_i, k) = \Phi(t_i, k)$, while X should vanish at the dynamical time $t_\Lambda(= 60)$, thus we set

$$X(t_i, k) = \frac{\pi}{\sqrt{b_c}}, \qquad X(t_\Lambda, k) = 0. \tag{10.21}$$

To evaluate time evolution of the field with momentum k, we have to solve (10.19) with these boundary conditions at k, together with those where k is replaced with λ.

Figures 10.2, 10.3 and 10.4 show the results of solving time evolution of the fluctuations numerically. The details of how to solve it are summarized in Appendix Sect. 10.4. Figure 10.2 shows the results for the largest value of Δ calculated, and the right is its enlarged view of the vicinity of the phase transition. Figure 10.3 summarizes time evolution of Φ for various Δ, including the case in Fig. 10.2, and the contribution from the h-component in Φ. We can see that as h contributes, the fluctuation amplitude turns to reduce. The $k = \lambda$ line at the right end of each figure showing time evolution is the result of solving only the latter two equations in (10.19) simultaneously, and each of the other k lines is the result of solving the four equations simultaneously.

The spectrum of fluctuations, which was scale-invariant before inflation begins, deforms over time and eventually reduces. Figure 10.4 shows the value of Φ at the

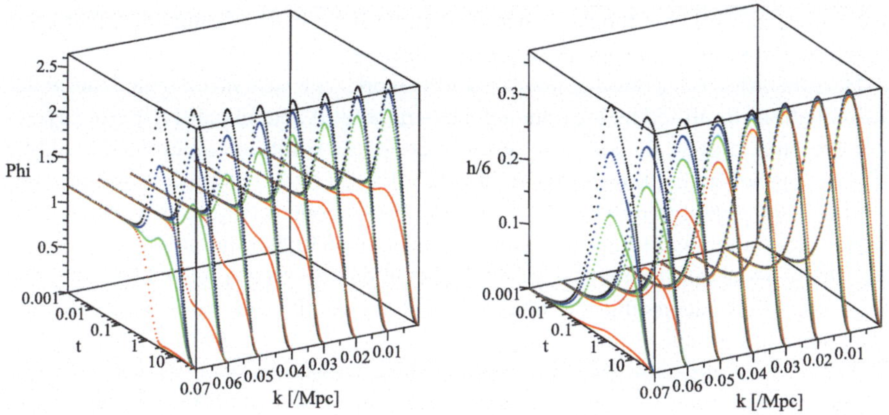

Fig. 10.3 The left shows time evolution of the fluctuation Φ for different values of Δ. The results are when $\Delta/2\pi$ is set as 1.3λ (black), λ (blue), and 0.7λ (green) from above, and black and red ($\Delta = 0$) are the same as in Fig. 10.2. The right shows time evolution of the h-component in the corresponding Φ. Reprinted from [Ham24a] under the Creative Commons Attribution (CC-BY) license

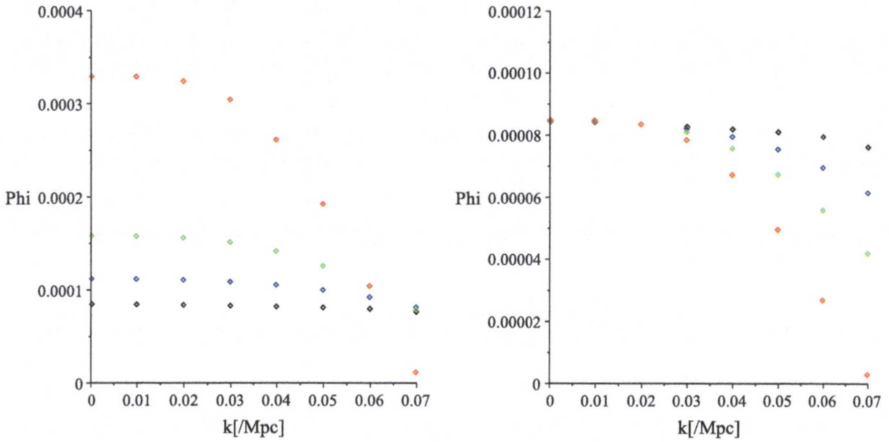

Fig. 10.4 Plots of the value of Φ at the time of the spacetime phase transition, providing the primordial spectrum (a line connecting the points and the amplitude is its square). The right is the plots normalized so that the others match the black at $k = m$ (10.16). It can be seen that the nonlinear term serves to maintain the scale invariance of the spectrum beyond the comoving Planck scale m. Reprinted from [Ham24a] under the Creative Commons Attribution (CC-BY) license

time of the phase transition, which gives the primordial spectrum that becomes the initial value for the subsequent cosmic evolution. The right normalized panel in Fig. 10.4 shows that the nonlinear term has effects of flattening the spectrum in the region $k > m$. Although the linear approximation calculation shows a large deformation, we can see that the deformation is suppressed by the nonlinear effect up to a high momentum region, exceeding three times m.

Here, in order to see how the nonlinear term works, we have investigated the evolution equation simplified by introducing the parameter Δ, but in reality, it is necessary to solve on a large scale a system of multi-coupled evolution equations derived by expressing the momentum integrals in (10.13) and (10.14) as piecewise integrals. Furthermore, although only nonlinear terms up to second order are considered here, higher-order nonlinear terms are likely to contribute in higher momentum regions. In fact, in the linear approximation, the spectrum undulates greatly and exhibits strange behavior, which means that the nonlinear effect given by the exponential becomes essentially important.

Then, what is the reason why the nonlinear term with an exponential of ϕ is thought to play an important role in maintaining scale invariance despite the presence of mass scale? The Einstein-Hilbert action with such an exponential is one of marginal conformal field in terms of conformal field theory. In general, it is believed that the power-law behavior of correlation functions is not lost even if such a conformal field is added to the system as a potential term. Therefore, it may be said that this behavior appears here.

The nonlinear effect contributes significantly to the early stage of inflation, but becomes negligibly small at the time of the spacetime phase transition. If the strength of the nonlinearity is expressed in terms of the non-Gaussianity parameter $f_{\rm NL}$ often used as its indicator [KS01], then it is on the order of unity.

In this way, we can expect that the scale invariance before inflation is maintained up to the relatively high momentum region of the CMB due to the nonlinear effects. But how far will it be maintained, and will complicated structures emerge beyond that? The results here are the first step toward clarifying this subject.

10.4 Appendix: How to Solve Simplified Evolution Equation

In order to examine time evolution of the fluctuations, it is necessary to simultaneously solve both the homogeneous equation that determines the inflationary background and the evolution equations of the fluctuations [Ham24a].

The homogeneous equation can be expressed as

$$\partial_t \bar{Z} + 7\bar{H}\bar{Z} + 4\bar{Y}^2 + 18\bar{H}^2\bar{Y} + 6\bar{H}^4 - 3\left[1 + \frac{\gamma_1}{\kappa}\bar{\alpha}_G(t)\right]^{\kappa}\left(\bar{Y} + 2\bar{H}^2\right) = 0, \tag{10.22}$$

using the dimensionless time $t = H_{\rm D}\tau$ and the normalized Hubble variables, $\bar{H} = \partial_t\bar{a}/\bar{a}$, $\bar{Y} = \partial_t\bar{H}$, and $\bar{Z} = \partial_t\bar{Y}$, defined before. Here, the time-dependent running coupling constant (7.9) is denoted as $\bar{\alpha}_G(t) = \bar{t}^2(t)/4\pi$. The dimensionless initial time $t_i = H_{\mathbf{D}}/E_i$ is set to 10^{-3}, with the energy scale E_i being sufficiently larger than the Planck scale.

In the following, the evolution equation of the fluctuations that we actually dealt with and the method for solving it are described in detail. We use the trace equation (10.8) with multiplied by $(8\pi^2/b_c) \times (1/\bar{B}) \times (1/H_{\mathbf{D}}^4 a^4)$ and simplified the nonlinear terms as in Sects. 10.2 and 10.3. Denoting it as $TR_{[\lambda]}(k) = 0$ yields:

$$
\begin{aligned}
TR_{[\lambda]}(k) &= -2D_t^4\Phi(k) - 4P_a(k)D_t^2\Phi(k) - 2P_a^2(k)\Phi(k) - \frac{4}{3}\bar{H}D_t^3\,h(k) \\
&-\frac{16}{3}\left[\bar{Y} + \bar{H}^2\right]D_t^2\,h(k) + \frac{4}{9}P_a(k)D_t^2\,h(k) - 8\left[\bar{Z} + 4\bar{H}\bar{Y} + 2\bar{H}^3\right]\partial_t h(k) \\
&-\frac{20}{9}\bar{H}P_a(k)\partial_t h(k) - \frac{32}{9}\left[\bar{Y} + \bar{H}^2\right]P_a(k)h(k) + \frac{4}{9}P_a^2(k)h(k) \\
&+\frac{1}{(2\pi)^2}\frac{\Delta^2}{\lambda^2}\Bigg\{-8\partial_t h(\lambda)D_t^3\Phi(k) - 8\partial_t h(k)D_t^3\Phi(\lambda) - \frac{16}{3}D_t^2\,h(\lambda)D_t^2\Phi(k) \\
&-\frac{16}{3}D_t^2\,h(k)D_t^2\Phi(\lambda) - \frac{4}{3}D_t^3\,h(\lambda)\partial_t\Phi(k) - \frac{4}{3}D_t^3\,h(k)\partial_t\Phi(\lambda) \\
&+\left[-\frac{20}{9}P_a(\lambda) + \frac{16}{3}P_a(k)\right]h(\lambda)D_t^2\Phi(k) + \left[-\frac{32}{9}P_a(k) + \frac{20}{3}P_a(\lambda)\right]h(k)D_t^2\Phi(\lambda) \\
&+\left[-\frac{20}{9}P_a(\lambda) - \frac{8}{3}P_a(k)\right]\partial_t h(\lambda)\partial_t\Phi(k) + \left[-\frac{20}{9}P_a(k) - \frac{8}{3}P_a(\lambda)\right]\partial_t h(k)\partial_t\Phi(\lambda) \\
&-\frac{2}{3}P_a(\lambda)D_t^2\,h(\lambda)\Phi(k) - \frac{2}{3}P_a(\lambda)D_t^2\,h(k)\Phi(\lambda) \\
&+\left[-\frac{2}{9}P_a^2(\lambda) + \frac{16}{3}P_a^2(k) + \frac{4}{9}P_a(\lambda)P_a(k)\right]h(\lambda)\Phi(k) \\
&+\left[4P_a^2(\lambda) + \frac{14}{9}P_a(\lambda)P_a(k)\right]h(k)\Phi(\lambda)\Bigg\} \\
&+\left[1 + \frac{\gamma_1}{\kappa}\bar{\alpha}_G(t)\right]^{\kappa}\Bigg\{6D_t^2\Phi(k) + 12\bar{H}\partial_t\Phi(k) + 12\left[\bar{Y} + 2\bar{H}^2\right]\Phi(k) + 6P_a(k)\Phi(k) \\
&+4\bar{H}\partial_t h(k) - 8\left[\bar{Y} + 2\bar{H}^2\right]h(k) - \frac{4}{3}P_a(k)h(k)\Bigg\} \\
&+\frac{1}{(2\pi)^2}\frac{\Lambda^2}{\lambda^2}\left[1 + \frac{\gamma_1}{\kappa}\bar{\alpha}_G(t)\right]^{\kappa}\Bigg\{12\Phi(\lambda)D_t^2\Phi(k) + 12\Phi(k)D_t^2\Phi(\lambda) + 12\partial_t\Phi(\lambda)\partial_t\Phi(k) \\
&+24\bar{H}\left[\Phi(\lambda)\partial_t\Phi(k) + \Phi(k)\partial_t\Phi(\lambda)\right] + 24\left[\bar{Y} + 2\bar{H}^2\right]\Phi(\lambda)\Phi(k) \\
&+\left[12P_a(k) + 6P_a(\lambda)\right]\Phi(\lambda)\Phi(k) - 8\,h(\lambda)D_t^2\Phi(k) - 8\,h(k)D_t^2\Phi(\lambda) \\
&+4\partial_t h(\lambda)\partial_t\Phi(k) + 4\partial_t h(k)\partial_t\Phi(\lambda) - 16\bar{H}\left[h(\lambda)\partial_t\Phi(k) + h(k)\partial_t\Phi(\lambda)\right] \\
&+8\bar{H}\left[\partial_t h(\lambda)\Phi(k) + \partial_t h(k)\Phi(\lambda)\right] - 16\left[\bar{Y} + 2\bar{H}^2\right]\left[h(\lambda)\Phi(k) + h(k)\Phi(\lambda)\right] \\
&-\left[\frac{2}{3}P_a(\lambda) + 16P_a(k)\right]h(\lambda)\Phi(k) - \left[\frac{8}{3}P_a(k) + 14P_a(\lambda)\right]h(k)\Phi(\lambda)\Bigg\}. \qquad (10.23)
\end{aligned}
$$

If writing the constraint equation (10.9) multiplied by $(3/16) \times \bar{t}^2 \times (1/H_{\mathbf{D}}^2 a^2)$ as $SP(k) = 0$, then we get

$$\begin{aligned} SP(k) &= D_t^2 h(k) + \frac{1}{3} P_a(k) h(k) + \frac{b_c}{8\pi} \bar{\alpha}_G(t) \bar{B}(t) \Big\{ D_t^2 \Phi(k) + P_a(k) \Phi(k) \\ &+ 2\bar{H} \partial_t \Phi(k) + 6\big[\bar{Y} + \bar{H}^2\big] \Phi(k) + \frac{2}{3} \bar{H} \partial_t h(k) + \frac{2}{3}\big[\bar{Y} - \bar{H}^2\big] h(k) \\ &- \frac{2}{9} P_a(k) h(k) \Big\} + \frac{b_c}{8\pi} \bar{\alpha}_G(t) \big\{ -3\Phi(k) + h(k) \big\}. \end{aligned} \tag{10.24}$$

Moreover, in order to carry out numerical calculations, the terms with $D_t^3 h \, (= \partial_\eta^3 h / a^3 H_{\mathbf{D}}^3)$ in (10.23) has to be replaced with terms up to the third time-derivative of Φ and up to the second time-derivative of h by using

$$\begin{aligned} D_t^3 h(k) = &- \frac{1}{3} P_a(t) \partial_t h(k) + \frac{\partial_t \bar{\alpha}_G(t)}{\bar{\alpha}_G(t)} \left(D_t^2 h + \frac{1}{3} P_a(k) h \right) \\ &- \frac{b_c}{8\pi} \bar{\alpha}_G(t) \Big\{ \bar{B}(t) \Big[D_t^3 \Phi(k) + 2\bar{H} D_t^2 \Phi(k) + 8\big(\bar{Y} + \bar{H}^2\big) \partial_t \Phi(k) \\ &\quad + 6\big(\bar{Z} + 4\bar{H}\bar{Y} + 2\bar{H}^3\big) \Phi(k) + P_a(k) \partial_t \Phi(k) + \frac{2}{3} \bar{H} D_t^2 h(k) \\ &\quad + \frac{4}{3} \bar{Y} \partial_t h(k) + \frac{2}{3} \big(\bar{Z} - 2\bar{H}^3\big) h(k) - \frac{2}{9} P_a(k) \partial_t h(k) \Big] \\ &\quad + \partial_t \bar{B}(t)) \Big[D_t^2 \Phi(k) + 2\bar{H} \partial_t \Phi(k) + 6\big(\bar{Y} + \bar{H}^2\big) \Phi(k) \\ &\quad + P_a(k) \Phi(k) + \frac{2}{3} \bar{H} \partial_t h(k) + \frac{2}{3} \big(\bar{Y} - \bar{H}^2\big) h(k) - \frac{2}{9} P_a(k) h(k) \Big] \Big\} \\ &+ \frac{b_c}{8\pi} \bar{\alpha}_G(t) \Big\{ 3\partial_t \Phi(k) + 6\bar{H} \Phi(k) - \partial_t h(k) - 2\bar{H} h(k) \Big\}, \end{aligned}$$

obtained by time-derivative of the constraint equation (10.9).

The time evolution of the field with momentum k, including the nonlinear effects, is obtained by simultaneously solving the four equations $TR_{[\lambda]}(k) = 0$, $SP(k) = 0$, $TR_{[\lambda]}(\lambda) = 0$, and $SP(\lambda) = 0$. Note here that the last two form a closed system of equations for the field with momentum λ that can be solved by them alone.

When solving (10.23) and (10.24), we also have to solve the homogeneous equation (10.22) simultaneously to determine the inflationary background, but the scale factor $\bar{a}$ appears only in the form $P_a(k)$ (10.15) in these equations. Since the solution of $\bar{a}$ in the early stage is given by $\bar{a}(t) = e^{\bar{H}_i(t - t_i)}$ with $\bar{H}_i = 1/\sqrt{\bar{B}(t_i)}$ and then $P_a(k)$ vanishes rapidly with time, using this function to reduce computational load does not affect the result. Here, in order to further reduce computational load, we fix the Hubble variables in the inflationary background as $\bar{H} = \bar{H}_i$, $\bar{Y} = 0$, and $\bar{Z} = 0$. If contributions from the nonlinear terms are small, the evolution equation can be solved using only the $\bar{a}$ function above without fixing the Hubble variables in this way. However, this manipulation only raises or lowers overall amplitude of the spectrum and does not affect the spectral pattern, which is almost determined in the early stage. Therefore, the background is here fixed as above so that the calculations can be performed with more aggressive parameter values.

If we write the ratio of the two physical scales as $H_{\mathbf{D}}/\Lambda_{\mathrm{QG}} = N$ (7.12), then the running coupling constant (7.9) is written as $\bar{\alpha}_G(t) = [4\pi\beta_0 \log(N^2/t^2)]^{-1}$. When performing the numerical calculation actually, we first convert this to $\bar{\alpha}_G^c(t) = [4\pi\beta_0 \log((N+c)^2/t^2)]^{-1}$ and find a solution of $c = 1$, then use it as an approximate solution to find the solution of $c = 0$. However, since the introduction of c also only raises or lowers the amplitude slightly as a whole and does not affect the spectral pattern, we here present only the results for c=1 to reduce the load, as mentioned above. In this case, ϵ, introduced in Footnote 1 of Chap. 7, in not necessary.

Special programs are needed to handle boundary value problems for differential equations where some variables that are not observable diverge at boundaries. Moreover, we need to solve simultaneous equations with multiple variables. The commercially available Maple software has a built-in program for such boundary value problems, and we use it here. Specifically, we use the "dsolve" command in Maple15, set it to "method=bvp[midrich], abserr=1e-2, initmesh=1000, maxmesh=8192 (maximum value)", and solve it numerically. In addition, as Δ gets larger, it becomes more difficult to converge initial Newton iterations. In that case, we use the "continuation" parameter to start from a smaller Δ where the initial iteration converges, leading to the desired result.

Figures 10.2, 10.3, and 10.4 are the results of solving the nonlinear evolution equation system with the boundary conditions (10.20) and (10.21), by simplifying some functions and calculation methods, as described above. In order to perform large-scale computations involving many fields with different momentums with higher precision, a special program such as the Fortran software, BVP_SOLVER [EM96], and a device that can run it will be necessary.

Chapter 11
Topology and Quantum Gravity

There are some researchers who believe that quantization of gravity means taking the sum over all topologies from the viewpoint of the path integral method. In two and three dimensions, there are no dynamical degrees of freedom in the gravitational field, thus taking the sum over topologies may be significant, but in four dimensions, that is not the main issue, the gravitational field dynamics is much more important. Still, spacetimes with special topologies may contribute to the physics of the early universe.

In general, configurations of the gravitational field with topologies different from S^4 are sometimes called gravitational instantons. They have been extensively studied mainly in the 1970s and summarized in a report by Eguchi, Gilkey, and Hanson [EGH80]. Most of them are given as solutions to the Einstein equation satisfying the self-duality condition for the Riemann tensor, and many of them have boundaries. Here we focus only on a solution that is closely related to the Weyl action and is extremely similar to the Yang-Mills instanton solution. It is a compact four-dimensional manifold with no boundary, satisfying both the self-duality condition for the Weyl tensor and the gravitational equation of motion derived from the Weyl action.

Compared to the study of Yang-Mills instantons [BPST75, tHo76, CDG78, CDG79, SVZ80, VZNS82], there has been little research on physics of gravitational instantons. Behind this is the problem of spin structure, i.e., whether fermions can exist on the manifold. Addressing this problem is one of the subjects of this chapter.

K. Hamada, *Trans-Planckian Physics and Inflation*, Fundamental Theories of Physics 223, https://doi.org/10.1007/978-981-96-3476-7_11

11.1 Differential Geometry and Fermions

First, briefly summarize the basic tools of differential geometry [GSW87, Nak90, PT09]. We consider locally Euclidean spacetime as with Yang-Mills instantons, and introduce the vierbein fields e^a_μ ($a = 0, 1, 2, 3$) at each point in spacetime, where unless otherwise specified, a, b, c, etc. denote the local Euclidean indices of $SO(4)$, while μ, ν, λ, etc. are the Einstein indices. They are orthogonal tangent vectors, such that $e^{a\mu} e^b_\mu = \delta^{ab}$. Equivalently, the metric field can be expressed as $g_{\mu\nu} = \delta_{ab}\, e^a_\mu e^b_\nu$ using the vierbein fields.

The covariant derivative of the vector field $V_a = e^\mu_a V_\mu$ with a local Euclidean index is expressed as $\nabla_\mu V_a = \partial_\mu V_a + \omega_{\mu a}{}^b\, V_b$, where ω^{ab}_μ is a spin connection. In order for this to match the ordinary covariant derivative $\nabla_\mu V_\nu = \partial_\mu V_\nu - \Gamma^\lambda_{\mu\nu} V_\lambda$ such that $\nabla_\mu V_\nu = e^a{}_\nu \nabla_\mu V_a$, the covariant derivative of the vierbein field must vanish as

$$\nabla_\mu e^a_\nu = \partial_\mu e^a_\nu + \omega_\mu{}^a{}_b e^b_\nu - \Gamma^\lambda_{\mu\nu} e^a_\lambda = 0.$$

From this, the spin connection can be expressed as $\omega_{\mu ab} = e^\nu_a(\partial_\mu e_{\nu b} - \Gamma^\lambda_{\mu\nu} e_{\lambda b})$, which satisfies an antisymmetric property $\omega^{ab}_\mu = -\omega^{ba}_\mu$.

To describe geometrical properties of spacetime, introduce various differential forms. If we write the vierbein 1-form as $e^a = e^a_\mu dx^\mu$, then its exterior derivative satisfies

$$de^a + \omega^a{}_b \wedge e^b = 0, \tag{11.1}$$

where $\omega^{ab} = \omega^{ab}_\mu dx^\mu$ is the connection 1-form. Using the connection 1-form, the curvature 2-form is defined by

$$\mathcal{R}_{ab} = d\omega_{ab} + \omega_a{}^c \wedge \omega_{cb} = \frac{1}{2} R_{ab\mu\nu}\, dx^\mu \wedge dx^\nu, \tag{11.2}$$

where the Riemann tensor $R_{ab\mu\nu}$ ($= R_{abcd}\, e^c_\mu e^d_\nu$) is expressed as

$$R_{ab\mu\nu} = \partial_\mu \omega_{\nu ab} - \partial_\nu \omega_{\mu ab} + \left[\omega_\mu, \omega_\nu\right]_{ab}. \tag{11.3}$$

It can be easily shown that the exterior derivative of the curvature 2-form satisfies $d\mathcal{R}_{ab} + \omega_{ac} \wedge \mathcal{R}^c{}_b - \mathcal{R}_{ac} \wedge \omega^c{}_b = 0$. Rewriting this gets $\nabla_\lambda R_{ab\mu\nu}\, dx^\lambda \wedge dx^\mu \wedge dx^\nu/2 = 0$, thus the Bianchi identity is derived.

There are many orthonormal frames e^a, which are connected by local orthogonal rotation. The transformation is expressed as

$$e^a \to e'^a = \Phi^a{}_b\, e^b. \tag{11.4}$$

In fact, the condition that the metric field remains unchanged as $g'_{\mu\nu} = g_{\mu\nu}$, i.e., $\delta_{ab}\, e'^a e'^b = \delta_{ab}\, \Phi^a{}_c \Phi^b{}_d\, e^c\, e^d = \delta_{cd}\, e^c\, e^d$, leads to

$$\delta_{ab}\,\Phi^a{}_c\,\Phi^b{}_d = \delta_{cd}.$$

This is a orthogonal transformation of $SO(4)$ that satisfies $\Phi^T\,\Phi = I$ $(\Phi^{-1} = \Phi^T)$. Here $\det\Phi = 1$ means that the orientation of the manifold is fixed, and $\det\Phi = -1$ represents the opposite orientation.

The dimension of the $SO(4)$ group is 6 . This coincides with the number obtained by subtracting 10, the number of components of the metric field $g_{\mu\nu}$, from $4^2 = 16$, that of the vierbein field e^a_μ.[1] In other words, the vierbein field has too many components to represent the metric field, thus the $SO(4)$ transformation serves to remove the extra degrees of freedom as gauge degrees of freedom. The connection 1-form then transforms as

$$\omega'^a{}_b = \Phi^a{}_c\,\omega^c{}_d\left(\Phi^{-1}\right)^d{}_b + \Phi^a{}_c\,d\left(\Phi^{-1}\right)^c{}_b, \tag{11.5}$$

and the curvature 2-form as $\mathcal{R}'^a{}_b = \Phi^a{}_c\,\mathcal{R}^c{}_d\left(\Phi^{-1}\right)^d{}_b$.

Let Σ^{ab} be the generator of the $SO(4)$ group and normalize its algebra as

$$\left[\Sigma^{ab}, \Sigma^{cd}\right] = -\delta^{ac}\,\Sigma^{bd} + \delta^{ad}\,\Sigma^{bc} + \delta^{bc}\,\Sigma^{ad} - \delta^{bd}\,\Sigma^{ac}. \tag{11.6}$$

The generator acting on a vector index is given by $\left(\Sigma^{ab}\right)_{cd} = \delta^a{}_c\delta^b{}_d - \delta^a{}_d\delta^b{}_c$, and if the transformation parameters are denoted as ε^{ab} $(= -\varepsilon^{ba})$, the elements of the $SO(4)$ group is expressed as

$$\Phi_{cd} = \exp\left(\frac{1}{2}\varepsilon_{ab}\,\Sigma^{ab}\right)_{cd}. \tag{11.7}$$

Now, we introduce the Clifford algebra to define fermions. $SO(4)$ is the group of transformations that preserve the quadratic form $p_0^2 + p_1^2 + p_2^2 + p_3^2$ of a vector p_a unchanged. Therefore, let us consider expressing this quadratic form as the square of a linear form $\gamma^a\,p_a$. Then it turns out that the coefficient γ^a satisfying this condition does not exist as a c-number, but is expressed as a matrix that satisfies the anti-commutative relation

$$\left\{\gamma^a, \gamma^b\right\} = -2\delta^{ab}. \tag{11.8}$$

This is the Clifford algebra.[2] The independent elements of this algebra are 1, γ^a, $\gamma^a\gamma^b$ $(a < b)$, $\gamma^a\gamma^b\gamma^c$ $(a < b < c)$, and $\gamma_5 = \gamma^0\gamma^1\gamma^2\gamma^3$ of 16 $(= 1+4+6+4+1)$, where γ_5 is the chirality operator satisfying $\gamma_5^2 = 1$ and $\gamma_5^\dagger = \gamma_5$. From this, γ^a

---

[1] Generally in d-dimensional space, the number of components of the d-bein is d^2, while that of the metric field is $d(d+1)/2$, thus the difference between them becomes $d(d-1)/2$, which is the number of degrees of freedom for $SO(d)$ gauge transformation.

[2] The overall sign is a convention. The negative sign is used because we are considering the case where the Clifford algebra $\{\gamma^a, \gamma^b\} = -2\eta^{ab}$ (B.1) with the Minkowski metric is Wick-rotated to the Euclidean one.

is described by a matrix of 4×4 to represent 16 elements.[3] The gamma matrix in curved space is defined by $\gamma^\mu = e^\mu_a \gamma^a$ or $\{\gamma^\mu, \gamma^\nu\} = -2g^{\mu\nu}$, where note that γ^a is a constant matrix, but γ^μ is a field.

The generator for fermions satisfying (11.6) and its group element are given by

$$\Sigma^{ab} = -\frac{1}{4}[\gamma^a, \gamma^b], \qquad S = \exp\left(\frac{1}{2}\varepsilon_{ab}\Sigma^{ab}\right), \tag{11.9}$$

respectively. According to Osterwalder and Schrader [OS73, OS75], a fermion ψ and its conjugate field $\bar{\psi}$ in Euclidean space are defined as independent fields that transform such as

$$\psi' = S\psi, \qquad \bar{\psi}' = \bar{\psi}S^{-1}. \tag{11.10}$$

This is a double cover of $SO(4)$ and is called the spinor group $Spin(4)$.[4] However, since both Lie algebras are the same, the $Spin(4)$ representation is often simply called the spinor representation of $SO(4)$ when its global group structure is not considered.

Considering an infinitesimal transformation ($\varepsilon_{ab} \ll 1$), the matrix elements of the local $SO(4)$ transformation are given by $\Phi^a{}_b = (I + \varepsilon_{cd}\Sigma^{cd}/2)^a{}_b = \delta^a{}_b + \varepsilon^a{}_b$. From (11.5), the spin connection then transforms as $\omega'_{\mu ab} = \omega_{\mu ab} - \nabla_\mu \varepsilon_{ab}$.

From the transformation law (11.10), it can be seen that $\bar{\psi}\psi$ is a scalar quantity as $\bar{\psi}'\psi' = \bar{\psi}S^{-1}S\psi = \bar{\psi}\psi$. In addition, we can find that the current $J^a = \bar{\psi}\gamma^a\psi$ is transformed as a vector such as $J^{a\prime} = \bar{\psi}'\gamma^a\psi^{a\prime} = \bar{\psi}S^{-1}\gamma^a S\psi = \Phi^a{}_b J^b$ by using

[3] In general, the dimension of the $SO(d)$ Clifford algebra is $\sum_{r=0}^{d} {}_dC_r = 2^d$. In particular, when d is an even number, γ^a is given by a $2^{d/2} \times 2^{d/2}$ matrix, and a fermion is represented as a vector of $2^{d/2}$ components.

[4] In general, $Spin(d)$ is a double cover of $SO(d)$, that is, $Spin(d)/Z_2 = SO(d)$. For simplicity, if we take the element of the $SO(3)$ group with $\varepsilon_{12} = \varepsilon_{23} = 0$ and $\varepsilon_{13} = \alpha$ as an example, then from $(\Sigma^{13})_{cd} = \delta^1_c\delta^3_d - \delta^1_d\delta^3_c$, it is given by

$$\Phi = \exp(\alpha\Sigma^{13}) = \begin{pmatrix} \cos\alpha & 0 & \sin\alpha \\ 0 & 0 & 0 \\ -\sin\alpha & 0 & \cos\alpha \end{pmatrix}.$$

On the other hand, if we choose a representation of the Clifford algebra (11.8) as $\gamma^a = i\lambda^a$ using the Pauli matrix λ^a, we find that from $\Sigma^{13} = -[\gamma^1, \gamma^3]/4 = -i\lambda^2/2$, the element of $Spin(3)$ is given by

$$S = \exp\left(-\alpha\frac{i}{2}\lambda^2\right) = \begin{pmatrix} \cos\frac{\alpha}{2} & -\sin\frac{\alpha}{2} \\ \sin\frac{\alpha}{2} & \cos\frac{\alpha}{2} \end{pmatrix}.$$

Thus, if we go around once in the $Spin(3)$ group, we go around twice in the $SO(3)$ group. As can be seen from this discussion, $Spin(3) = SU(2)$. In addition, $Spin(4) = SU(2) \times SU(2)$ and $Spin(6) = SU(4)$.

$$S^{-1}\gamma^a S = \left(1 - \frac{1}{2}\varepsilon_{cd}\Sigma^{cd}\right)\gamma^a\left(1 + \frac{1}{2}\varepsilon_{cd}\Sigma^{cd}\right)$$
$$= \gamma^a - \frac{1}{2}\varepsilon_{cd}[\Sigma^{cd}, \gamma^a] = \gamma^a + \varepsilon^a{}_b\gamma^b = \Phi^a{}_b\gamma^b. \qquad (11.11)$$

This relation generally holds and gives a surjective homomorphism from $Spin(4)$ to $SO(4)$. Since it holds even if we change $S \to -S$, we can see that the kernel is Z_2 and $Spin(4)$ is a double cover of $SO(4)$.

Next, consider the covariant derivative for fermions, which is defined by

$$\nabla_\mu\psi = \partial_\mu\psi + \frac{1}{2}\omega_{\mu cd}\Sigma^{cd}\psi,$$
$$\nabla_\mu\bar{\psi} = \partial_\mu\bar{\psi} - \frac{1}{2}\omega_{\mu cd}\bar{\psi}\Sigma^{cd},$$

and since the vierbein field satisfies $\nabla_\mu e_a^\nu = 0$, $\nabla_\mu\gamma^\nu = \gamma^\nu\nabla_\mu$ holds true. According to the transformation laws (11.4), (11.5), and (11.10), $\nabla_a\psi = e_a^\mu\nabla_\mu\psi$ transforms as

$$(\nabla_a\psi)' = \Phi_a{}^b e_b{}^\mu S\left[S^{-1}\partial_\mu S + \frac{1}{2}(\omega_{\mu cd} + 2\,\varepsilon_c{}^e\,\omega_{\mu ed} - \partial_\mu\varepsilon_{cd})S^{-1}\Sigma^{cd}S\,\psi\right]$$
$$= S\Phi_a{}^b\nabla_b\psi,$$

where $S^{-1}\partial_\mu S = \partial_\mu\varepsilon_{cd}\,\Sigma^{cd}/2$, $S^{-1}\Sigma^{cd}S = \Sigma^{cd} + \varepsilon_{ef}\,[\Sigma^{cd}, \Sigma^{ef}]/2$, and the $SO(4)$ algebra (11.6) are used. Using this transformation law and the relation (11.11), we can show that

$$(\bar{\psi}\gamma^a\nabla_a\psi)' = \bar{\psi}'\gamma^a(\nabla_a\psi)' = \bar{\psi}S^{-1}\gamma^a S\,\Phi_a{}^b\nabla_b\psi = \bar{\psi}\gamma^a\nabla_a\psi.$$

A fermion action has to be invariant under both diffeomorphism and local $SO(4)$ transformation. By adding $\sqrt{g}$ to make it diffeomorphism invariant, it is given as

$$I_\psi = \int d^4x\sqrt{g}\left[\frac{i}{2}(\bar{\psi}\gamma^a\nabla_a\psi - \nabla_a\bar{\psi}\gamma^a\psi) - m\bar{\psi}\psi\right]. \qquad (11.12)$$

Since the Hermitian property of fermions in Euclidean space is obscure, each coefficient has been determined so that the Feynman amplitude in flat spacetime derived using this Euclidean action agrees with that given by Wick-rotating the amplitude defined in Minkowski spacetime.[5] For differences with the definition of fermions in Minkowski spacetime, see Appendix B.

[5] In flat spacetime, the equation of motion becomes $(i\gamma^a\partial_a - m)\psi = 0$. In momentum space we get $(\gamma^a k_a + m)\psi = 0$. In this case, the propagator is correctly given by $-1/(\gamma^a k_a + m) = (\gamma^a k_a - m)/(k^2 + m^2)$. [Note that when using the sign convention $\{\gamma^a, \gamma^b\} = 2\delta^{ab}$ as the Clifford algebra, the propagator is not correct unless the imaginary unit is introduced in the mass term.].

11.2 $\mathbb{CP}^2$ Gravitational Instanton

As a gravitational instanton solution, we here consider a 4-dimensional manifold without boundaries that satisfies both the self-duality condition for the Weyl tensor,

$$C_{abcd} = \frac{1}{2}\epsilon_{abef}\, C^{ef}{}_{cd}, \tag{11.13}$$

and the equation of motion for the Weyl action, $T^{\rm W}_{ab} = 0$, where

$$\begin{aligned} T^{\rm W}_{ab} &= -4R_{acde}R_b{}^{cde} + 8R_{ac}R_b{}^c - \frac{4}{3}R_{ab}R - 4\nabla^2 R_{ab} + \frac{4}{3}\nabla_a\nabla_b R \\ &\quad + \delta_{ab}C_{cdef}C^{cdef} + \frac{2}{3}\delta_{ab}\nabla^2 R \end{aligned} \tag{11.14}$$

is a rewrite of the energy-momentum tensor $T^{\mu\nu}_{\rm W} = (2/\sqrt{g}) \times \delta\int d^4x\sqrt{g}\, C^2_{\alpha\beta\lambda\sigma}/\delta g_{\mu\nu}$ into Euclidean indices. The solution currently known that satisfies these two conditions is only the complex projective space $\mathbb{CP}^2$ [EF76].

The $\mathbb{CP}^2$ space is a real four-dimensional space that satisfies $\sum_j |z_j|^2 = 1/\mathfrak{a}$ and equivalence relation $z_j \simeq e^{i\alpha}z_j$ when expressed by complex coordinates z_1, z_2, z_3, where $\mathfrak{a}$ is a constant with mass-squared dimension and α is a real number. The characteristic of the $\mathbb{CP}^2$ space that will be important below is that there is $\mathbb{CP}^1 = S^2$ as a subspace, for example, $z_3 = 0$.

The metric of $\mathbb{CP}^2$ found by Fubini [Fub1904] and Study [Stu1905] in the 1900s can be written in real space as

$$ds^2_{\mathbb{CP}^2} = \frac{dr^2 + r^2\sigma_3^2}{(1+\mathfrak{a}r^2)^2} + \frac{r^2(\sigma_1^2+\sigma_2^2)}{1+\mathfrak{a}r^2}, \tag{11.15}$$

where σ_A $(A = 1, 2, 3)$ is a 1-form satisfying $d\sigma_A = \epsilon_{ABC}\,\sigma_B \wedge \sigma_C$ and see Appendix Sect. 11.6.1 for its expression. As we will see in detail later, $r \to \infty$ in this metric becomes S^2 instead of a point, which represents the subspace of $z_3 = 0$ mentioned above. In this way, while S^4 is a closed space by adding a point to infinity of $\mathbb{R}^4$, $\mathbb{CP}^2$ may be thought of as $\mathbb{R}^4$ made into a closed space by adding S^2.

If we express the Fubini-Study metric (11.15) as $ds^2_{\mathbb{CP}^2} = \delta_{ab}\, e^a e^b$, then the vierbein 1-form $e^a = e^a_\mu dx^\mu$ is written as

$$e^0 = \frac{dr}{1+\mathfrak{a}r^2}, \quad e^1 = \frac{r\sigma_1}{(1+\mathfrak{a}r^2)^{\frac{1}{2}}}, \quad e^2 = \frac{r\sigma_2}{(1+\mathfrak{a}r^2)^{\frac{1}{2}}}, \quad e^3 = \frac{r\sigma_3}{1+\mathfrak{a}r^2}.$$

Each exterior derivative of e^a is given by $de^0 = 0$ and

$$
\begin{aligned}
de^1 &= \frac{1}{(1+ar^2)^{\frac{3}{2}}} dr \wedge \sigma_1 + \frac{2r}{(1+ar^2)^{\frac{1}{2}}} \sigma_2 \wedge \sigma_3 \\
&= \frac{1}{r} e^0 \wedge e^1 + \frac{2(1+ar^2)}{r} e^2 \wedge e^3, \\
de^2 &= \frac{1}{(1+ar^2)^{\frac{3}{2}}} dr \wedge \sigma_2 + \frac{2r}{(1+ar^2)^{\frac{1}{2}}} \sigma_3 \wedge \sigma_1 \\
&= \frac{1}{r} e^0 \wedge e^2 + \frac{2(1+ar^2)}{r} e^3 \wedge e^1, \\
de^3 &= \frac{1-ar^2}{(1+ar^2)^2} dr \wedge \sigma_3 + \frac{2r}{1+ar^2} \sigma_1 \wedge \sigma_2 \\
&= \frac{1-ar^2}{r} e^0 \wedge e^3 + \frac{2}{r} e^1 \wedge e^2.
\end{aligned}
$$

From (11.1), the connection 1-forms $\omega^{ab} = \omega^{ab}_{\mu} dx^{\mu}$ are given by

$$
\begin{aligned}
\omega^{01} &= -\frac{1}{r} e^1, & \omega^{23} &= \frac{1}{r} e^1, \\
\omega^{02} &= -\frac{1}{r} e^2, & \omega^{31} &= \frac{1}{r} e^2, \\
\omega^{03} &= -\frac{1-ar^2}{r} e^3, & \omega^{12} &= \frac{1+2ar^2}{r} e^3,
\end{aligned} \tag{11.16}
$$

which satisfy $\omega^{0A} \propto e^A$ and $\omega^{AB} \propto \epsilon^{ABC} e^C$. Note here that there is an anti-self-dual relationship between ω^{01} and ω^{23} and between ω^{02} and ω^{31}, while there are both self-dual and anti-self-dual components between ω^{03} and ω^{12}, as seen by decomposing them into

$$
\omega^{03} = -\frac{1+\frac{1}{2}ar^2}{r} e^3 + \frac{3ar}{2} e^3, \quad \omega^{12} = \frac{1+\frac{1}{2}ar^2}{r} e^3 + \frac{3ar}{2} e^3.
$$

Thus, the spin connection does not become self-dual, though the Weyl tensor is self-dual. This is a significant difference from self-dual solutions of the Riemann tensor in which case the spin connection also becomes self-dual.

Substituting these connection 1-forms into (11.2) yields the curvature 2-form of $\mathbb{CP}^2$ as

$$
\begin{aligned}
\mathcal{R}_{01} &= a\, e^0 \wedge e^1 - a\, e^2 \wedge e^3, & \mathcal{R}_{23} &= -a\, e^0 \wedge e^1 + a\, e^2 \wedge e^3, \\
\mathcal{R}_{02} &= a\, e^0 \wedge e^2 - a\, e^3 \wedge e^1, & \mathcal{R}_{31} &= -a\, e^0 \wedge e^2 + a\, e^3 \wedge e^1, \\
\mathcal{R}_{03} &= 4a\, e^0 \wedge e^3 + 2a\, e^1 \wedge e^2, & \mathcal{R}_{12} &= 2a\, e^0 \wedge e^3 + 4a\, e^1 \wedge e^2.
\end{aligned}
$$

Since the curvature 2-form can be written as $\mathcal{R}_{ab} = R_{abcd}\, e^c \wedge e^d/2$ using the vierbein 1-form, the non-zero components of the Riemann tensor are $R_{0101} = R_{0202} = R_{2323} = R_{3131} = a$, $R_{0123} = R_{0231} = -a$, $R_{0312} = 2a$, and $R_{0303} = R_{1212} = 4a$. The Ricci tensor $R_{ab} = R^c{}_{acb}$ and the scalar curvature $R = R^c{}_c$ are given by

$$R_{ab} = 6\mathfrak{a}\,\delta_{ab}, \qquad R = 24\mathfrak{a},$$

respectively.

The Weyl tensor is given by $C_{abcd} = R_{abcd} - 2\mathfrak{a}(\delta_{ac}\delta_{bd} - \delta_{ad}\delta_{bc})$. Writing down each component of the Weyl tensor gives

$$C_{0101} = C_{2301} = -\mathfrak{a}, \quad C_{0202} = C_{3102} = -\mathfrak{a}, \quad C_{0303} = C_{1203} = 2\mathfrak{a},$$
$$C_{0123} = C_{2323} = -\mathfrak{a}, \quad C_{0231} = C_{3131} = -\mathfrak{a}, \quad C_{0312} = C_{1212} = 2\mathfrak{a},$$

which satisfies the Weyl self-duality condition (11.13). Using the self-dual 't Hooft symbol (11.29) summarized in Appendix Sect. 11.6.1, it can be written as $C_{abcd} = -\mathfrak{a}\,(\eta^1_{ab}\eta^1_{cd} + \eta^2_{ab}\eta^2_{cd}) + 2\mathfrak{a}\,\eta^3_{ab}\eta^3_{cd}$.

Topological properties The volume element of $\mathbb{CP}^2$ is defined by

$$\begin{aligned} e^0 \wedge e^1 \wedge e^2 \wedge e^3 &= \frac{r^3}{(1+\mathfrak{a}r^2)^3}\,dr \wedge \sigma_1 \wedge \sigma_2 \wedge \sigma_3 \\ &= \frac{r^3 \sin\theta}{8(1+\mathfrak{a}r^2)^3}\,dr \wedge d\varphi \wedge d\theta \wedge d\psi, \end{aligned}$$

and thus its volume $\int e^0 \wedge e^1 \wedge e^2 \wedge e^3 = \int d^4x\sqrt{g}$ is given by

$$\mathrm{Vol}(\mathbb{CP}^2) = \int_0^{\infty_0} dr \int_0^{2\pi} d\varphi \int_0^{\pi} d\theta \int_0^{4\pi} d\psi\, \frac{r^3 \sin\theta}{8(1+\mathfrak{a}r^2)^3} = \frac{\pi^2}{2\mathfrak{a}^2}$$

From this, the Weyl action is calculated as $\int d^4x\sqrt{g}\,C^2_{abcd} = 48\pi^2$.

Now, we introduce various topological invariants as characteristics for classifying topology. In the following, $\mathcal{R}$ is a 4×4 matrix representing the curvature 2-form $\mathcal{R}_{ab}$, and $\mathrm{Tr}(\mathcal{R} \wedge \mathcal{R}) = \mathcal{R}_a^{\ b} \wedge \mathcal{R}_b^{\ a}$.

Letting M be a manifold without boundary, the first Pontrjagin number is defined by

$$\begin{aligned} P_1(M) &= -\frac{1}{8\pi^2}\int_M \mathrm{Tr}(\mathcal{R} \wedge \mathcal{R}) \\ &= \frac{1}{32\pi^2}\int_M d^4x\sqrt{g}\,\epsilon^{abcd} R_{abef} R_{cd}^{\ \ ef}, \end{aligned}$$

where $e^a \wedge e^b \wedge e^c \wedge e^d = d^4x\sqrt{g}\,\epsilon^{abcd}$ is used. This quantity does not change even if R_{abcd} is replaced with C_{abcd} due to the cyclic property of curvature $\epsilon^{abcd} R_{abce} = 0$ and so on. The Hirzebruch signature, generally called the signature, is given by one-third of the first Pontrjagin number,

$$\tau(M) = \frac{1}{3} P_1(M).$$

The Euler characteristic is defined by

$$\chi(M) = \frac{1}{32\pi^2} \int_M \epsilon^{abcd} \mathcal{R}_{ab} \wedge \mathcal{R}_{cd}$$
$$= \frac{1}{32\pi^2} \int_M d^4x \sqrt{g}\, G_4,$$

where $G_4 = \epsilon^{abcd} \epsilon^{efgh} R_{abef} R_{cdgh}/4$ is used, which holds only in four dimensions. These of S^4 are given by $\tau = 0$ and $\chi = 2$.

Substituting the $\mathbb{CP}^2$ metric into each definition yields

$$P_1(\mathbb{CP}^2) = 3, \qquad \tau(\mathbb{CP}^2) = 1, \qquad \chi(\mathbb{CP}^2) = 3. \tag{11.17}$$

If $\mathbb{CP}^2$ is the anti-self dual, then $\tau = -1$. Using the Betti number b_n, the Euler characteristic of $\mathbb{CP}^2$ is represented by $\chi = b_0 + b_2 + b_4 = 1 + 1 + 1 = 3$, whereas that of S^4 is $\chi = 2$ with $b_0 = b_4 = 1$ and $b_2 = 0$. Thus, $\mathbb{CP}^2$ has $b_2 = 1$, which number indicates that there is an incontractible S^2 subspace and that there is a closed but not exact 2-form.

Lastly, we give a geometric expression of the Atiyah-Singer spin 1/2 Dirac index [AS68] for a manifold M without boundary, defined by

$$\mathcal{I}_{1/2} = \nu_+ - \nu_-, \tag{11.18}$$

where $\nu_\pm$ is the number of fermion zero-modes with chirality $\pm$ for the γ_5 operator. Consider the general case where there is a gauge field associated with a group G on M, and denote the gauge field as $\mathcal{A}_\mu$ ($\mathcal{A}_\mu^\dagger = -\mathcal{A}_\mu$), then a fermion satisfies the equation of motion

$$\gamma^\mu \left(\partial_\mu + \frac{1}{2} \omega_{\mu ab} \Sigma^{ab} + \mathcal{A}_\mu \right) \psi = 0.$$

If $G = U(1)$, $\mathcal{A}_\mu$ is $-ixA_\mu/2$ with charge x, and if $SU(2)$, it is given by $-i\lambda^A A_\mu^A/2$, where $[\lambda^A, \lambda^B] = 2i\epsilon^{ABC}\lambda^C$. The curvature 2-form (field strength) of the gauge field is defined by

$$\Omega = d\mathcal{A} + \mathcal{A} \wedge \mathcal{A} = \frac{1}{2} \mathcal{F}_{\mu\nu}\, dx^\mu \wedge dx^\nu,$$

where $\mathcal{A} = \mathcal{A}_\mu dx^\mu$.

Letting V be a group representation to which the fermion belongs and $\dim V$ be its dimension, the spin 1/2 Dirac index (11.18) is then given by [AS68, EGH80, Nak90]

$$
\begin{aligned}
\mathcal{I}_{\frac{1}{2}} &= \int_M \hat{A}(M) \wedge ch(\Omega) \\
&= -\frac{\dim V}{24} P_1(M) + \int_M \left[\frac{1}{2} c_1^2(\Omega) - c_2(\Omega)\right], \qquad (11.19)
\end{aligned}
$$

where $\hat{A}(M)$ is the A-roof genus, and $ch(\Omega)$ is the Chern character, which are defined by (11.44) in Appendix Sect. 11.6.3. When integrated over a four-dimensional manifold, only the 4-form part remains, as shown on the right-hand side. The expression in square brackets is a rewrite of the second Chern character ch_2, and c_1 and c_2 are the first and second Chern classes for the curvature 2-form Ω defined by[6]

$$
c_1(\Omega) = i\frac{1}{2\pi} \mathrm{Tr}\, \Omega, \quad c_2(\Omega) = \frac{1}{8\pi^2}\left[\mathrm{Tr}\,(\Omega \wedge \Omega) - \mathrm{Tr}\, \Omega \wedge \mathrm{Tr}\, \Omega\right].
$$

From this, the spin $1/2$ Dirac index (11.19) can be expressed as

$$
\mathcal{I}_{\frac{1}{2}} = -\frac{\dim V}{8}\tau(M) - \frac{1}{8\pi^2}\int_M \mathrm{Tr}\, \Omega \wedge \Omega. \qquad (11.20)
$$

For $U(1)$ gauge field, substitute $\Omega = -ixF/2$ into this expression, and for $SU(2)$ gauge field, $\Omega = -i\lambda^A F^A/2$, where F and F^A are the 2-forms of their respective gauge fields.

The spin $1/2$ Dirac index on $\mathbb{CP}^2$ in the absence of a gauge field is

$$
\mathcal{I}_{\frac{1}{2}} = -\frac{\dim V}{8} \qquad (11.21)
$$

due to $\tau = 1$. If there is only one fermion ($\dim V = 1$), $\mathcal{I}_{1/2} = -1/8$, which is a non-integer. On the other hand, the index must be an integer because it counts the number of zero modes. This indicates that a fermion is not correctly defined on $\mathbb{CP}^2$.

11.3 Fermions on $\mathbb{CP}^2$ and Spin Structures

It is not obvious whether fermions can be correctly defined on a manifold. A manifold for which fermions can be defined globally is called a spin manifold. S^4 is a spin manifold, but $\mathbb{CP}^2$ is not so, that is called, it does not admit a spin structure. Introducing fermions whose number is a multiple of 8 can make the index (11.21) an integer, but this is a necessary condition, not a sufficient condition, for $\mathbb{CP}^2$ to have a spin structure [HP78, GP78].

6 The Chern classes are defined by $c(\Omega) = \det(1 + i\Omega/2\pi) = 1 + c_1(\Omega) + c_2(\Omega) + \cdots$.

The cause of this problem lies in the S^2 subspace existing in $\mathbb{CP}^2$. The $r \to \infty$ limit of the Fubini-Study metric (11.15) is not a point, but S^2 as

$$ds^2_{\mathbb{CP}^2}\Big|_{r\to\infty} = \frac{1}{\mathfrak{a}}(\sigma_1^2 + \sigma_2^2) = \frac{1}{4\mathfrak{a}}(d\theta^2 + \sin^2\theta\, d\varphi^2).$$

Consider the parallel transport of a fermion that goes around once along a certain curve C_v on S^2. Since the Dirac equation can be written as $\psi(x^\mu + \Delta x^\mu) = (1 - \omega_{\mu ab}\Sigma^{ab}\Delta x^\mu/2)\psi(x^\mu)$ for an infinitesimal movement Δx^μ, we obtain a formula

$$\psi_v = \exp\left(-\int_{C_v}\omega\right)\psi, \qquad \omega = \frac{1}{2}\omega_{\mu ab}\,\Sigma^{ab}dx^\mu. \tag{11.22}$$

Let C_v be the S^1 curve that goes around the sphere from a certain point x_0 on S^2, and v is a parameter that runs from 0 to 1. The curves C_0 and C_1 of $v = 0$ and $v = 1$ are trivial curves, i.e., the point x_0 itself, then the bundle of continuous curves of $0 \le v \le 1$ becomes S^2 (see Fig. 11.1).

Consider two non-trivial curves C_v and $C_{v'}$, and let Y be the region on S^2 surrounded by them. We then obtain

$$\int_{C_{v'}}\omega - \int_{C_v}\omega = \int_Y d\omega \tag{11.23}$$

for the 1-form ω according to the Stokes theorem. Using this to eliminate ψ introduced as a reference from (11.22), we get

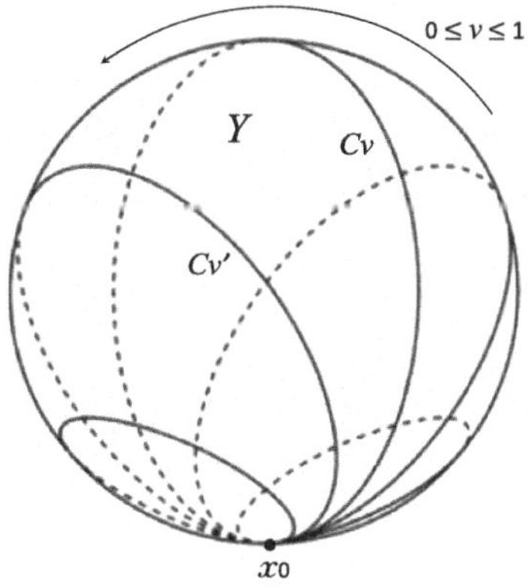

Fig. 11.1 How to construct S^2 from the smash product of S^1 and S^1, written as $S^1 \wedge S^1 \cong S^2$. Y denotes the surface surrounded by two curves C_v and $C_{v'}$

$$\psi_{v'} = \exp\left(-\int_{C_{v'}} \omega + \int_{C_v} \omega\right)\psi_v = \exp\left(-\int_Y d\omega\right)\psi_v.$$

Let us now consider the case of $v' = 1$ and $v = 0$ with trivial curves $C_1 = C_0 = x_0$ and thus $Y = S^2$. This case is expressed as

$$\psi_1 = \exp\left(-\int_{S^2} d\omega\right)\psi_0. \tag{11.24}$$

Since ψ_0 and ψ_1 are defined at the same point x_0, they must be the same. Therefore, the condition that fermions can be consistently defined on the manifold, i.e., the condition for spin structure to exist, is given by

$$\psi_1 = \psi_0. \tag{11.25}$$

Note here that this condition holds regardless of the sign arbitrariness associated with the double covering.

Unfortunately, it turns out that $\mathbb{CP}^2$ does not admit a spin structure. To confirm this, let us specifically calculate the phase of (11.24). Taking $r \to \infty$ in (11.16), the connection 1-form ω on S^2 becomes

$$\omega = \frac{1}{2}\omega_{ab}\,\Sigma^{ab} \to \omega^{03}\,\Sigma^{03} + \omega^{12}\,\Sigma^{12},$$

and $\omega^{03} \to \sigma_3$ and $\omega^{12} \to 2\sigma_3$, respectively. If the coordinates unrelated to S^2 are fixed, then $\sigma_3 = \cos\theta\, d\varphi/2$, thus at $r \to \infty$, we obtain

$$\int_{S^2} d\omega^{03} = \int_{S^2} \frac{1}{2}\sin\theta\, d\varphi \wedge d\theta = 2\pi, \qquad \int_{S^2} d\omega^{12} = 4\pi.$$

Using this value and $[\Sigma^{03}, \Sigma^{12}] = 0$ and $(\Sigma^{03})^2 = (\Sigma^{12})^2 = -1/4$, the phase of (11.24) can be easily calculated as

$$\exp\left(-\int_{S^2} d\omega\right) = \exp(-2\pi\Sigma^{03})\exp(-4\pi\Sigma^{12}) = -1, \tag{11.26}$$

so that $\psi_1 = -\psi_0$. Thus, since the condition (11.25) is not satisfied, we find that $\mathbb{CP}^2$ is not a spin manifold.

As seen above, the condition for whether manifolds have a spin structure is decided by whether a non-exact 2-form $d\omega$ exists in the two-dimensional closed subspace and it generates a non-trivial phase like (11.26).[7]

11.4 Generalized Spin Structures

Fermions are also coupled with gauge fields, therefore by introducing a background gauge field, we can cancel the phase from the spin connection with that from the gauge field, so that the manifold has a spin structure. That is called generalized spin structure [HP78, GP78, BFF78, AI80, IP82]. However, when there are multiple fermions, some restrictions arises on gauge groups and representations.

$U(1)$ **background gauge field** Consider a $U(1)$ background gauge field on $\mathbb{CP}^2$. It is known that there exists a solution where the equation of motion $\nabla^\mu F_{\mu\nu} = 0$ is satisfied and the energy-momentum tensor $T^{U(1)}_{\mu\nu} = -F_{\mu\lambda}F_\nu{}^\lambda + g_{\mu\nu}F_{\lambda\sigma}F^{\lambda\sigma}/4$ vanishes. The existence of this solution does not change the equation of motion (11.14) for the gravitational field due to $T^{U(1)}_{\mu\nu} = 0$, therefore the $\mathbb{CP}^2$ manifold is preserved.

Expressed in terms of the gauge-field curvature 2-form $F = F_{\mu\nu}\,dx^\mu \wedge dx^\nu/2 = F_{ab}\,e^a \wedge e^b/2$, the solution is given by

$$\begin{aligned} F &= 2\mathfrak{a}\left(e^0 \wedge e^3 + e^1 \wedge e^2\right) \\ &= 2\mathfrak{a}\left(\frac{r}{(1+\mathfrak{a}r^2)^2}dr \wedge \sigma_3 + \frac{r^2}{1+\mathfrak{a}r^2}\sigma_1 \wedge \sigma_2\right). \end{aligned} \tag{11.27}$$

This is a self-dual solution satisfying $F_{ab} = \epsilon_{abcd}F^{cd}/2$, and changing the sign in the middle yields an anti-self-dual solution. The 1-form $A = A_\mu dx^\mu$ defined by $F = dA$ can be expressed as

$$A = \frac{\mathfrak{a}r^2}{1+\mathfrak{a}r^2}\sigma_3 = -\bar{\eta}_{3\mu\nu}\frac{\mathfrak{a}x^\nu}{1+\mathfrak{a}r^2}dx^\mu.$$

Let us consider S^2 in $\mathbb{CP}^2$, which causes a problem with spin structure. At $r \to \infty$ where S^2 appears, the scale $\mathfrak{a}$ disappears and the gauge field becomes $F \to 2\sigma_1 \wedge \sigma_2 = d\sigma_3$. From this, we obtain

$$\int_{S^2} F = \int_{S^2} d\sigma_3 = \int_{S^2} \frac{1}{2}\sin\theta\, d\varphi \wedge \theta = 2\pi.$$

[7] In mathematical terms, the necessary and sufficient condition for spin structure to exist is that the integral of the second Stiefel-Whitney class $w_2 \in H^2(M, Z_2)$ over the closed subspace becomes trivial (0 mod 2) [EGH80, IP82]. $M = \mathbb{CP}^2$ does not admit it because $\int_{S^2} w_2 = (i/\pi)\int_{S^2} d\omega = 1$.

The equation of motion of a fermion on $\mathbb{CP}^2$ in the presence of the $U(1)$ gauge field is changed to $\gamma^\mu(\partial_\mu + \omega_{\mu ab}\,\Sigma^{ab}/2 - ixA_\mu/2)\psi = 0$, where x is a charge that the fermion has, and the unit e is absorbed into the gauge field. As with the spin connection, the phase contribution from the gauge field when considering parallel transport on S^2 is given as

$$\psi_v = \exp\left(i\frac{x}{2}\int_{C_v} A\right)\psi.$$

Considering parallel transport on two different curves $C_{v'}$ and C_v and eliminating ψ, we get

$$\psi_{v'} = \exp\left(i\frac{x}{2}\int_{C_{v'}} A - i\frac{x}{2}\int_{C_v} A\right)\psi_v = \exp\left(i\frac{x}{2}\int_{Y} F\right)\psi_v.$$

If we choose the trivial curves $v' = 1$ and $v = 0$ as before, they represent the same point $C_1 = C_0 = x_0$, thus Y is S^2. From this, we obtain

$$\psi_1 = \exp\left(i\frac{x}{2}\int_{S^2} F\right)\psi_0 = e^{i\pi x}\,\psi_0.$$

Therefore, if x is an odd number, the negative sign appears and can be canceled out that from the spin connection.

This generalized spin structure is represented as $Spin_c(4) = [Spin(4) \times U(1)]/Z_2$, where Z_2 refers to choosing a representation such that the negative sign from the spin connection and that from the $U(1)$ gauge field cancel out.

In this way, by introducing the background $U(1)$ gauge field and assigning an "odd" charge to fermions, we can introduce the spin structure into $\mathbb{CP}^2$. On the other hand, there is no contribution from the spin connection for bosons, thus their charges must be "even" so that the negative sign does not appear. This condition shows that whether $\mathbb{CP}^2$ admits generalized spin structures depends on the content of matter coupled to the gravitational field.

$U(1)$ **background gauge field in** $SU(2)$ Next, we extend the discussion to a larger group G, and consider $U(1)$ subgroups within it. Here, considering $SU(2)$ and letting F^3 of its $U(1)$ subgroup be (11.27), introduce the background gauge field $-(i/2)\lambda^3 F^3$. The equation of motion is then given by $\gamma^\mu(\partial_\mu + \omega_{\mu ab}\,\Sigma^{ab}/2 - i\lambda^3 A^3_\mu/2)\,\psi = 0$. Considering the doublet of $SU(2)$, since $\lambda^3 = \mathrm{diag}(1, -1)$, the upper and lower components each satisfy the equation of motion with the $U(1)$ charge $x = \pm 1$. Therefore, there is a spin structure because the doublet entirely produces a negative sign, and cancels out that of $\mathbb{CP}^2$.

In the general $2j+1$ representation, $\lambda^3 = \text{diag}(2j, 2j-2, \cdots, -2j)$ and the sign from the background gauge field is $e^{i2\pi j}$. Therefore, when j is a half-integer, there is a spin structure, but when it is an integer, the negative sign of $\mathbb{CP}^2$ remains, thus there is no spin structure. On the other hand, bosons must belong to a representation that j is an integer.

$SU(2)$ **background gauge field** On the $\mathbb{CP}^2$ manifold, there exists a solution for the $SU(2)$ gauge field where the equation of motion is satisfied and the energy-momentum tensor vanishes, rather than the $U(1)$ part of $SU(2)$. Using the spin connection (11.16) and the 't Hooft symbol (11.29), it is expressed as

$$A_\mu^{\mathrm{A}} = -\frac{1}{2}\bar{\eta}^{\mathrm{A}}{}_{ab}\,\omega_\mu^{ab}, \tag{11.28}$$

and the curvature is given by

$$\begin{aligned} F_{\mu\nu}^{\mathrm{A}} &= \partial_\mu A_\nu^{\mathrm{A}} - \partial_\nu A_\mu^{\mathrm{A}} + \epsilon^{\mathrm{ABC}} A_\mu^{\mathrm{B}} A_\nu^{\mathrm{C}} \\ &= -\frac{1}{2}\bar{\eta}^{\mathrm{A}}{}_{ab}\left(\partial_\mu \omega_\nu^{ab} - \partial_\nu \omega_\mu^{ab} + [\omega_\mu, \omega_\nu]^{ab}\right) = -\frac{1}{2}\bar{\eta}^{\mathrm{A}}{}_{ab} R_{\mu\nu}{}^{ab}, \end{aligned}$$

where $\epsilon^{\mathrm{ABC}}\,\bar{\eta}^{\mathrm{B}}_{ab}\,\bar{\eta}^{\mathrm{C}}_{cd} = \delta_{ac}\,\bar{\eta}^{\mathrm{A}}{}_{bd} - \delta_{ad}\,\bar{\eta}^{\mathrm{A}}{}_{bc} - \delta_{bc}\,\bar{\eta}^{\mathrm{A}}{}_{ad} + \delta_{bd}\,\bar{\eta}^{\mathrm{A}}{}_{ac}$ is used. In fact, on $\mathbb{CP}^2$, this satisfies

$$\begin{aligned} D^\mu F_{\mu\nu}^{\mathrm{A}} &= \nabla^\mu F_{\mu\nu}^{\mathrm{A}} + \epsilon^{\mathrm{ABC}} A^{\mathrm{B}\mu} F_{\mu\nu}^{\mathrm{C}} \\ &= -\frac{1}{2}\bar{\eta}^{\mathrm{A}}{}_{ab}\, g^{\mu\lambda}\left(\partial_\lambda R_{\mu\nu}{}^{ab} - \Gamma^\sigma_{\lambda\mu} R_{\sigma\nu}{}^{ab} - \Gamma^\sigma_{\lambda\nu} R_{\mu\sigma}{}^{ab} + \omega_\lambda^{ac} R_{\mu\nu c}{}^{b} + \omega_\lambda^{bc} R_{\mu\nu}{}^{a}{}_{c}\right) \\ &= -\frac{1}{2}\bar{\eta}^{\mathrm{A}}{}_{ab}\,\nabla^\mu R_{\mu\nu ab} = 0 \end{aligned}$$

and $T_{\mu\nu}^{SU(2)} = -F_{\mu\lambda}^{\mathrm{A}}\,F_\nu^{\mathrm{A}\lambda} + g_{\mu\nu}\,F_{\lambda\sigma}^{\mathrm{A}}\,F^{\mathrm{A}\lambda\sigma}/4 = 0$. Since the existence of this solution does not change the equation of motion of the gravitational field (11.14), the $\mathbb{CP}^2$ solution remains unchanged.

This $F_{\mu\nu}^{\mathrm{A}}$ is neither self-dual nor anti-self-dual, but has both components. In addition, the expression in which 't Hooft symbol $\bar{\eta}^{\mathrm{A}}{}_{ab}$ is replaced with $\eta^{\mathrm{A}}{}_{ab}$ is also a solution and has similar properties.

If this $SU(2)$ background gauge field exists, we can introduce spin structure for fermion doublets, generally half-integer j representations. In fact, since the behavior of the gauge field on S^2 at $r \to \infty$ is $A_\mu^{\mathrm{A}} \to \delta^{\mathrm{A}3}\,\omega_\mu^{03} - \epsilon^{\mathrm{A}12}\,\omega_\mu^{12}$, the gauge field 1-form $A^{\mathrm{A}} = A_\mu^{\mathrm{A}}\,dx^\mu$ is given by A^1, $A^2 \to 0$, and $A^3 \to \omega^{03} - \omega^{12} = -\sigma_3$. Thus, the curvature becomes

$$F^1,\ F^2 \to 0, \qquad F^3 \to -d\sigma_3.$$

Therefore, since the discussion of generalized spin structure in the case of $U(1)$ subgroup can be used as it is, fermions with half-integer j representation can be defined on $\mathbb{CP}^2$.

11.5 Particle Models with Generalized Spin Structures

Finally, consider realistic particle models in which multiple fermions exist. First of all, we find that on $\mathbb{CP}^2$, it is not possible to introduce spin structure into the Standard Model with $G = SU(3)_c \times SU(2)_L \times U(1)_Y$. In fact, in order to make the spin 1/2 Dirac index $\mathcal{I}_{1/2}$ an integer, we first need to add right-handed neutrinos to increase the number of Weyl fermions to 16 in each generation. However, right-handed neutrinos do not have any gauge charge, thus the sign from the $\mathbb{CP}^2$ manifold cannot be erased by introducing a background gauge field. In addition, for other fermions and boson fields, charges are not properly assigned, so that there is no spin structure.[8]

In addition, there is no spin structure in the $SU(5)$ GUT model because the number of Weyl fermions per generation is 15 and thus $\mathcal{I}_{1/2}$ is not an integer.

Among the widely-known GUT models that include the Standard Model as a part, the simplest gauge group that can introduce spin structure on $\mathbb{CP}^2$ is $SO(10)$ (strictly $Spin(10)$) [BFF78, CS96, WWW19]. There are several options for how to introduce the background gauge field to admit a generalized spin structure. First, consider the case where spontaneous symmetry breaking occurs such as $SO(10) \to SU(5) \times U(1)_X$. Then, Weyl fermions of the 16 representation of $SO(10)$ constituting one generation are decomposed as

$$16 \quad \to \quad 10_{-5} + \bar{5}_3 + 1_{-1},$$

where the subscripts are $U(1)_X$ charges, all of which are odd numbers. On the other hand, bosons have even charges as

$$\begin{aligned} 45 &\to 24_0 + 10_{-4} + \bar{10}_4 + 1_0, \\ 10 &\to 5_{-2} + \bar{5}_2. \end{aligned}$$

Therefore, if we introduce the background gauge field (11.27) to $U(1)_X$, then we can make $\mathbb{CP}^2$ a spin manifold.

Then we can also see that the spin 1/2 Dirac index (11.20) is an integer as follows. For each fermion with charge x, $\Omega = i(x/2)F$, thus using (11.27) we find

$$\frac{1}{8\pi^2}\int_{\mathbb{CP}^2} \mathrm{Tr}\, \Omega \wedge \Omega = -\frac{x^2}{4}\frac{2(2\mathfrak{a})^2}{8\pi^2}\mathrm{Vol}(\mathbb{CP}^2) = -\frac{x^2}{8}.$$

[8] There is an idea to modify the Standard Model by adding an extra $U(1)$ gauge field and assigning its charges so that spin structure is allowed.

Since dimV $= 16$ and the sum of squared charges is $N_X = 10 \times (-5)^2 + 5 \times 3^2 + 1 \times (-1)^2 = 296$, the index becomes

$$\mathcal{I}_{\frac{1}{2}} = -\frac{\dim V}{8}\tau(\mathbb{CP}^2) - \left(-\frac{N_X}{8}\right) = -2 + 37 = 35.$$

If F is an anti-self-dual solution, the sign in the middle becomes negative and becomes -39.

Thus, we can introduce spin structures in these cases, but unfortunately the index becomes non-zero. This implies that fermion zero-modes exist and the partition function disappears.[9]

Next, consider $SO(10) \to SU(4) \times SU(2)_L \times SU(2)_R$ as another pattern of symmetry breaking. The fermion representation is then decomposed as

$$16 \;\to\; (4, 2, 1) + (\bar{4}, 1, 2)$$

On the other hand, the boson representations are decomposed as

$$\begin{aligned} 45 &\to (15, 1, 1) + (6, 2, 2) + (1, 3, 1) + (1, 1, 3), \\ 10 &\to (6, 1, 1) + (1, 2, 2). \end{aligned}$$

Hence, introduce $-(i/2)\lambda^3 F^3_{L,R}$ for both $U(1)$ subgroups of $SU(2)_L$ and $SU(2)_R$, where $F^3_{L,R}$ is the 2-form F (11.27). The equation of motion is given by $\gamma^\mu(\partial_\mu + \omega_{\mu ab}\,\Sigma^{ab}/2 - i\lambda^3 A^3_{L\mu}/2 - i\lambda^3 A^3_{R\mu}/2\,)\,\psi = 0$. The fermion with the 16 representation is decomposed into 4 doublets each of L and R, for a total of 8 doublets. Since representations with integer j are not included, a negative sign appears and cancels with the sign from the spin connection of $\mathbb{CP}^2$. On the other hand, the bosons have representations that do not produce the sign. Therefore, it is possible to introduce a generalized spin structure.

Further calculating the spin $1/2$ Dirac index in this case yields

$$\begin{aligned} \mathcal{I}_{\frac{1}{2}} &= -\frac{\dim V}{8}\tau(\mathbb{CP}^2) + \frac{4}{16\pi^2}\int_{\mathbb{CP}^2} F^3_L \wedge F^3_L + \frac{4}{16\pi^2}\int_{\mathbb{CP}^2} F^3_R \wedge F^3_R \\ &= -2 + \frac{1}{2\pi^2}\int_{\mathbb{CP}^2} F \wedge F = -2 + 2 = 0, \end{aligned}$$

due to $\dim V = 8 \times 2$, $\Omega = -i\lambda^3 F^3/2$, and $\mathrm{Tr}\,(\lambda^3\lambda^3) = 2$. Since the index vanishes, there is a possibility that the partition function exists. On the other hand, if we bring an

[9] Since the kinetic term of the action does not contain the zero mode, the Grassmann path integral for this mode disappears. If there is a mass term $m\bar{\psi}\psi$, it contains a zero mode, thus the Grassmann integral has a value, and the partition function is proportional to the mass m [tHo76, SVZ80]. If there are multiple zero modes, it is the product of the masses related to them. However, here we are considering a high-energy limit where mass can be considered to be zero.

anti-self-dual solution as F, the sign is reversed and becomes $\mathcal{I}_{1/2} = -2 - 2 = -4$, thus the partition function disappears.

Similarly to above, consider the case where the symmetry is broken as $SO(10) \to SU(4) \times SU(2)_L \times SU(2)_R$ and apply the background gauge field (11.28) to the left and right $SU(2)$, we can introduce a generalized spin structure. The difference from the above discussion is that the spin 1/2 Dirac index does not vanish. In fact, since $\Omega = -i\lambda^{\mathrm{A}} F^{\mathrm{A}}/2$, $\tau(\mathbb{CP}^2) = 1$, and $\chi(\mathbb{CP}^2) = 3$, the gauge field contribution to the index is given for each fermion doublet by

$$\begin{aligned}
\frac{1}{8\pi^2}\int_{\mathbb{CP}^2} \mathrm{Tr}\,\Omega \wedge \Omega &= -\frac{1}{64\pi^2}\int_{\mathbb{CP}^2} d^4x\sqrt{g}\,\epsilon^{\mu\nu\lambda\sigma} F^{\mathrm{A}}_{\mu\nu} F^{\mathrm{A}}_{\lambda\sigma} \\
&= -\frac{1}{64\pi^2}\int_{\mathbb{CP}^2} d^4x\sqrt{g}\left(\frac{1}{2}\epsilon^{\mu\nu\lambda\sigma} R_{\mu\nu ab} R_{\lambda\sigma}{}^{ab}\right. \\
&\qquad\qquad \left. -\frac{1}{4}\epsilon^{\mu\nu\lambda\sigma}\epsilon^{abcd} R_{\mu\nu ab} R_{\lambda\sigma cd}\right) \\
&= -\frac{3}{4}\tau(\mathbb{CP}^2) + \frac{1}{2}\chi(\mathbb{CP}^2) = \frac{3}{4}.
\end{aligned}$$

If we change $\bar{\eta}^{\mathrm{A}}_{ab}$ to η^{A}_{ab}, it becomes $-9/4$ because the sign in the middle is reversed. Since there are 8 doublets, when applying the same $\bar{\eta}^{\mathrm{A}}_{ab}$ solution to both $SU(2)$, we get

$$\begin{aligned}
\mathcal{I}_{\frac{1}{2}} &= -\frac{\dim V}{8}\tau(\mathbb{CP}^2) - \frac{8}{8\pi^2}\int_{\mathbb{CP}^2} \mathrm{Tr}\,\Omega \wedge \Omega \\
&= -\frac{16}{8} - 8 \times \frac{3}{4} = -8.
\end{aligned}$$

Therefore, the index becomes an integer, but does not vanish. In addition, if we apply the solution of η^{A}_{ab} to one $SU(2)$, we get $\mathcal{I}_{1/2} = -2 - 4 \times (3/4) - 4 \times (-9/4) = 4$. Similarly, if we change both to the η^{A}_{ab} solution, we get $\mathcal{I}_{1/2} = -2 - 8 \times (-9/4) = 16$. In each case, since the fermion zero-mode exists, the partition function disappears.

Moreover, even when applying a combination of the $U(1)$ and $SU(2)$ solutions to the $SU(2)_L \times SU(2)_R$ part, we can introduce a generalized spin structure. However, the spin 1/2 Dirac index vanishes only when the self-dual background gauge field (11.27) is applied to both $U(1)$ subgroups of $SU(2)$.

In any case, if $\mathbb{CP}^2$ space is physically required, the $SO(10)$ GUT model is preferable. The inclusion of 16 Weyl fermions in one generation is a favorable property since mass has been found in neutrinos [Kaj+98] and this model has not yet been rejected by the Kamioka's proton decay experiment [SKDE].

Finally, from the perspective of the path integral, it would seem to be necessary to take the sum over all possible topologies, but the existence of fermions should be a condition for adding to the sum. However, it is not sure about the physical significance of doing the sum. The gravitational instanton is the topology of spacetime itself that is the vessel that contains us, therefore the situation is different from that of Yang-Mills instantons [tHo76, CDG78, CDG79, SVZ80]. Thinking about topology is fun in itself, but we still do not understand anything about it in terms of physics.

11.6 Appendix: Addenda to Differential Geometry

11.6.1 Tools for Instantons

First, introduce the 't Hooft symbol [tHo76, VZNS82], which is often used in instanton physics. The indices of $SU(2)$ are A, B $= 1, 2, 3$, the spacetime indices of $\mathbb{R}^4$ are $a, b = 0, 1, 2, 3$, and the totally antisymmetric tensor is normalized to $\epsilon^{0123} = 1$. By convention, the same index means to take the sum over it. The symbol is then defined by

$$\begin{aligned}\eta^{\mathrm{A}}{}_{ab} &= \epsilon_{0\mathrm{A}ab} + \frac{1}{2}\epsilon_{\mathrm{ABC}}\,\epsilon_{\mathrm{BC}ab},\\ \bar{\eta}^{\mathrm{A}}{}_{ab} &= \epsilon_{0\mathrm{A}ab} - \frac{1}{2}\epsilon_{\mathrm{ABC}}\,\epsilon_{\mathrm{BC}ab},\end{aligned} \tag{11.29}$$

where $\eta^{\mathrm{A}}{}_{ab}$ and $\bar{\eta}^{\mathrm{A}}{}_{ab}$ are self-dual and anti-self-dual about indices of a and b, respectively, satisfying

$$\eta^{\mathrm{A}}{}_{ab} = \frac{1}{2}\epsilon_{abcd}\,\eta^{\mathrm{A}}{}_{cd}, \qquad \bar{\eta}^{\mathrm{A}}{}_{ab} = -\frac{1}{2}\epsilon_{abcd}\,\bar{\eta}^{\mathrm{A}}{}_{cd}.$$

In terms of components, they are written as

$$\begin{aligned}\eta^{\mathrm{A}}{}_{0b} &= \delta_{\mathrm{A}b}, \quad \eta^{\mathrm{A}}{}_{\mathrm{BC}} = \epsilon_{\mathrm{ABC}},\\ \bar{\eta}^{\mathrm{A}}{}_{0b} &= -\delta_{\mathrm{A}b}, \quad \bar{\eta}^{\mathrm{A}}{}_{\mathrm{BC}} = \epsilon_{\mathrm{ABC}}.\end{aligned}$$

A differential 1-form constructed using the anti-self-dual 't Hooft symbol is introduced as

$$\sigma_{\mathrm{A}} = \bar{\eta}^{\mathrm{A}}{}_{ab}\frac{x^a dx^b}{r^2},$$

where $x^a = (t, x, y, z)$ and $r^2 = t^2 + x^2 + y^2 + z^2$. Expressing dr and σ_{A} in coordinate components become

$$dr = \frac{1}{r}(tdt + xdx + ydy + zdz),$$
$$\sigma_1 = \frac{1}{r^2}(xdt - tdx - zdy + ydz),$$
$$\sigma_2 = \frac{1}{r^2}(ydt + zdx - tdy - xdz),$$
$$\sigma_3 = \frac{1}{r^2}(zdt - ydx + xdy - tdz).$$

The exterior derivative of the 1-form σ_A satisfies a relation $d\sigma_A = \epsilon_{ABC}\,\sigma_B \wedge \sigma_C$, that is,

$$d\sigma_1 = 2\sigma_2 \wedge \sigma_3, \quad d\sigma_2 = 2\sigma_3 \wedge \sigma_1, \quad d\sigma_3 = 2\sigma_1 \wedge \sigma_2.$$

Using σ_A, the $\mathbb{R}^4$ space can be written as

$$\begin{aligned} ds^2_{\mathbb{R}^4} &= dt^2 + dx^2 + dy^2 + dz^2 \\ &= dr^2 + r^2\left(\sigma_1^2 + \sigma_2^2 + \sigma_3^2\right). \end{aligned}$$

Furthermore, if we introduce coordinates

$$x + iy = r\cos\frac{\theta}{2}e^{\frac{i}{2}(\psi+\varphi)},$$
$$z + it = r\sin\frac{\theta}{2}e^{\frac{i}{2}(\psi-\varphi)},$$

using the Euler angle $0 \le \theta \le \pi$, $0 \le \varphi \le 2\pi$, $0 \le \psi \le 4\pi$, which is often used in this research area, we obtain

$$\sigma_1 = \frac{1}{2}(\sin\psi d\theta - \sin\theta\cos\psi d\varphi),$$
$$\sigma_2 = \frac{1}{2}(-\cos\psi d\theta - \sin\theta\sin\psi d\varphi),$$
$$\sigma_3 = \frac{1}{2}(d\psi + \cos\theta d\varphi).$$

The $\mathbb{R}^4$ space is then expressed as

$$ds^2_{\mathbb{R}^4} = dr^2 + \frac{r^2}{4}\left(d\varphi^2 + d\theta^2 + d\psi^2 + 2\cos\theta d\varphi d\psi\right).$$

Here, we mention why the range of the Euler angle ψ should be set above. Considering a two-dimensional surface with fixed the coordinates θ and φ of S^2, its metric becomes $dr^2 + (r^2/4)d\psi^2$. In order for this to be identified with a $\mathbb{R}^2$ plane with metric $ds_{R^2} = dr^2 + r^2 d\omega^2$ $(0 \le \omega \le 2\pi)$ without conical singularity, it is necessary that $\omega = \psi/2$, so that $0 \leq \psi \leq 4\pi$. This range is the same as when the unit S^3 is expressed using Euler angles.

't Hooft symbol formulas Various formulas that η^{A}_{ab} satisfies are summarized below:

$$
\begin{aligned}
\eta^{\mathrm{A}}_{ab} &= -\eta^{\mathrm{A}}_{ba}, \qquad \eta^{\mathrm{A}}_{ab}\,\eta^{\mathrm{B}}_{ab} = 4\delta^{\mathrm{AB}}, \\
\eta^{\mathrm{A}}_{ab}\,\eta^{\mathrm{A}}_{ac} &= 3\delta_{bc}, \qquad \eta^{\mathrm{A}}_{ab}\,\eta^{\mathrm{A}}_{ab} = 12, \\
\eta^{\mathrm{A}}_{ab}\,\eta^{\mathrm{A}}_{cd} &= \delta_{ac}\delta_{bd} - \delta_{ad}\delta_{bc} + \epsilon_{abcd}, \\
\eta^{\mathrm{A}}_{ab}\,\eta^{\mathrm{B}}_{ac} &= \delta^{\mathrm{AB}}\delta_{bc} + \epsilon^{\mathrm{ABC}}\eta^{\mathrm{C}}_{bc}, \\
\epsilon^{\mathrm{ABC}}\,\eta^{\mathrm{B}}_{ac}\,\eta^{\mathrm{C}}_{bd} &= \delta_{ab}\eta^{\mathrm{A}}_{cd} - \delta_{ad}\eta^{\mathrm{A}}_{cb} - \delta_{cb}\eta^{\mathrm{A}}_{ad} + \delta_{cd}\eta^{\mathrm{A}}_{ab}, \\
\epsilon_{abcd}\,\eta^{\mathrm{A}}_{de} &= -\delta_{ce}\eta^{\mathrm{A}}_{ab} - \delta_{be}\eta^{\mathrm{A}}_{ca} - \delta_{ae}\eta^{\mathrm{A}}_{bc}.
\end{aligned}
$$

Similar formulas for $\bar{\eta}^{\mathrm{A}}_{ab}$ can be obtained by replacing above as

$$
\eta^{\mathrm{A}}_{ab}, \quad \epsilon_{abcd} \quad \longrightarrow \quad \bar{\eta}^{\mathrm{A}}_{ab}, \quad -\epsilon_{abcd}
$$

In adition, as formulas for the product of the two, there are

$$
\eta^{\mathrm{A}}_{ab}\,\bar{\eta}^{\mathrm{B}}_{ab} = 0, \qquad \eta^{\mathrm{A}}_{ab}\,\bar{\eta}^{\mathrm{B}}_{ac} = \eta^{\mathrm{A}}_{ac}\,\bar{\eta}^{\mathrm{B}}_{ab}.
$$

11.6.2 Eguchi-Hanson Metric

The Eguchi-Hanson metric [EH79] is widely known as a gravitational instanton solution, which has been found as a solution that satisfies the (anti-) self-duality condition for the Riemann tensor,

$$
R_{abcd} = \pm\frac{1}{2}\epsilon_{abef}\,R^{ef}{}_{cd},
$$

where the positive and negative signs represent self-duality and anti-self-duality, respectively.

First, assume that the metric field representing the gravitational instanton has a form

$$
ds^2 = f^2(r)dr^2 + r^2\left(\sigma_1^2 + \sigma_2^2\right) + r^2 g^2(r)\sigma_3^2.
$$

The vierbein 1-form $e^a = e^a_\mu dx^\mu$ is then written as $e^0 = f dr$, $e^1 = r\sigma_1$, $e^2 = r\sigma_2$, $e^3 = rg\sigma_3$, and the spin connection 1-form $\omega^{ab} = \omega^{ab}_\mu dx^\mu$ is given by

$$\omega^{01} = -\frac{1}{rf}e^1, \quad \omega^{02} - \frac{1}{rf}e^2, \quad \omega^{03} = -\frac{1}{rfg}\left(g + r\frac{\partial g}{\partial r}\right)e^3,$$

$$\omega^{23} = \frac{g}{r}e^1, \quad \omega^{31} = \frac{g}{r}e^2, \quad \omega^{12} = \left(\frac{2}{rg} - \frac{g}{r}\right)e^3.$$

If the Riemann tensor is (anti-) self-dual, it can be shown that the spin connection is also (anti-) self-dual such that

$$\omega_{\mu ab} = \pm\frac{1}{2}\epsilon_{abcd}\,\omega_\mu^{cd}.$$

Choosing the anti-self-duality condition according to the original paper, we obtain

$$fg = 1, \qquad \frac{1}{f}\left(g + r\frac{\partial g}{\partial r}\right) = 2 - g^2.$$

Eliminating f from the second equation yields

$$r\frac{\partial g^2}{\partial r} = 4(1 - g^2),$$

thus by solving this equation, we get $g^2 = 1 - a^4/r^4$, where a (> 0) is an integration constant. Here, although we can consider $1 + a^4/r^4$ as a solution to g^2, it has a singularity at $r = 0$. On the other hand, for the former with a negative sign, the singularity can be removed by setting $r \geq a$ as shown below. In this way, the anti-self-dual Eguchi-Hanson metric is found as

$$ds_{\mathrm{EH}}^2 = \frac{1}{1 - \frac{a^4}{r^4}}dr^2 + r^2(\sigma_1^2 + \sigma_2^2) + r^2\left(1 - \frac{a^4}{r^4}\right)\sigma_3^2. \tag{11.30}$$

The curvature 2-form of this space is given by

$$\mathcal{R}_{01} = -\mathcal{R}_{23} = \frac{2a^4}{r^6}(-e^0 \wedge e^1 + e^2 \wedge e^3),$$

$$\mathcal{R}_{02} = -\mathcal{R}_{31} = \frac{2a^4}{r^6}(-e^0 \wedge e^2 + e^3 \wedge e^1),$$

$$\mathcal{R}_{03} = -\mathcal{R}_{12} = \frac{4a^4}{r^6}(e^0 \wedge e^3 - e^1 \wedge e^2).$$

Now, let us remove the conical singularity that appears at $r = a$. To examine its neighborhood, convert the coordinate variable to $u^2 = r^2(1 - a^4/r^4)$, then $dr = du\sqrt{1 - a^4/r^4}/(1 + a^4/r^4)$ is obtained, and the metric can be written as

$$ds_{\mathrm{EH}}^2 = \frac{1}{(1 + \frac{a^4}{r^4})^2}du^2 + u^2\sigma_3^2 + r^2(\sigma_1^2 + \sigma_2^2).$$

If we take the limit of $u \to 0$ $(r \to a)$, we get

$$ds_{\text{EH}}^2 \to \frac{1}{4}du^2 + \frac{1}{4}u^2(d\psi + \cos\theta\, d\varphi)^2 + \frac{a^2}{4}(d\theta^2 + \sin^2\theta\, d\varphi^2).$$

Considering a two-dimensional surface with fixed θ and φ of the S^2 coordinates, the metric near $u = 0$ is expressed as

$$\frac{1}{4}(du^2 + u^2 d\psi^2).$$

If this can be identified with the metric of the $\mathbb{R}^2$ plane, then there is no conical singularity. Since the metric of $\mathbb{R}^2$ is $d\rho^2 + \rho^2 d\omega^2$ with $\rho \geq 0$ and $0 \leq \omega \leq 2\pi$, we can see that the conical singularity can be removed by restricting the range of ψ to

$$0 \leq \psi \leq 2\pi.$$

In this case, the topology near $r = a$ becomes $\mathbb{R}^2 \times S^2$.

The Eguchi-Hanson metric (11.30) is a solution in which $r \to \infty$ is a boundary. Lastly, let us see the topological structure of the boundary. If we consider the originally defined region $0 \leq \psi \leq 4\pi$, $r \to \infty$ becomes a boundary whose topology is S^3. Since ψ is now restricted to half, the boundary is $\mathbb{RP}^3 = S^3/Z_2$.

11.6.3 Spin 1/2 Dirac Index (Atiyah-Singer Index)

Consider a massless fermion action $I_\psi = \int d^4x \sqrt{g}\, i\bar{\psi}\gamma^\mu D_\mu \psi$ coupled to a background gauge field $\mathcal{A}_\mu$ in curved spacetime. The covariant derivative is given by $D_\mu = \nabla_\mu + \mathcal{A}_\mu$, where $\mathcal{A}_\mu = A_\mu^{\text{A}} T^{\text{A}}$ $(\mathcal{A}_\mu^\dagger = -\mathcal{A}_\mu)$. In the main text, it is defined for $SU(2)$ as $T^{\text{A}} = -i\lambda^{\text{A}}/2$ with $[\lambda^A, \lambda^B] = 2i\epsilon^{ABC}\lambda^C$, and for $U(1)$ as $\mathcal{A}_\mu = -ixA_\mu/2$, where x is the charge. The curvature 2-form (field strength) of the gauge field is defined by $\Omega = d\mathcal{A} + \mathcal{A} \wedge \mathcal{A} = \mathcal{F}_{\mu\nu}\, dx^\mu \wedge dx^\nu/2$, where $\mathcal{A} = \mathcal{A}_\mu dx^\mu$.

Denoting the partition function as $Z = \int d\psi d\bar{\psi} e^{-I_\psi}$, the vacuum expectation value of an axial current $J_\mu^5 = i\bar{\psi}\gamma_\mu\gamma_5\psi$ is given by

$$\begin{aligned}\langle J_\mu^5(x)\rangle &= \frac{1}{Z}\int d\psi d\bar{\psi}\, e^{-I_\psi}\, i\bar{\psi}(x)\gamma_\mu\gamma_5\psi(x)\\ &= -i\text{Tr}[\gamma_\mu\gamma_5\, H(x,x)],\end{aligned}$$

where the Green function H can be expressed by introducing an eigenequation $i\gamma^\mu D_\mu\psi_n = E_n\psi_n$ as $H(x,x') = \sum_n \psi_n(x)\bar{\psi}_n(x')/E_n$. In fact, when the derivative is applied, $i\gamma^\mu D_\mu H(x,0) = \sum_n \psi_n(x)\bar{\psi}_n(0) = \delta^4(x)/\sqrt{g}$ holds. Therefore, we obtain

$$\langle J_\mu^5(x)\rangle = \sum_n \frac{1}{E_n} i\bar{\psi}_n(x)\gamma_\mu\gamma_5\psi_n(x).$$

From the eigenequation $i\gamma^\mu\nabla_\mu\psi_n = E_n\psi_n - i\mathcal{A}_\mu\gamma^\mu\psi_n$ and its conjugate equation $i\nabla_\mu\bar{\psi}_n\gamma^\mu = -E_n\bar{\psi}_n + i\mathcal{A}_\mu\bar{\psi}_n\gamma^\mu$, the divergence of the axial current is expressed as

$$\begin{aligned}
\nabla^\mu\langle J_\mu^5(x)\rangle &= \sum_n \frac{1}{E_n} i\nabla^\mu(\bar{\psi}_n\gamma_\mu\gamma_5\psi_n) \\
&= \sum_n \frac{1}{E_n}\left(i\nabla^\mu\bar{\psi}_n\gamma_\mu\gamma_5\psi_n - \bar{\psi}_n\gamma_5 i\gamma_\mu\nabla^\mu\psi_n\right) \\
&= -2\sum_n \bar{\psi}_n\gamma_5\psi_n = 2\,\mathrm{Tr}\langle x|\gamma_5|x\rangle,
\end{aligned}$$

where $\{\gamma_5, \gamma_\mu\} = 0$ is used and Tr is a general term for various traces required. The Atiyah-Singer spin $1/2$ Dirac index is then defined as

$$\mathcal{I}_{\frac{1}{2}} = \int d^4x\sqrt{g}\sum_n \bar{\psi}_n\gamma_5\psi_n = -\int d^4x\sqrt{g}\,\mathrm{Tr}\langle x|\gamma_5|x\rangle. \tag{11.31}$$

In the following, the trace part involving γ_5 is calculated by regularizing as

$$\mathrm{Tr}\langle x|\gamma_5|x\rangle = \lim_{s\to 0}\mathrm{Tr}\,\langle x|\gamma_5\, e^{-sK}|x\rangle, \tag{11.32}$$

where $K = (\gamma^\mu D_\mu)^2 = \gamma^\mu\gamma^\nu D_\mu D_\nu$ is a positive-definite operator, rewritten by using $\nabla_\mu\gamma_\nu = \gamma_\nu\nabla_\mu$.

The operator K is expressed using $\gamma^\mu\gamma^\nu = \{\gamma^\mu, \gamma^\nu\}/2 + [\gamma^\mu, \gamma^\nu]/2 = -g^{\mu\nu} - 2\Sigma^{\mu\nu}$ as

$$\begin{aligned}
K &= \left(-g^{\mu\nu} - 2\Sigma^{\mu\nu}\right)D_\mu D_\nu = -D^\mu D_\mu - \Sigma^{\mu\nu}[D_\mu, D_\nu] \\
&= D^\mu D_\mu - \Sigma^{\mu\nu}\left(\frac{1}{2}R_{\mu\nu ab}\Sigma^{ab} + \mathcal{F}_{\mu\nu}\right) \\
&= -D^\mu D_\mu + \frac{1}{4}R - \mathcal{F}_{ab}\Sigma^{ab},
\end{aligned} \tag{11.33}$$

where the last equality is from $R_{abcd}\Sigma^{ab}\Sigma^{cd} = R_{abcd}\gamma^{ab}\gamma^{cd}/4 = -R/2$, in which we use the cyclic property of the Riemann tensor and a formula

$$\gamma^{ab}\gamma^{cd} = \gamma^{abcd} + \delta^{ac}\gamma^{bd} - \delta^{ad}\gamma^{bc} - \delta^{bc}\gamma^{ad} + \delta^{bd}\gamma^{ac} - \delta^{ac}\delta^{bd} + \delta^{ad}\delta^{bc}$$

for antisymmetric products of the gamma matrices, $\gamma^{ab} = [\gamma^a, \gamma^b]/2!$ and $\gamma^{abcd} = (\gamma^a\gamma^b\gamma^c\gamma^d + \text{anti-sym.})/4!$.

In order to calculate (11.32), introduce the heat kernel $G^{(s)}(x, y) = \langle x|e^{-sK}|y\rangle$. This is given as a solution to the heat conduction equation

$$\frac{\partial}{\partial s} G^{(s)}(x, y) = -K G^{(s)}(x, y), \tag{11.34}$$

satisfying the initial condition $\lim_{s \to 0} G^{(s)}(x, y) = \delta^4(x - y)/\sqrt{g}$. When the parameter s is small, the heat kernel at the same point is expanded as

$$G^{(s)}(x, x) = \frac{1}{(4\pi)^2} \frac{1}{s^2} \left[1 + s a_1(x) + s^2 a_2(x) + \cdots \right].$$

First, consider the case in which there is a background gauge field in flat spacetime. Decomposing the regularization operator as

$$K = -\partial^2 - V, \tag{11.35}$$

then $V = \mathcal{F}_{ab} \Sigma^{ab} + \cdots$, where the dots are gauge field terms that do not contain the gamma matrix. The heat kernel satisfying the above initial condition in flat spacetime is given by

$$G_0^{(s)}(x, y) = \langle x | e^{s\partial^2} | y \rangle = \frac{1}{(4\pi)^2 s^2} e^{-(x-y)^2/4s}, \tag{11.36}$$

thus using this we obtain

$$\begin{aligned} G^{(s)}(x, x) &= G_0^{(s)}(x, x) \exp\left(s \mathcal{F}_{ab} \Sigma^{ab}\right) \\ &= \frac{1}{(4\pi)^2} \left[\frac{1}{s^2} + \frac{1}{s} \mathcal{F}_{ab} \Sigma^{ab} + \frac{1}{2} \mathcal{F}_{ab} \mathcal{F}_{cd} \Sigma^{ab} \Sigma^{cd} + \cdots \right]. \end{aligned}$$

Multiplying this by γ_5 and taking a trace yields (11.32). Due to $\mathrm{Tr}(\gamma_5) = \mathrm{Tr}(\gamma_5 \Sigma^{ab}) = 0$, the term that diverges at $s \to 0$ disappears, and from $\mathrm{Tr}(\gamma_5 \Sigma^{ab} \Sigma^{cd}) = \epsilon^{abcd}$, we get a finite value

$$\lim_{s \to 0} \mathrm{Tr}\left[\gamma_5 G^{(s)}(x, x)\right] = \frac{1}{32\pi^2} \epsilon^{abcd} \, \mathrm{Tr}\left(\mathcal{F}_{ab} \mathcal{F}_{cd}\right). \tag{11.37}$$

Next, in order to calculate (11.32) in curved spacetime in the same way, we take a coordinate system that is locally flat at the origin, i.e., $\Gamma^{\lambda}_{\mu\nu}(0) = 0$, and expand the operator K around the origin. If the metric field is expanded assuming that the Riemann tensor at the origin is a constant, we get[10]

[10] In fact, $\Gamma^{\lambda}_{\mu\nu}(x)$ is found from $g_{\mu\nu}(x)$ in the first line, and substituting it into the definition of the Riemann tensor yields $R^{\lambda}{}_{\mu\sigma\nu}(x) = \partial_\sigma \Gamma^{\lambda}_{\mu\nu}(x) - \partial_\nu \Gamma^{\lambda}_{\mu\sigma}(x) + o(x^2) = 2R^{\lambda}{}_{\mu\sigma\nu}(0)/3 - [R^{\lambda}{}_{\nu\mu\sigma}(0) + R^{\lambda}{}_{\sigma\nu\mu}(0)]/3 + o(x^2) = R^{\lambda}{}_{\mu\sigma\nu}(0) + o(x^2)$ using the cyclic property. Similarly, $R_{\mu\nu ab}(0) + o(x^2)$ can be found by substituting the last spin connection into (11.3).

$$g_{\mu\nu}(x) = \delta_{\mu\nu} - \frac{1}{3}x^\lambda x^\sigma R_{\mu\lambda\nu\sigma}(0) + o(x^3),$$
$$\sqrt{g(x)} = 1 - \frac{1}{6}x^\lambda x^\sigma R_{\lambda\sigma}(0) + o(x^3),$$
$$\Gamma^\lambda_{\mu\nu}(x) = \frac{1}{3}x^\sigma\left[R^\lambda{}_{\mu\sigma\nu}(0) + R^\lambda{}_{\nu\sigma\mu}(0)\right] + o(x^2),$$
$$\omega_{\mu ab}(x) = \frac{1}{2}x^\lambda R_{\lambda\mu ab}(0) + o(x^2).$$

In the following, the subscript (0) representing the origin is omitted. The Laplacian in curved spacetime is expressed under this expansion as

$$\begin{aligned}\nabla^\mu\nabla_\mu &= g^{\mu\nu}\left(\partial_\mu\nabla_\nu - \Gamma^\lambda_{\mu\nu}\nabla_\lambda + \frac{1}{2}\omega_{\mu ab}\Sigma^{ab}\nabla_\nu\right)\\ &= \partial^2 + \frac{1}{3}x^\lambda x^\sigma R^{\mu}{}_{\lambda}{}^{\nu}{}_{\sigma}\partial_\mu\partial_\nu - \frac{2}{3}x^\lambda R_\lambda{}^\mu\partial_\mu + \frac{1}{2}x^\lambda R_\lambda{}^\mu{}_{ab}\Sigma^{ab}\partial_\mu\\ &\quad -\frac{1}{6}x^\lambda x^\sigma R_\lambda{}^\mu R_{\sigma\mu ab}\,\Sigma^{ab} + \frac{1}{16}x^\lambda x^\sigma R_{\lambda\mu ab}R_\sigma{}^\mu{}_{cd}\,\Sigma^{ab}\,\Sigma^{cd}.\end{aligned} \quad (11.38)$$

Therefore, the operator K (11.33) can be decomposed into the Laplacian in flat spacetime and a potential term V involving the curvature at the origin as in (11.35). As the V part, extract only the terms containing Σ^{ab} that remain after multiplying by γ_5 and taking the trace, which are written as $V = V_1 + V_2$, so that

$$\begin{aligned}V_1 &= \frac{1}{2}x^\lambda R_{\lambda\mu ab}\,\Sigma^{ab}\partial^\mu + \mathcal{F}_{ab}\,\Sigma^{ab},\\ V_2 &= -\frac{1}{6}x^\lambda x^\sigma R_\lambda{}^\mu R_{\sigma\mu ab}\,\Sigma^{ab} + \frac{1}{16}x^\lambda x^\sigma R^\mu{}_{\lambda ab}\,R_{\mu\sigma cd}\,\Sigma^{ab}\Sigma^{cd}.\end{aligned} \quad (11.39)$$

Using the heat kernel (11.36) in flat spacetime, the solution to the heat conduction equation is given by the integral equation $G = G_0 + G_0VG$. Solving this iteratively get a solution as

$$\begin{aligned}G^{(s)}(0,0) &= G_0^{(s)}(0,0) + \int_0^s dt \int d^4x\, G_0^{(s-t)}(0,x)V(x)\,G_0^{(t)}(x,0)\\ &+ \int_0^s dt \int d^4x\, G_0^{(s-t)}(0,x)V(x)\int_0^t du \int d^4y\, G_0^{(t-u)}(x,y)V(y)\,G_0^{(u)}(y,0)\\ &+\cdots.\end{aligned}$$

From this, the parts that remains after calculating $\mathrm{Tr}[\gamma_5\, G^{(s)}(0,0)]$ are the following two terms:

$$\mathrm{Tr}\left[\gamma_5 \int_0^s dt \int d^4x\, G_0^{(s-t)}(0,x)\, V_1(x) \int_0^t du \int d^4y\, G_0^{(t-u)}(x,y) V_1(y)\, G_0^{(u)}(y,0)\right]$$
$$+\mathrm{Tr}\left[\gamma_5 \int_0^s dt \int d^4x\, G_0^{(s-t)}(0,x) V_2(x)\, G_0^{(t)}(x,0)\right]. \tag{11.40}$$

Note here that K is expanded assuming that x and y are small, but since contributions from these integrals are concentrated near the origin when s is infinitesimal, the result does not change in the end even if the integration domain is $(-\infty, \infty)$. Therefore, it can be calculated using the integral formulas $\int d^4x\, e^{-ax^2} = \pi^2/a^2$, $\int d^4x\, x^a e^{-ax^2} = 0$, $\int d^4x\, x^2 e^{-ax^2} = 2\pi^2/a^3$, and $\int d^4x\, x^a x^b e^{-ax^2} = \delta^{ab}\pi^2/2a^3$.

First, calculating the second integral in (11.40), the contribution from the first term of V_2 in (11.39) disappears when we take the trace, and from the second term we obtain a term that does not depend on s such that

$$\mathrm{Tr}[\gamma_5\, G^{(s)}(0,0)] = \frac{1}{768\pi^2}\epsilon^{abcd} R_{\mu\nu ab} R^{\mu\nu}{}_{cd}.$$

Next, when we perform the integration of the first term in (11.40), the gravitational part of V_1 disappears due to $R^{\mu\nu} R_{\mu\nu ab} = 0$, and only the gauge field part (11.37) remains. Since the origin can be chosen freely, if we return $x = 0$ to general x, we find that the spin 1/2 Dirac index (11.31) is given by

$$\begin{aligned}\mathcal{I}_{\frac{1}{2}} &= -\frac{1}{768\pi^2}\int d^4x \sqrt{g}\, \epsilon^{abcd} R_{\mu\nu ab} R^{\mu\nu}{}_{cd} - \frac{1}{32\pi^2}\int d^4x \sqrt{g}\, \epsilon^{abcd}\, \mathrm{Tr}\big(\mathcal{F}_{ab}\mathcal{F}_{cd}\big)\\ &= -\frac{1}{8}\tau - \frac{1}{8\pi^2}\int \mathrm{Tr}\, \Omega \wedge \Omega,\end{aligned}$$

where $\Omega = \mathcal{F}_{ab}\, e^a \wedge e^b/2$.

General formula for even dimensions The method solving iteratively as described above is easy to understand, but it becomes complicated as spacetime dimension increases. Here we derive the $2n$ even-dimensional heat kernel using a more general method. At this time, the calculation is simplified by assuming that the origin is Ricci flat. In fact, as can be inferred from the above calculation, in this type of calculation, terms containing the Ricci tensor is generally ignored as they disappear when Tr is taken. The expansion formula (11.38) of the Laplacian can be used as it is for general dimensions by setting $R_{\mu\nu} = 0$.

In order to solve the heat conduction equation (11.34), assume the following form as a solution:

$$G^{(s)}(x,0) = \exp\left(\frac{1}{2}P_{\mu\nu}(s)x^\mu x^\nu + Q(s)\right), \tag{11.41}$$

where $P_{\mu\nu} = P_{\nu\mu}$. The quantity we want to derive is $Q(s)$ that remains at $x = 0$. In this case, the left-hand side of the heat conduction equation (11.34) becomes

$$\frac{\partial}{\partial s}G^{(s)}(x,0) = \left(\frac{1}{2}\frac{\partial P_{\mu\nu}}{\partial s}x^\mu x^\nu + \frac{\partial Q}{\partial s}\right)G^{(s)}(x,0).$$

On the other hand, the right-hand side is given by

$$\begin{aligned} -\ KG^{(s)}(x,0) &= g^{\mu\nu}\nabla_\mu\nabla_\nu G^{(s)}(x,0) \\ &= \left[P^\mu_{\ \mu} + \left(P_\mu^{\ \lambda}P_{\lambda\nu} + \frac{1}{3}R^\lambda{}_\mu{}^\sigma{}_\nu P_{\lambda\sigma} + \frac{1}{2}\bar{R}_\mu^{\ \lambda}P_{\lambda\nu} - \frac{1}{16}\bar{R}_\mu^{\ \lambda}\bar{R}_{\lambda\nu}\right)x^\mu x^\nu\right]G^{(s)}(x,0), \end{aligned}$$

where $\bar{R}_{\mu\nu} = R_{\mu\nu ab}\Sigma^{ab}$. If denoting a trace for the indices μ, ν as tr, then $\mathrm{tr}(\bar{R}P) = 0$ because $\bar{R}_{\mu\nu}$ is an antisymmetric matrix. Furthermore, noting that the Ricci tensor is now set to zero, we can find the following equation:

$$\frac{1}{2}\frac{\partial}{\partial s}\mathrm{tr}\,P = \mathrm{tr}\,P^2 - \frac{1}{16}\mathrm{tr}\,\bar{R}^2, \qquad \frac{\partial}{\partial s}Q = \mathrm{tr}\,P.$$

Solving the first equation reveals that

$$P = -\frac{1}{4}\left(\bar{R}\coth\frac{\bar{R}s}{2}\right) = -\frac{1}{2s}I - \frac{1}{24}\bar{R}^2 s + \frac{1}{1440}\bar{R}^4 s^3 + \cdots \tag{11.42}$$

is a solution, where I represents the Kronecker delta $\delta_{\mu\nu}$. Using this to solve the second equation, we get

$$\begin{aligned} Q &= \mathrm{tr}\left[-\frac{1}{2}\log\left(\sinh\frac{\bar{R}s}{2}\right) + C\right] \\ &= \mathrm{tr}\left(C - \frac{1}{2}\log\frac{\bar{R}s}{2} - \frac{1}{48}\bar{R}^2 s^2 + \frac{1}{5760}\bar{R}^4 s^4 + o(s^6)\right). \end{aligned} \tag{11.43}$$

Since the initial condition when substituting (11.42) into (11.41) and taking the limit of $s \to 0$ should be $\lim_{s\to 0} G^{(s)}(x,0) = \exp(-x^2/2s)/(4\pi s)^n = \delta^{2n}(x)$, a condition for determining the unknown constant C is given by $\lim_{s\to 0} Q(s) = -\log(4\pi s)^n = \mathrm{tr}[-\log(4\pi s I)/2]$. Applying this condition to (11.43), we obtain $C = \log(\bar{R}/8\pi)/2$. Therefore, using $\exp(\mathrm{tr}\log X) = \det X$, the heat kernel at $x = 0$ is found as

$$\begin{aligned} G^{(s)}(0,0) = e^{Q(s)} &= \frac{1}{(4\pi s)^n}\sqrt{\det\frac{\bar{R}s/2}{\sinh\bar{R}s/2}} \\ &= \frac{1}{(4\pi)^n}\frac{1}{s^n}\left(1 - \frac{s^2}{48}\mathrm{tr}\,\bar{R}^2 + o(s^4)\right). \end{aligned}$$

If the $2n$-dimensional chirality operator that anticommutes with all γ^a is defined as

$$\gamma_{2n+1} = -i^n \gamma^0 \gamma^1 \cdots \gamma^{2n-1},$$

so that the Hermitian property $\gamma_{2n+1}^\dagger = \gamma_{2n+1}$ and $\gamma_{2n+1}^2 = 1$ are satisfied, then the spin $1/2$ Dirac index is given by

$$\begin{aligned}\mathcal{I}_{\frac{1}{2}} &= -\lim_{s\to 0} \int d^{2n}x \sqrt{g}\, \mathrm{Tr}\big[\gamma_{2n+1} \exp(s\bar{\mathcal{F}}) G^{(s)}(x,x)\big] \\ &= \lim_{s\to 0} \frac{-1}{(4\pi s)^n} \int d^{2n}x \sqrt{g}\, \mathrm{Tr}\left[\gamma_{2n+1} \sqrt{\det \frac{\bar{R}s/2}{\sinh \bar{R}s/2}} \exp(s\bar{\mathcal{F}})\right],\end{aligned}$$

where the contribution from the gauge field $\bar{\mathcal{F}} = \mathcal{F}_{ab}\, \Sigma^{ab}$ is also added. The trace can be taken with $\Sigma^{ab} = -[\gamma^a, \gamma^b]/4$ and $\mathrm{Tr}(1) = 2^n$ in mind, so that only the part with n Σ^{ab} remains as

$$\begin{aligned}\mathrm{Tr}\big(\gamma_{2n+1} \Sigma^{a_1 b_1} \Sigma^{a_2 b_2} \ldots \Sigma^{a_n b_n}\big) &= \left(-\frac{1}{2}\right)^n \mathrm{Tr}\big(\gamma_{2n+1} \gamma^{a_1} \gamma^{b_1} \gamma^{a_2} \gamma^{b_2} \ldots \gamma^{a_n} \gamma^{b_n}\big) \\ &= -i^n \epsilon^{a_1 b_1 a_2 b_2 \cdots a_n b_n},\end{aligned}$$

thus a completely antisymmetric tensor appears. The s-dependence then cancels out and disappears.

Paying attention to this fact and $e^{a_1} \wedge e^{b_1} \wedge \cdots \wedge e^{a_n} \wedge e^{b_n} = d^{2n}x \sqrt{g}\, \epsilon^{a_1 b_1 a_2 b_2 \cdots a_n b_n}$, we find that instead of taking $-\mathrm{Tr}\, \gamma_{2n+1}$, it is sufficient to carry out a substitution such as $\Sigma^{ab} \to e^a \wedge e^b$ and then perform the $2n$-dimensional volume integration. In addition, the coefficient $(i/4\pi)^n$ remaining after integration is absorbed into the curvatures, that is, the following substitution is done:

$$\bar{R}^\mu{}_\nu \to \frac{i}{4\pi} R^\mu{}_{\nu ab}\, e^a \wedge e^b = \frac{i}{2\pi} \mathcal{R}^\mu{}_\nu, \qquad \bar{\mathcal{F}} \to \frac{i}{4\pi} \mathcal{F}_{ab}\, e^a \wedge e^b = \frac{i}{2\pi} \Omega.$$

In this way, the integrand is expressed in terms of the curvature 2-forms, $\mathcal{R}^\mu{}_\nu$ and Ω, and the Atiyah-Singer spin $1/2$ Dirac index can be written in the integral using the differential form as [EGH80, Nak90, PT09]

$$\mathcal{I}_{\frac{1}{2}} = \int_M \hat{A}(\mathcal{R}) \wedge ch(\Omega),$$

where

$$\hat{A}(\mathcal{R}) = \sqrt{\det \frac{i\mathcal{R}/4\pi}{\sinh i\mathcal{R}/4\pi}}, \qquad ch(\Omega) = \mathrm{Tr} \exp\left(\frac{i}{2\pi} \Omega\right) \tag{11.44}$$

are called the A-roof genus and the Chern character, respectively. M represents a $2n$-dimensional manifold with no boundary.

When $\hat{A}$ is expanded, it is expressed by a polynomial of the Pontrjagin class p_n as

$$\hat{A} = 1 - \frac{1}{24} p_1 + \frac{1}{5760} \left(7 p_1^2 - 4 p_2 \right) + \cdots .$$

In the case of four dimensions ($n = 2$), when volume integration is performed, only the second term remains and $\mathcal{I}_{1/2} = -P_1/24 = -\tau/8$ is obtained, where $P_1 = \int p_1$ is the first Pontrjagin number.

Chapter 12
Simplicial Quantum Gravity

The asymptotically background-free quantum gravity is a renormalizable quantum field theory formulated in a hybrid manner that combines conformal field theory, a typical nonperturbative method, and perturbation theory. As an approach to formulate this theory in a completely nonperturbative manner, a method based on dynamical triangulations has been proposed. This is a method of considering Euclidean spacetime as in lattice gauge theory and dividing it into random lattices for examination.

12.1 Dynamical Triangulation Method

The method of formulating the theory of gravity by discretizing spacetime was started by Regge in the 1960s [Reg61]. The initial idea was to express a variety of spacetime by changing the length of each link while fixing the connection pattern of the lattice. However, it is recognized that this method cannot sufficiently express all possible spacetime configurations required for the path integral over the gravitational field. In addition, there is a problem with what integral measure to use when performing it.

As an alternative to the Regge calculus, "simplicial quantum gravity" has been proposed, in which spacetime is divided into a random lattice, rather than changing the length of each link. An element that composes a lattice is given by a simplex as shown in Fig. 12.1, and n-dimensional spacetime is made by gluing a number of n-simplices. For example, when triangles that are typical of 2-simplices are glued, a two-dimensional surface as shown in Fig. 12.2 can be constructed. Here, we consider a spacetime with no boundary and fix its topology.

What is significant here is a variety of gluing patterns of simplices, not their individual shapes. Simplices are taken to be equivalent with the same volume. Spacetime exists inside simplices on average, and it is thought that a smooth spacetime can be obtained by taking a continuum limit. Quantization is done in the spirit of the path

K. Hamada, *Trans-Planckian Physics and Inflation*, Fundamental Theories of Physics 223, https://doi.org/10.1007/978-981-96-3476-7_12

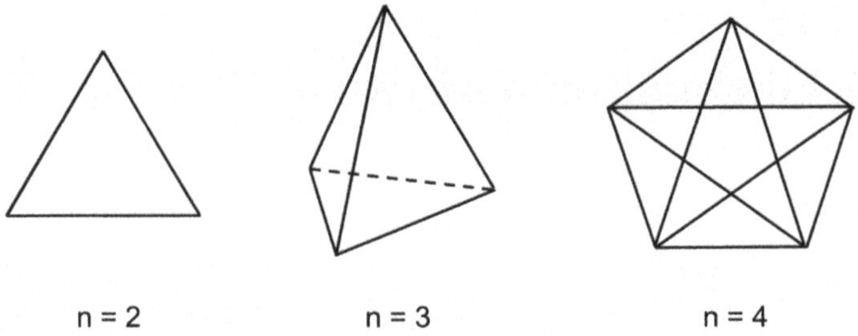

Fig. 12.1 Elements of n-dimensional simplicial quantum gravity

Fig. 12.2 Two-dimensional random surface (dotted line) and Feynman diagram dual to it (double line with arrows)

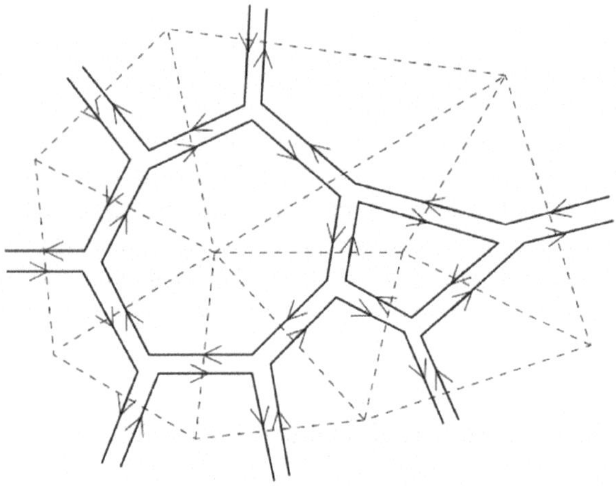

integral. Different gluing patterns represent locally different spacetime configurations, and by taking the sum over all combinatorial configurations, i.e., all possible triangulations T, the partition function of quantum gravity is expressed as

$$Z_{\text{SG}} = \sum_T e^{-\mathcal{I}_{\text{G}} - \mathcal{I}_{\text{M}}}, \tag{12.1}$$

where $\mathcal{I}_{\text{G}}$ is an action of gravity determined by spacetime configurations, and $\mathcal{I}_{\text{M}}$ is a matter action defined on the simplicial manifold. For topology, we mainly consider the sphere.

12.1.1 Two-Dimensional Simplicial Quantum Gravity

This method was first implemented successfully in two dimensions [Weig82, Dav85, ADF85, KKM85, BKKM86]. Let N_2 be the number of triangles forming a surface, and let T_2 be the set of all possible triangulations, i.e., the possible gluing patterns of N_2 pieces. Letting N_0 be the number of vertices (0-simplices) and N_1 be the number of links (1-simplices), the constructed surfaces satisfy the following conditions:

$$3N_2 = 2N_1,$$
$$N_0 - N_1 + N_2 = \chi,$$

where χ is the Euler characteristic. Since there are two equations for three variables N_i $(i = 0, 1, 2)$, there is only one independent variable, and let it be N_2.

The action is given by $\mathcal{I}_G = \lambda N_2$ of the area element only, and the partition function is defined by

$$Z_{SG_2}(\lambda) = \sum_{N_2=0}^{\infty} \sum_{T_2} e^{-\lambda N_2} = \sum_{N_2=0}^{\infty} \Omega(N_2) e^{-\lambda N_2}, \tag{12.2}$$

where $\Omega(N_2)$ is the number of states in T_2. This is known as a statistical model in which a second-order phase transition occurs. Denoting its critical point by $\lambda = \lambda^c$, the state number factor behaves in the large N_2 as

$$\Omega(N_2) \sim N_2^{\gamma_{st}^{(2)}-3} e^{\lambda^c N_2}, \tag{12.3}$$

where the exponent $\gamma_{st}^{(2)}$ is called the string susceptibility. Letting area of each triangle be 1, N_2 becomes area of the surface, and thus λ acts as a cosmological constant that controls it. Introducing a scale a corresponding to lattice spacing, two-dimensional quantum gravity is then derived by taking a continuum limit that $N_2 \to \infty$ at the same time as $a \to 0$ while retaining $(\lambda - \lambda^c)/a^2 \to \lambda_{cos}$ at the critical point, where λ_{cos} is a physical cosmological constant.

It is known that two-dimensional random surfaces have a one-to-one correspondence with Feynman diagrams when considering dual diagrams as shown in Fig. 12.2. The partition function (12.2) is then expressed by a matrix model, the path integral over a zero-dimensional $N \times N$ Hermitian matrix M with the kinetic term $\mathrm{tr}(M^2)/2$ and an interaction term $g\,\mathrm{tr}(M^3)/\sqrt{N}$, as

$$F_N(g) = \frac{1}{N^2} \log \int d^{N^2}M \, e^{-\mathrm{tr}\left(\frac{1}{2}M^2 + \frac{g}{\sqrt{N}}M^3\right)}.$$

Each three-point interaction in Feynman diagrams corresponds to a triangle, and a vacuum diagram with N_2 interactions is identified with a compact random surface with area N_2. Various vacuum diagrams represent how triangles are connected, namely, T_2. Topologies are classified according to 't Hooft's $1/N$-expansion [tHo74] as $F_N \propto N^{\chi-2}$. For large N, only planar Feynman diagrams corresponding to random surfaces of S^2 with $\chi = 2$ remain.

Since this path integral is zero-dimensional, it can be evaluated exactly and vacuum Feynman diagrams can be summed over at all orders [BIPZ78, BIZ80]. The second-order phase transition point g_c is defined as a point where analyticity of g is broken. In general, for finite N, the path integral is not well-defined because the potential term is not bounded below. However, taking the limit $N \to \infty$, contributions from the integral measure of the $N \times N$ matrix dominate,[1] so that the analyticity extends to $g \neq 0$ and the breaking point appears at a value $g_c = -e^{\lambda^c}$. This negativity is essential when identifying $\lim_{N\to\infty} F_N(g = -e^{\lambda})$ with the simplicial gravity partition function (12.2),[2] and its continuum limit can be taken at g_c [BK90, DS90, GM90].

On the other hand, in two-dimensional quantum gravity [DK89, Dav88], we can find the behavior of the partition function in a large area by focusing on the zero-mode ϕ_0 of the Liouville field. The ϕ_0-dependence of the Liouville action (6.25), except the cosmological term, is given by $S_{\mathrm{L}} = b_{\mathrm{L}}\chi\phi_0$, and when fixed the value of $v_2 = e^{\gamma_0\phi_0}$, the partition function has the following v_2-dependence:

$$\int d\phi_0 e^{-b_{\mathrm{L}}\chi\phi_0}\delta(e^{\gamma_0\phi_0} - v_2) \sim v_2^{-1-b_{\mathrm{L}}\chi/\gamma_0},$$

where γ_0 is the Liouville charge (6.31) of the cosmological term. Paying attention to $v_2 \sim N_2$, we can find that the string susceptibility is expressed as $\gamma_{\mathrm{st}}^{(2)} = 2 - b_{\mathrm{L}}\chi/\gamma_0$. If we consider spherical toporogy of $\chi = 2$ and let c be the central charge of conformal field theory [BPZ84, FQS84] coupled with gravity, then the coefficient of the Liouville action is $b_{\mathrm{L}} = (25 - c)/6$, and thus we get

$$\gamma_{\mathrm{st}}^{(2)} = 2 + \frac{1}{12}\Big[c - 25 - \sqrt{(25 - c)(1 - c)}\Big]. \tag{12.4}$$

This is consistent with the analytical results of the matrix model. At this time, matrix models corresponding to various central charges $c\ (\leq 1)$ are constructed by introducing spin degrees of freedom onto random lattices by increasing the number of matrix types [Kaz86, BK87].

In addition, it has been shown that correlation functions derived from the matrix model just on the phase transition point coincide with those of two-dimensional quantum gravity [GL91].

---

1 The path integral can be rewritten in terms of N real eigenvalues λ_i of M [Meh67]. The measure is then expressed as $\prod_i d\lambda_i \Delta^2(\lambda)$, where $\Delta(\lambda) = \prod_{1\leq i<j\leq N}(\lambda_i - \lambda_j)$ is the Vandermonde determinant. This determinant gives a logarithmic interaction, when it is rewritten in exponential form.

2 A four-point interaction $g\,\mathrm{tr}(M^4)/N$ which corresponds to a quadrangle can be considered as well. Since it is bounded below, the path integral is analytic for $g \geq 0$. However, as in the case of the three-point interaction, the analyticity extends to the negative region at $N \to \infty$. It has been confirmed that just on the critical point, this model shows the same universality as the three-point interaction [BK87].

12.1.2 Four-Dimensional Simplicial Quantum Gravity

Following the success in two dimensions, extension to higher dimensions was actively examined in the 1990s [AM92a, AM92b, AJ92, CKR94, EHI+97]. A four-dimensional manifold by the dynamical triangulation method is represented by the number of i-simplices, N_i $(i = 0, 1, \ldots, 4)$.

These five variables are not independent due to the following conditions for the manifold without boundaries.[3] First, two 4-simplices are attached through one 3-simplex, and inversely, any 3-simplex is shared by two 4-simplices. This condition gives

$$5N_4 = 2N_3$$

In addition, from the condition that 4-simplices sharing a vertex form 4-ball, we get

$$-2N_1 + 3N_2 - 4N_3 + 5N_4 = 0.$$

Furthermore, Euler relation yields

$$N_0 - N_1 + N_2 - N_3 + N_4 = \chi.$$

For the S^4 topology, $\chi = 2$.

Since three relationships hold for five variables, there are two independent variables. We choose them as N_2 and N_4, and express the others as

$$\begin{aligned} N_0 &= \chi + \frac{1}{2}N_2 - N_4, \\ N_1 &= \frac{3}{2}N_2 - \frac{5}{2}N_4, \\ N_3 &= \frac{5}{2}N_4. \end{aligned}$$

The action of gravity is then expressed in terms of N_2 and N_4 as

$$\mathcal{I}_G = -\kappa_2 N_2 + \lambda_4 N_4, \tag{12.5}$$

where κ_2 corresponds to the square of the Planck mass, and λ_4 corresponds to the cosmological constant.

Letting T_4 be all possible triangulations with volume N_4, the partition function is defined from the action (12.5) as

[3] There should be no two d-simplices $(d = 1, 2, 3, 4)$ having the same set of vertices.

$$Z_{\mathrm{SG}_4}(\kappa_2, \lambda_4) = \sum_{N_4=0}^{\infty} \sum_{T_4} e^{\kappa_2 N_2 - \lambda_4 N_4} = \sum_{N_4=0}^{\infty} \Omega(\kappa_2, N_4) e^{-\lambda_4 N_4}. \tag{12.6}$$

If the second-order phase transition occurs at $\lambda_4 = \lambda_4^c(\kappa)$, then the number of states at this point is written for large N_4 as

$$\Omega(\kappa_2, N_4) \sim N_4^{\gamma_{\mathrm{st}}^{(4)} - 3} e^{\lambda_4^c(\kappa_2) N_4}, \tag{12.7}$$

where $\gamma_{\mathrm{st}}^{(4)}$ is a four-dimensional version of the string susceptibility.

The four-dimensional simplicial quantum gravity defined in this way is considered to represent spacetime states in the ultraviolet limit of the asymptotically background-free quantum gravity. Predictions from quantum gravity can be obtained from the behavior of the zero-mode of the conformal-factor field, just as in the two-dimensional case. The action of the conformal-factor field in the ultraviolet limit where the coupling constant t disappears becomes the Riegert action plus the Einstein-Hilbert term as a potential given by the quantum scalar curvature $e^{\gamma_2 \phi} \mathcal{R}$ with the Riegert charge γ_2 (6.21). Therefore, the zero-mode dependence, except the cosmological term, is provided by $S_{\mathrm{QG}} = 2b_1 \chi \phi_0 - \kappa_2 e^{\gamma_2 \phi_0} \mathcal{R}$, and the $v_4 = e^{\gamma_0 \phi_0}$ dependence of the partition function can be found as

$$\begin{aligned} &\int d\phi_0 \exp\left\{-2b_1 \chi \phi_0 + \kappa_2 e^{\gamma_2 \phi_0} \mathcal{R}\right\} \delta(e^{\gamma_0 \phi_0} - v_4) \\ &\sim v_4^{-1-2b_1\chi/\gamma_0} f(\kappa_2 v_4^{\gamma_2/\gamma_0}), \end{aligned}$$

where b_1 is the coefficient (2.22) of the Riegert action S_{R}, and γ_0 is the Riegert charge (6.21) of the cosmological term operator. From this, the susceptibility is expressed as $\gamma_{\mathrm{st}}^{(4)} = 2 - 2b_1 \chi / \gamma_0$. For $\chi = 2$, we obtain

$$\gamma_{\mathrm{st}}^{(4)} = 2 - \frac{1}{2}\left(b_1 + \sqrt{b_1^2 - 4b_1} \right). \tag{12.8}$$

12.2 Grand-Canonical Monte-Carlo Quantization

Quantization of the simplicial quantum gravity is carried out by taking the sum over all possible triangulations. The sum is performed by the Monte Carlo method creating a Markov chain by the Pachner moves [Pac91, GV92].

In two dimensions, any two configurations in T_2 are connected by combinations of the three Pachner moves, $(p, q) = (1, 3)$, $(2, 2)$, and $(3, 1)$. Figure 12.3 shows these moves, and the parameters p and q represent the number of 2-simplices annihilated and created by each move, respectively.

In four dimensions, any two configurations in T_4 are connected by combinations of the five Pachner moves, $(p, q) = (1, 5)$, $(2, 4)$, $(3, 3)$, $(4, 2)$, and $(5, 1)$, where p and

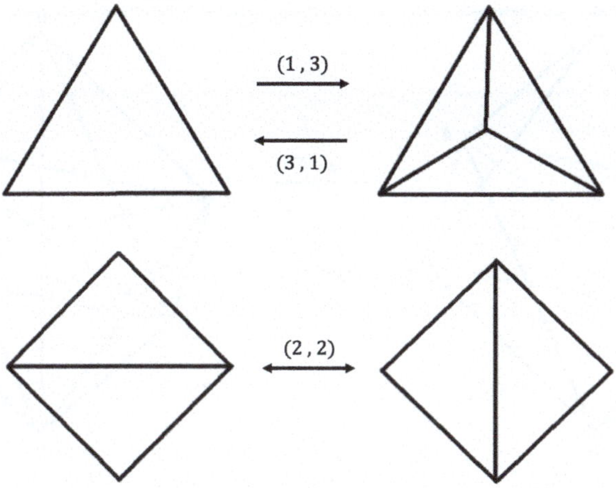

Fig. 12.3 Three types of the Pachner moves in two dimensions

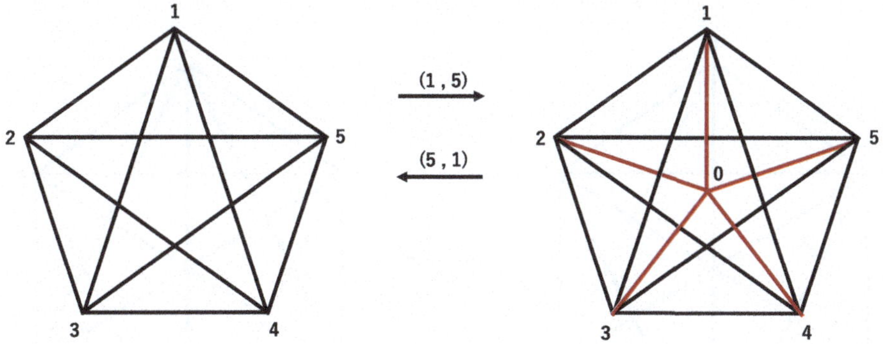

Fig. 12.4 The Pachner (1, 5) move splits a 4-simplex (12345) into five 4-simplices (01234), (01235), (01245), (01345), and (02345) by introducing a vertex 0. The (5, 1) move is the reverse

q represent the number of 4-simplices annihilated and created by each move, respectively. Figures 12.4, 12.5, and 12.6 show these five moves, and Table 12.1 describes changes in variables by each move. These moves conserve the topology of the manifold.

Let us describe how to calculate the partition function (12.6) of the four-dimensional simplicial quantum gravity, and the two dimensional case as well. Since the moves except (3, 3) involve changes of N_4, we should be employ the grand-canonical Monte-Carlo method.[4] Let $w_{a\to b}$ and $w_{b\to a}$ be the transition probabilities of the Markov process between neighboring configurations a and b. In order to

[4] To simplify the calculation, a quasi-canonical Monte-Carlo method is still widely used, in which N_4 is constrained to fluctuate around some target volume $\bar{N}_4$. This is done by adding a volume fixing term $\delta\mathcal{I} = \delta\kappa_4(N_4 - \bar{N}_4)^2$ to the action (12.5), where $\delta\kappa_4$ is a free parameter. However, this method has limitations on finite-size effects and ergodic properties. Furthermore, the size-dependence of

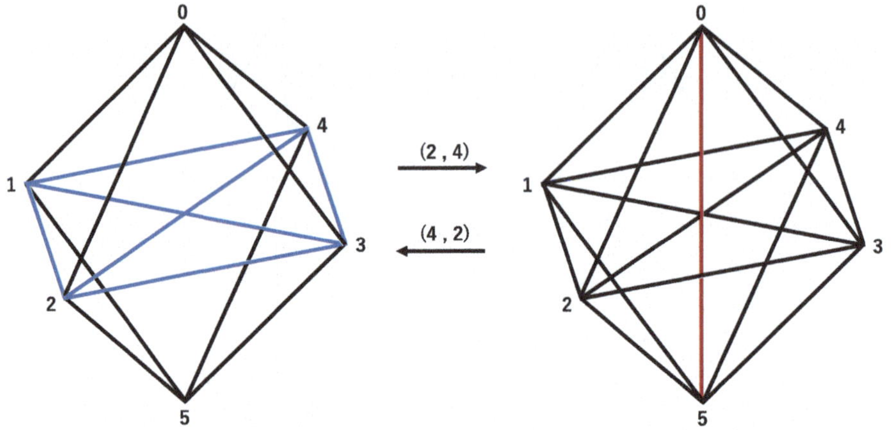

Fig. 12.5 The Pachner (2, 4) move changes two 4-simplices (01234) and (12345), which share the tetrahedron (1234), to four 4-simplices (01235), (01245), (01345), and (02345) by introducing a edge (05). The (4, 2) move is the reverse

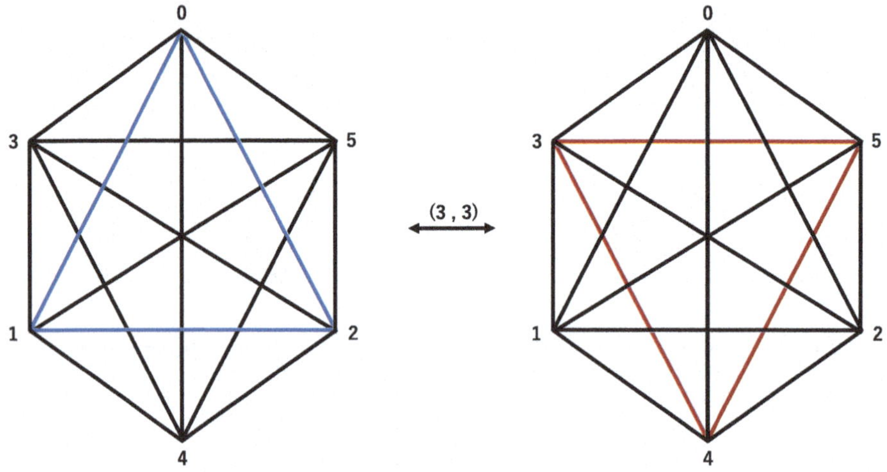

Fig. 12.6 The Pachner (3, 3) move changes a configuration of three 4-simplices (01234), (01245), and (01235) with a triangle (012) to a different configuration of three 4-simplices (01345), (12345), and (02345) with a triangle (345) and vice-versa

achieve the equilibrium distribution $P_a \sim e^{-\mathcal{I}_a}$, the detailed balance condition

$$\frac{1}{n_a} P_a w_{a \to b} = \frac{1}{n_b} P_b w_{b \to a}$$

---

the partition function becomes unclear. To improve numerical accuracy, it is necessary to perform the grand-canonical method with varying the volume [HEY03].

Table 12.1 Changes of variables by each Pachner move in four dimensions

(p, q)	δN_0	δN_1	δN_2	δN_3	δN_4
$(1, 5)$	1	5	10	10	4
$(2, 4)$	0	1	4	5	2
$(3, 3)$	0	0	0	0	0
$(4, 2)$	0	-1	-4	-5	-2
$(5, 1)$	-1	-5	-10	-10	-4

should be satisfied. Here, the additional factor n_a takes care of changes in number of states caused by the change of degrees of freedom, which is given by the number of all possible moves starting from a configuration a,

$$n_a = n_a^{(1,5)} + n_a^{(2,4)} + n_a^{(3,3)} + n_a^{(4,2)} + n_a^{(5,1)},$$

where $n_a^{(p,q)}$ is the number of (p, q)-moves therein. This calculation is performed using the Metropolis method.

Moreover, let us add matter fields on the manifold. Consider N_X scalar fields $\{X_i^{(\alpha)}\}$ $(\alpha = 1, \cdots, N_X)$ on vertices, and N_A vector fields $\{A_{ij}^{(\beta)}\}$ $(\beta = 1, \cdots, N_A)$ on links with $A_{ji}^{(\beta)} = -A_{ij}^{(\beta)}$ [BBK+98, HETY01]. Their actions are given by

$$\mathcal{I}_X = \frac{1}{2} \sum_{\alpha=1}^{N_X} \sum_{i,j} \mathcal{N}(l_{ij}) \left\{ X_i^{(\alpha)} - X_j^{(\alpha)} \right\}^2,$$

$$\mathcal{I}_A = \frac{1}{2} \sum_{\beta=1}^{N_A} \sum_{i,j,k} \mathcal{N}(t_{ijk}) \left\{ A_{ij}^{(\beta)} + A_{jk}^{(\beta)} + A_{ki}^{(\beta)} \right\}^2, \tag{12.9}$$

where the factors $\mathcal{N}(l_{ij})$ and $\mathcal{N}(t_{ijk})$ are numbers of 4-simplices sharing the link l_{ij} and the triangle t_{ijk}, respectively. These actions have a correct thermodynamical limit for large N_4 because they are proportional to N_4. The path integral of the matter field is performed using the heat-bath method together with the over-relaxation technique.

12.2.1 Numerical Calculation Results

We first present the results of numerical simulations using the grand-canonical method for the two-dimensional simplicial quantum gravity [OTY98, HHY06b]. Considering a system with n $(= 1, 2)$ Ising spins whose central charge becomes $c = n/2$ or a system with m $(= 1, 2)$ scalar fields such that $c = m$, the string susceptibility $\gamma_{\text{st}}^{(2)}$ is extracted from logarithm of state numbers (12.3) for them.

Fig. 12.7 Numerical results of the string susceptibility $\gamma_{st}^{(2)}$ for two-dimensional simplicial quantum gravity

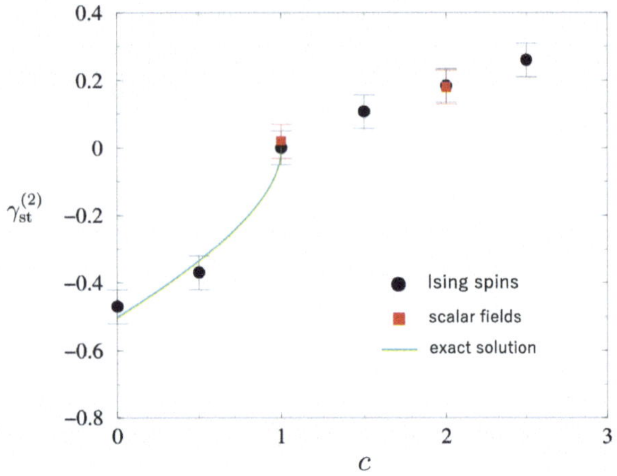

Figure 12.7 shows the results, and as far as $c \leq 1$ is concerned, they are in good agreement with the exact solution (12.4).

For $1 < c < 25$, the string susceptibility (12.4) is predicted to be a complex number reflecting an unstable configuration, while the numerical simulation gives positive real numbers, indicating the emergence of a different phase involving surface instability specific to the simplicial model.

In four-dimensional simplicial quantum gravity, the existence of a second-order phase transition necessary to take the continuum limit is not as obvious as in the case of two-dimensional surfaces. Horata, Egawa, and Yukawa constructed a manifold with the S^4 topology by gluing together more than 10,000 4-simplices, and conducted numerical simulations by incorporating matter fields into the manifold [EHY02, HEY03, HHY06b]. As a result, it was shown that although the phase transition is first-order in a system consisting of simplices only, a second-order phase transition occurs when matter fields are added (see Fig. 12.8). The change of order of the phase transition is also expected from the sign of the susceptibility $\gamma_{st}^{(4)}$ (12.7) changing to negative.

Figure 12.9 shows the numerical results of $\gamma_{st}^{(4)}$, evaluated just on the critical line, where the corresponding b_1 values are plotted using the formula (12.8). The straight line drawn in it is a regression curve, given by

$$b_1 = 0.0030(3)\,(N_X + 62N_A) + 3.98(3). \tag{12.10}$$

On the other hand, the exact value of the coefficient b_1 of the Riegert action in the ultraviolet limit is given by (2.23), which is expressed numerically as $b_1 = 0.00278\,(N_X + 62N_A) + 4.27$, where the breakdown of the last term is 4.35 from the Weyl action and -0.08 from the Riegert action, derived by quantizing each. Thus, we can find that the numerical result is consistent with this value derived from the asymptotically background-free quantum gravity.

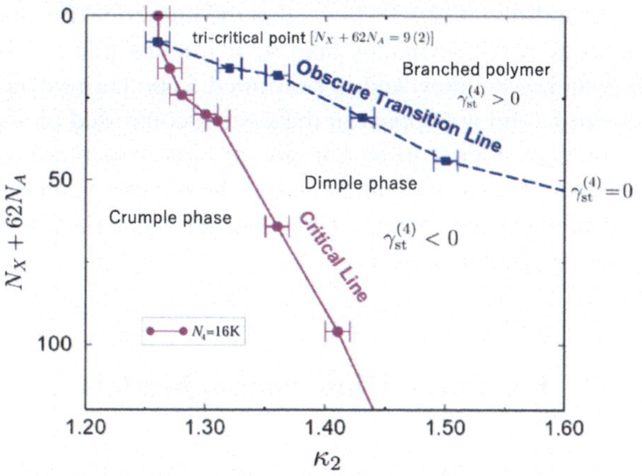

Fig. 12.8 Phase diagram of four-dimensional simplicial quantum gravity as a function of κ_2 and $N_X + 62N_A$, where N_X and N_A represent the number of scalar fields and $U(1)$ gauge fields, respectively. The critical line of the second-order phase transition appears where the susceptibility $\gamma_{st}^{(4)}$ becomes negative

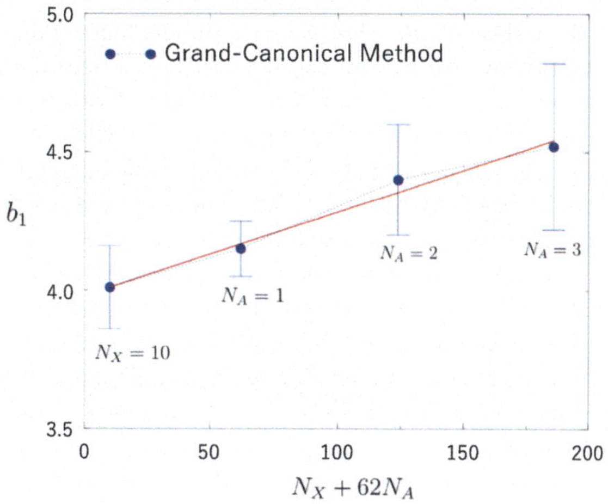

Fig. 12.9 Numerical results for the susceptibility $\gamma_{st}^{(4)}$. Using the formula $\gamma_{st}^{(4)} = 2 - \left(b_1 + \sqrt{b_1^2 - 4b_1}\right)/2$ (12.8) derived from the continuum theory, the results are displayed in terms of b_1. The straight line is an estimate of b_1 as a linear function of the combination $N_X + 62N_A$

When comparing the two approaches, it should be noted that the Boltzmann weight $e^{\kappa_2 N_2 - \lambda_4 N_4}$ in four-dimensional simplicial quantum gravity only has parts corresponding to the Einstein-Hilbert action and the cosmological term in the continuum theory. In other words, information on the dimensionless fourth-derivative gravitational action is contained in the possible gluing patterns T_4. As mentioned in Chap. 2, this corresponds to the fact that the fourth-derivative part $I^{(4)}$ of the action I (2.5) and the Wess-Zumino action S are purely quantum quantities that do not depend on $\hbar$, and if the continuum theory of (12.1) is written as

$$Z_{QG} = \int \mathcal{D}g_{\mu\nu} e^{-\mathcal{I}_{EG}},$$

then the fourth-derivative part can be considered as part of the path integral measure $\mathcal{D}g_{\mu\nu}$ as in (2.31).

In addition, (12.10) shows that the numerical result of b_1 in the absence of matter fields is slightly smaller than 4. It implies that the susceptibility (12.8) becomes a complex number, and is presumed to be the reason why the second-order phase transition did not appear in the system composed of 4-simplices only. However, this can be considered to be a temporary phenomenon that occurs because the number of 4-simplices is still small, and if the number is increased and the lattice spacing is substantially narrowed, a second-order phase transition point may appear under the grand-canonical method.

12.3 On Three-Dimensional Models

Lastly, we briefly discuss three-dimensional quantum gravity. Unlike even dimensions, there is no gravitational scalar quantity whose volume integral is dimensionless in three dimensions. Therefore, a dimensionful parameter has to be introduced into the gravitational action, and the conformal field theory technique developed in two and four dimensions cannot be applied to three dimensions as it is.

As one of the methods using continuum fields, Witten's approach [Wit88] describing the gravity theory without the cosmological term and matter fields as a Chern-Simons theory [CS74] is widely known. It claims that three-dimensional gravity is a topological field theory with no local degrees of freedom. Corresponding to this, there are some models expressing topological invariants as a simplicial gravity such as the model by Ponzano and Regge [PR68] representing 6 sides of a tetrahedron by using the Racah-Wigner $6j$ symbol and the model by Turaev and Viro [TV92] that is a more mathematical development of it. Research on the relationship between these and the Chern-Simons theory is still ongoing.

On the other hand, Boulatov's model [Bou92] proposed as a three-dimensional simplicial quantum gravity is well known, which is defined by performing the sum over all possible gluing patterns of 3-simplices under a fixed topology, with adding the Boltzmann weights due to the cosmological term and the Einstein-Hilbert term as in (12.2) and (12.6). It aims to generalize the matrix model that succeeded in 2-simplices to a model with tensor legs, and to execute the sum in 3-simplices. However, since it is a fairly complicated model, efforts to analyse it have been made [Oog93] but they cannot be said to be progressing well. In addition, perhaps because there are no proper physical quantities that can be used for comparison, three-dimensional quantum gravity is primarily studied as a mathematical object, although numerical calculations of three-dimensional simplicial gravity have been performed [AV92, HTY98, Tho99].

Correction to: Trans-Planckian Physics and Inflation

Correction to:
K. Hamada, *Trans-Planckian Physics and Inflation*,
Fundamental Theories of Physics 223,
https://doi.org/10.1007/978-981-96-3476-7

The original version of the book was published with an incorrect volume number which has now been changed from "26" to "223". The book has been updated with the change.

The updated version of this book can be found at
https://doi.org/10.1007/978-981-96-3476-7

K. Hamada, *Trans-Planckian Physics and Inflation*, Fundamental Theories of Physics 223, https://doi.org/10.1007/978-981-96-3476-7_13

Finale

There is an old scientific thought called "Ockham's razor". That is a principle advocated by William of Ockham, a philosopher believed to have been born in the village of Ockham, England in the fourteenth century, stating that you should not make more assumptions than necessary to explain an event. A razor is used to mean cutting off unnecessary things.

The Ptolemaic theory, for instance, can be made as precise as you like by adding circular motions one after another if all the planets are on the same disk. This is, however, typical of parameter physics which introduces variables one after another to match observations, thus is against the thought of Occam's razor. In fact, when Pluto and comets with oblique orbits are added, it can no longer deal with them, and if you consider outside the solar system, it breaks down. Occam's razor does not judge authenticity, but it makes sense to think that it is better to be able to explain by simpler principles.

One of the guiding principles that have been respected throughout this book is diffeomorphism invariance. However, it seems that many people in the world put the Einstein equation first. Since the Einstein equation is beautiful and so clear that it can be used as a T-shirt design, there is a certain idolatry aspect, and there is a tendency that denying this equation is denying Einstein himself. That is, of course, wrong, and Einstein's greatest achievement was undoubtedly his discovery of diffeomorphism invariance, one of the fundamental principle of physics. That is never in conflict with quantum theory. The Einstein equation is one of equations derived from the principle, but it is just still incomplete.

Diffeomorphism invariance demands the existence of unphysical ghost modes. They are constantly involved in gravitational systems we consider, although many researchers, especially particle physicists, turn a blind eye to this fact. The expansion and structure formation of the universe are caused by the ghost mode that prevents the boundedness of the Einstein-Hilbert action. Gravitational objects would also not be formed without the ghost mode. Nevertheless, when considering quantum theory of gravity, ones start by removing the ghost mode. In fact, it is possible in Einstein's theory of gravity if we reduce the gravitational system to a system of

K. Hamada, *Trans-Planckian Physics and Inflation*, Fundamental Theories of Physics 223, https://doi.org/10.1007/978-981-96-3476-7

special relativity by applying the weak-field approximation, where physical modes are called gravitons. For this reason, many researchers consider that there is no problem with unitarity as long as we consider Einstein's theory of gravity. However, this is only true within the scope of the weak-field approximation, in which the effect of gravitons is assumed to be local, weak, and not to affect the entire spacetime. When gravity becomes stronger, singularities arise and unitarity breaks down.

When constructing quantum theory of gravity, diffeomorphism-invariant systems cannot be grasped as an extension of Lorentz-invariant ones, which is because the latter considers eigenstates of the Hamiltonian, while the former considers states where the Hamiltonian vanishes. In the Lorentz-invariant system, we only deal with physical eigenstates as quantum states, but when considered such a state in gravitational systems, there is only the vacuum state and nothing remains. In other words, ghost modes are essential to any gravitational system, although they are invisible. Describing a gravitational system as a Lorentz-invariant system is an approximation that can only be applied when gravity is so weak that back-reactions can be neglected. String theory has not escaped the curse of this method either. To construct a quantum theory of gravity, the gravitational field must be quantized as a field that satisfies the Hamiltonian constraint. This means quantizing spacetime itself, away from the perspective of particle physics.

In order to define quantum gravity, the path integral method is employed because it is possible to preserve diffeomorphism invariance. On the other hand, care must be taken when using the canonical quantization method as this point is unclear. In order for the path integral to be correctly defined, renormalizability is also necessary. Diffeomorphism invariance and renormalizability can be trusted as guiding principles due to the numerous achievements that have been derived from these. Renormalizablility of quantum gravity has long been ignored, but now is the time to consider deeply its physical implications.

In addition, renormalizability provides a theoretical foundation for spacetime being four-dimensional and continuous. We sometimes view the universe from outside the fourth dimension without realizing it. The idea that the universe was born at a certain moment is like looking at that moment from outside four-dimensional spacetime. However, we are in a four-dimensional world, and there is no way we can know whether there is a world outside the universe. An extra dimension is like an unknown continent, which adds an infinitely more extra degree of freedom than what needs to be explained.

Nobody knows how vast the universe is. The observed region is only a part of it. In the scenario of quantum gravity inflation, if we go back to the beginning of the universe, all of the currently visible range would be inside the correlation length ξ_Λ. In other words, it suggests that the universe began from a highly concentrated bubble of quantum gravity, inside which scale invariance was realized, remnants of which remain to this day. The gravitational interaction that we feel is extremely weak compared to other interactions, but its weakness can also be said to be the flip side of the fact that gravity has the ability to confine the entire universe in a small bubble. And now, it is amazing that we are in an age where such effects of quantum gravity can be observed.

This book is a specialized book focusing on the progress in the trans-Planckian physics that could not discussed in my book *Quantum Gravity and Cosmology Based on Conformal Field Theory* (Cambridge Scholar Publishing, Newcastle, 2018), and also serves as an introductory book to renormalizable and background-free quantum gravity. In publishing this book, I would like to thank Tetsuyuki Yukawa, and my family, Nonn and Kyouka, for supporting me. I also wish to thank Springer Nature Publishing for giving me the opportunity for publication.

Tsukuba, Japan
April 2025

Appendix A
Useful Gravitational Formulas

A.1 Formulas on Curvatures

The Lorentz signature employed in this book is $(-1, 1, \cdots, 1)$.[1] In this appendix, unless otherwise noted, dimensions of spacetime are arbitrary D. The Christoffel symbol and the Riemann tensor are defined by

$$\Gamma^{\lambda}_{\mu\nu} = \frac{1}{2} g^{\lambda\sigma} \left(\partial_\mu g_{\nu\sigma} + \partial_\nu g_{\mu\sigma} - \partial_\sigma g_{\mu\nu} \right),$$
$$R^{\lambda}{}_{\mu\sigma\nu} = \partial_\sigma \Gamma^{\lambda}{}_{\mu\nu} - \partial_\nu \Gamma^{\lambda}{}_{\mu\sigma} + \Gamma^{\lambda}_{\rho\sigma} \Gamma^{\rho}_{\mu\nu} - \Gamma^{\lambda}_{\rho\nu} \Gamma^{\rho}_{\mu\sigma},$$

respectively. The Ricci tensor is defined by $R_{\mu\nu} = R^{\lambda}{}_{\mu\lambda\nu}$ and the scalar curvature (Ricci scalar) is $R = R^{\mu}{}_{\mu}$. A covariant derivative is expressed using the Christoffel symbol as

$$\nabla_\mu A^{\sigma_1 \cdots \sigma_m}_{\lambda_1 \cdots \lambda_n} = \partial_\mu A^{\sigma_1 \cdots \sigma_m}_{\lambda_1 \cdots \lambda_n} - \sum_{j=1}^{n} \Gamma^{\nu_j}_{\mu\lambda_j} A^{\sigma_1 \cdots \sigma_m}_{\lambda_1 \cdots \nu_j \cdots \lambda_n} + \sum_{j=1}^{m} \Gamma^{\sigma_j}_{\mu\nu_j} A^{\sigma_1 \cdots \nu_j \cdots \sigma_m}_{\lambda_1 \cdots \lambda_n},$$

which satisfies

$$\left[\nabla_\mu, \nabla_\nu \right] A_{\lambda_1 \cdots \lambda_n} = \sum_{j=1}^{n} R_{\mu\nu\lambda_j}{}^{\sigma_j} A_{\lambda_1 \cdots \sigma_j \cdots \lambda_n}.$$

The Riemann tensor satisfies an antisymmetric property $R_{\mu\nu\lambda\sigma} = -R_{\nu\mu\lambda\sigma} = -R_{\mu\nu\sigma\lambda}$ and

[1] In Euclidean space, rewrite $\sqrt{-g}$ to $\sqrt{g}$, and also $\eta_{\mu\nu}$ to $\delta_{\mu\nu}$ for the flat metric.

K. Hamada, *Trans-Planckian Physics and Inflation*, Fundamental Theories of Physics 223, https://doi.org/10.1007/978-981-96-3476-7

$$R^{\mu}{}_{\nu\lambda\sigma} + R^{\mu}{}_{\lambda\sigma\nu} + R^{\mu}{}_{\sigma\nu\lambda} = 0,$$
$$\nabla_{\rho} R^{\mu}{}_{\nu\lambda\sigma} + \nabla_{\lambda} R^{\mu}{}_{\nu\sigma\rho} + \nabla_{\sigma} R^{\mu}{}_{\nu\rho\lambda} = 0,$$

where the last is the Bianchi identity, from which $\nabla_{\mu} R^{\mu}{}_{\lambda\nu\sigma} = \nabla_{\nu} R_{\lambda\sigma} - \nabla_{\sigma} R_{\lambda\nu}$ and $\nabla_{\mu} R^{\mu}_{\nu} = \nabla_{\nu} R/2$ are obtained.

The Weyl tensor is defined by

$$C_{\mu\nu\lambda\sigma} = R_{\mu\nu\lambda\sigma} - \frac{1}{D-2}\left(g_{\mu\lambda} R_{\nu\sigma} - g_{\mu\sigma} R_{\nu\lambda} - g_{\nu\lambda} R_{\mu\sigma} + g_{\nu\sigma} R_{\mu\lambda}\right) + \frac{1}{(D-1)(D-2)}\left(g_{\mu\lambda} g_{\nu\sigma} - g_{\mu\sigma} g_{\nu\lambda}\right) R. \tag{A.1}$$

This tensor vanishes no matter which index is contracted such as $C^{\mu}{}_{\mu\lambda\sigma} = C^{\mu}{}_{\nu\mu\sigma} = 0$. The number of independent components is $(D-3)D(D+1)(D+2)/12$, which is zero in three dimensions and ten in four dimensions.

The square of the Weyl tensor (A.1) is given by

$$C^2_{\mu\nu\lambda\sigma} = R^2_{\mu\nu\lambda\sigma} - \frac{4}{D-2} R^2_{\mu\nu} + \frac{2}{(D-1)(D-2)} R^2, \tag{A.2}$$

where $R^2_{\mu\nu\lambda\sigma} = R^{\mu\nu\lambda\sigma} R_{\mu\nu\lambda\sigma}$ and so on. For gauge fields, $F^2_{\mu\nu} = F_{\mu\nu} F^{\mu\nu}$ as well. The Euler density (Gauss-Bonnet combination) is defined by

$$G_4 = R^2_{\mu\nu\lambda\sigma} - 4R^2_{\mu\nu} + R^2 \tag{A.3}$$

as a formula that can be used in any dimension. In four dimensions, this can be written as $G_4 = {}^*R_{\mu\nu\lambda\sigma} {}^*R^{\mu\nu\lambda\sigma}$ using ${}^*R_{\mu\nu\lambda\sigma} = \epsilon_{\mu\nu}{}^{\alpha\beta} R_{\alpha\beta\lambda\sigma}/2$. In Euclidean space, the Euler characteristic is then expressed as

$$\chi = \frac{1}{32\pi^2} \int d^4x \sqrt{g}\, G_4.$$

Variational formulas Variations of the curvatures are given by

$$\delta g^{\mu\nu} = -g^{\mu\lambda} g^{\nu\sigma} \delta g_{\lambda\sigma}, \qquad \delta\sqrt{-g} = \frac{1}{2}\sqrt{-g}\, g^{\mu\nu} \delta g_{\mu\nu},$$
$$\delta\Gamma^{\lambda}_{\mu\nu} = \frac{1}{2} g^{\lambda\sigma}\left(\nabla_{\mu}\delta g_{\nu\sigma} + \nabla_{\nu}\delta g_{\mu\sigma} - \nabla_{\sigma}\delta g_{\mu\nu}\right),$$
$$\begin{aligned} \delta R^{\lambda}{}_{\mu\sigma\nu} &= \nabla_{\sigma}\delta\Gamma^{\lambda}_{\mu\nu} - \nabla_{\nu}\delta\Gamma^{\lambda}_{\mu\sigma} \\ &= \frac{1}{2} g^{\lambda\rho}\Big\{\nabla_{\sigma}\nabla_{\mu}\delta g_{\nu\rho} + \nabla_{\sigma}\nabla_{\nu}\delta g_{\mu\rho} - \nabla_{\sigma}\nabla_{\rho}\delta g_{\mu\nu} - \nabla_{\nu}\nabla_{\mu}\delta g_{\sigma\rho} \\ &\quad -\nabla_{\nu}\nabla_{\sigma}\delta g_{\mu\rho} + \nabla_{\nu}\nabla_{\rho}\delta g_{\mu\sigma}\Big\}, \end{aligned}$$

$$\begin{aligned}
\delta R_{\mu\nu} &= \delta R^{\lambda}{}_{\mu\lambda\nu} \\
&= \frac{1}{2}\left\{\nabla_\mu \nabla^\lambda \delta g_{\lambda\nu} + \nabla_\nu \nabla^\lambda \delta g_{\lambda\mu} - \nabla_\mu \nabla_\nu \left(g^{\lambda\sigma}\delta g_{\lambda\sigma}\right) - \nabla^2 \delta g_{\mu\nu}\right\} \\
&\quad -R^{\lambda}{}_{\mu}{}^{\sigma}{}_{\nu}\delta g_{\lambda\sigma} + \frac{1}{2}\left(R_\mu{}^\lambda \delta g_{\lambda\nu} + R_\nu{}^\lambda \delta g_{\lambda\mu}\right), \\
\delta R &= \delta g^{\mu\nu} R_{\mu\nu} + g^{\mu\nu}\delta R_{\mu\nu} \\
&= -R^{\mu\nu}\delta g_{\mu\nu} + \nabla^\mu \nabla^\nu \delta g_{\mu\nu} - \nabla^2\left(g^{\mu\nu}\delta g_{\mu\nu}\right).
\end{aligned}$$

In addition, variation formulas including covariant derivatives are

$$\begin{aligned}
\delta(\nabla_\mu A) &= \nabla_\mu \delta A, \\
\delta(\nabla_\mu \nabla_\nu A) &= \nabla_\mu \nabla_\nu \delta A - \frac{1}{2}\nabla^\lambda A\left(\nabla_\mu \delta g_{\nu\lambda} + \nabla_\nu \delta g_{\mu\lambda} - \nabla_\lambda \delta g_{\mu\nu}\right), \\
\delta(\nabla^2 A) &= \nabla^2 \delta A - \delta g_{\mu\nu}\nabla^\mu \nabla^\nu A - \nabla^\mu A \nabla^\nu \delta g_{\mu\nu} + \frac{1}{2}\nabla^\lambda A \nabla_\lambda (g^{\mu\nu}\delta g_{\mu\nu}),
\end{aligned}$$

where A is an arbitrary scalar.

Conformal variations of curvatures From the above variational formulas, a conformal (Weyl) variation $\delta_\omega g_{\mu\nu} = 2\omega g_{\mu\nu}$ of the scalar curvature is given by

$$\delta_\omega \sqrt{-g}R = (D-2)\omega\sqrt{-g}R - 2(D-1)\sqrt{-g}\nabla^2\omega.$$

Conformal variations of various curvatures squared are given by

$$\begin{aligned}
\delta_\omega \sqrt{-g}R^2_{\mu\nu\lambda\sigma} &= (D-4)\omega\sqrt{-g}R^2_{\mu\nu\lambda\sigma} - 8\sqrt{-g}R^{\mu\nu}\nabla_\mu\nabla_\nu\omega, \\
\delta_\omega \sqrt{-g}R^2_{\mu\nu} &= (D-4)\omega\sqrt{-g}R^2_{\mu\nu} - 2\sqrt{-g}R\nabla^2\omega - 2(D-2)\sqrt{-g}R^{\mu\nu}\nabla_\mu\nabla_\nu\omega, \\
\delta_\omega \sqrt{-g}R^2 &= (D-4)\omega\sqrt{-g}R^2 - 4(D-1)\sqrt{-g}R\nabla^2\omega, \\
\delta_\omega \sqrt{-g}\nabla^2 R &= (D-4)\omega\sqrt{-g}\nabla^2 R + (D-6)\sqrt{-g}\nabla^\lambda R\nabla_\lambda\omega \\
&\quad -2\sqrt{-g}R\nabla^2\omega - 2(D-1)\sqrt{-g}\nabla^4\omega, \\
\delta_\omega \sqrt{-g}\,C^2_{\mu\nu\lambda\sigma} &= (D-4)\omega\sqrt{-g}\,C^2_{\mu\nu\lambda\sigma}, \\
\delta_\omega \sqrt{-g}\,\mathrm{Tr}\left(F^2_{\mu\nu}\right) &= (D-4)\omega\sqrt{-g}\,\mathrm{Tr}\left(F^2_{\mu\nu}\right).
\end{aligned}$$

A.2 Mode Expansion Formulas

When the metric field is decomposed into the conformal-factor field and others as $g_{\mu\nu} = e^{2\phi}\bar{g}_{\mu\nu}$ (2.3), the curvatures are expressed as

$$
\begin{aligned}
\Gamma^\lambda{}_{\mu\nu} &= \bar{\Gamma}^\lambda{}_{\mu\nu} + \bar{g}^\lambda{}_\mu \bar{\nabla}_\nu \phi + \bar{g}^\lambda{}_\nu \bar{\nabla}_\mu \phi - \bar{g}_{\mu\nu} \bar{\nabla}^\lambda \phi, \\
R^\lambda{}_{\mu\sigma\nu} &= \bar{R}^\lambda{}_{\mu\sigma\nu} + \bar{g}^\lambda{}_\nu \bar{\Delta}_{\mu\sigma} - \bar{g}^\lambda{}_\sigma \bar{\Delta}_{\mu\nu} + \bar{g}_{\mu\sigma} \bar{\Delta}^\lambda{}_\nu - \bar{g}_{\mu\nu} \bar{\Delta}^\lambda{}_\sigma \\
&\quad + (\bar{g}^\lambda{}_\nu \bar{g}_{\mu\sigma} - \bar{g}^\lambda{}_\sigma \bar{g}_{\mu\nu}) \bar{\nabla}_\rho \phi \bar{\nabla}^\rho \phi, \\
R_{\mu\nu} &= \bar{R}_{\mu\nu} - (D-2)\bar{\Delta}_{\mu\nu} - \bar{g}_{\mu\nu} \left\{ \bar{\nabla}^2 \phi + (D-2) \bar{\nabla}_\lambda \phi \bar{\nabla}^\lambda \phi \right\}, \\
R &= e^{-2\phi} \left\{ \bar{R} - 2(D-1)\bar{\nabla}^2 \phi - (D-1)(D-2) \bar{\nabla}_\lambda \phi \bar{\nabla}^\lambda \phi \right\}, \\
C^\lambda{}_{\mu\sigma\nu} &= \bar{C}^\lambda{}_{\mu\sigma\nu}, \\
F_{\mu\nu} &= \bar{F}_{\mu\nu},
\end{aligned}
$$

where $\bar{\Delta}_{\mu\nu} = \bar{\nabla}_\mu \bar{\nabla}_\nu \phi - \bar{\nabla}_\mu \phi \bar{\nabla}_\nu \phi$. The quantities with the bar are defined using the metric field $\bar{g}_{\mu\nu}$, and raising and lowering of the indices of these quantities are also done using that. The Weyl tensor squared (A.2) and the gauge field strength squared are then expressed as

$$
\begin{aligned}
\sqrt{-g}\, C^2_{\mu\nu\lambda\sigma} &= \sqrt{-\bar{g}}\, e^{(D-4)\phi}\, \bar{C}^2_{\mu\nu\lambda\sigma}, \\
\sqrt{-g}\, \mathrm{Tr}(F^2_{\mu\nu}) &= \sqrt{-\bar{g}}\, e^{(D-4)\phi}\, \mathrm{Tr}(\bar{F}^2_{\mu\nu}),
\end{aligned}
\tag{A.4}
$$

respectively, and the Euler density (A.3) is

$$
\sqrt{-g}\, G_4 = \sqrt{-\bar{g}}\, e^{(D-4)\phi} \left[\bar{G}_4 + (D-3)\bar{\nabla}_\mu J^\mu + (D-3)(D-4) K \right], \tag{A.5}
$$

where

$$
\begin{aligned}
J^\mu &= 8\bar{R}^{\mu\nu}\bar{\nabla}_\nu \phi - 4\bar{R}\bar{\nabla}^\mu \phi + 4(D-2)\big(\bar{\nabla}^\mu \phi \bar{\nabla}^2 \phi - \bar{\nabla}^\mu \bar{\nabla}^\nu \phi \bar{\nabla}_\nu \phi \\
&\quad + \bar{\nabla}^\mu \phi \bar{\nabla}_\lambda \phi \bar{\nabla}^\lambda \phi\big), \\
K &= 4\bar{R}^{\mu\nu}\bar{\nabla}_\mu \phi \bar{\nabla}_\nu \phi - 2\bar{R}\bar{\nabla}_\lambda \phi \bar{\nabla}^\lambda \phi + 4(D-2)\bar{\nabla}^2 \phi \bar{\nabla}_\lambda \phi \bar{\nabla}^\lambda \phi \\
&\quad + (D-1)(D-2)(\bar{\nabla}_\lambda \phi \bar{\nabla}^\lambda \phi)^2.
\end{aligned}
$$

Thus, the ϕ-dependent part of the Euler density multiplied by $\sqrt{-g}$ becomes a total-divergence form in four dimensions.

Further expanding the metric field with bar in terms of the traceless tensor field $h_{\mu\nu}$ as $\bar{g}_{\mu\nu} = (\hat{g}e^h)_{\mu\nu}$ up to $o(h^2)$ yields

$$
\begin{aligned}
\sqrt{-\bar{g}} &= \sqrt{-\hat{g}}, \\
\bar{\Gamma}^{\lambda}_{\mu\nu} &= \hat{\Gamma}^{\lambda}_{\mu\nu} + \hat{\nabla}_{(\mu} h^{\lambda}{}_{\nu)} - \frac{1}{2}\hat{\nabla}^{\lambda} h_{\mu\nu} + \frac{1}{2}\hat{\nabla}_{(\mu}(h^2)^{\lambda}{}_{\nu)} - \frac{1}{4}\hat{\nabla}^{\lambda}(h^2)_{\mu\nu} \\
&\quad - h^{\lambda}{}_{\sigma}\hat{\nabla}_{(\mu} h^{\sigma}{}_{\nu)} + \frac{1}{2} h^{\lambda}{}_{\sigma}\hat{\nabla}^{\sigma} h_{\mu\nu}, \\
\bar{R} &= \hat{R} - \hat{R}_{\mu\nu} h^{\mu\nu} + \hat{\nabla}_{\mu}\hat{\nabla}_{\nu} h^{\mu\nu} + \frac{1}{2}\hat{R}^{\sigma}{}_{\mu\lambda\nu} h^{\lambda}{}_{\sigma} h^{\mu\nu} - \frac{1}{4}\hat{\nabla}^{\lambda} h^{\mu}{}_{\nu}\hat{\nabla}_{\lambda} h^{\nu}{}_{\mu} \\
&\quad + \frac{1}{2}\hat{\nabla}_{\nu} h^{\nu}{}_{\mu}\hat{\nabla}_{\lambda} h^{\lambda\mu} - \hat{\nabla}_{\mu}(h^{\mu}{}_{\nu}\hat{\nabla}^{\lambda} h^{\nu}{}_{\lambda}), \\
\bar{R}_{\mu\nu} &= \hat{R}_{\mu\nu} - \hat{R}^{\sigma}{}_{\mu\lambda\nu} h^{\lambda}{}_{\sigma} + \hat{R}^{\lambda}_{(\mu} h_{\nu)\lambda} + \hat{\nabla}_{(\mu}\hat{\nabla}^{\lambda} h_{\nu)\lambda} - \frac{1}{2}\hat{\nabla}^2 h_{\mu\nu} \\
&\quad - \frac{1}{2} h^{\lambda}_{(\mu}\hat{\nabla}^2 h_{\nu)\lambda} - \frac{1}{2}\hat{\nabla}^{\lambda} h^{\sigma}{}_{\mu}\hat{\nabla}_{\sigma} h_{\nu\lambda} - \frac{1}{4}\hat{\nabla}_{\mu} h^{\lambda}{}_{\sigma}\hat{\nabla}_{\nu} h^{\sigma}{}_{\lambda} \\
&\quad - \frac{1}{2}\hat{\nabla}_{\lambda}(h^{\lambda}{}_{\sigma}\hat{\nabla}_{(\mu} h^{\sigma}{}_{\nu)}) + \frac{1}{2}\hat{\nabla}_{\lambda}(h^{\sigma}_{(\mu}\hat{\nabla}_{\nu)} h^{\lambda}{}_{\sigma}) + \frac{1}{2}\hat{\nabla}_{\lambda}(h^{\lambda}{}_{\sigma}\hat{\nabla}^{\sigma} h_{\mu\nu}),
\end{aligned}
$$

where raising and lowering of the indices on the right-hand sides are performed using the background metric $\hat{g}_{\mu\nu}$, and the traceless condition is $h^{\mu}{}_{\mu} = \hat{g}^{\mu\nu} h_{\mu\nu} = 0$. The symmetric product is defined by $a_{(\mu} b_{\nu)} = (a_{\mu} b_{\nu} + a_{\nu} b_{\mu})/2$. With attention to $\bar{R} = \bar{g}^{\mu\nu}\bar{R}_{\mu\nu}$ and $\bar{g}^{\mu\nu} = (\hat{g} e^{-h})^{\mu\nu} = \hat{g}^{\mu\nu} - h^{\mu\nu} + \cdots$, we can derive $\bar{R}$ from $\bar{R}_{\mu\nu}$, where $[\hat{\nabla}_{\lambda}, \hat{\nabla}_{\nu}] h^{\lambda}{}_{\mu} = h^{\lambda}{}_{\sigma}\hat{R}^{\sigma}{}_{\mu\nu\lambda} + h_{\mu\sigma}\hat{R}^{\sigma}{}_{\nu}$ is used.

If the flat background metric $\hat{g}_{\mu\nu} = \eta_{\mu\nu}$ is employed, then the square of each curvature with bar, and so on, is expanded up to $o(h^2)$ as

$$
\begin{aligned}
\bar{R}^2_{\mu\nu\lambda\sigma} &= \partial_{\lambda}\partial_{\sigma} h_{\mu\nu}\partial^{\lambda}\partial^{\sigma} h^{\mu\nu} - 2\partial_{\nu}\partial_{\lambda} h_{\mu\sigma}\partial^{\mu}\partial^{\lambda} h^{\nu\sigma} + \partial_{\lambda}\partial_{\sigma} h_{\mu\nu}\partial^{\mu}\partial^{\nu} h^{\lambda\sigma}, \\
\bar{R}^2_{\mu\nu} &= \frac{1}{2}\partial_{\mu}\chi_{\nu}\partial^{\mu}\chi^{\nu} - \partial^2 h_{\mu\nu}\partial^{\mu}\chi^{\nu} + \frac{1}{2}\partial_{\mu}\chi_{\nu}\partial^{\nu}\chi^{\mu} + \frac{1}{4}\partial^2 h_{\mu\nu}\partial^2 h^{\mu\nu}, \\
\bar{R}^2 &= \partial_{\mu}\chi^{\mu}\partial_{\nu}\chi^{\nu}, \\
\bar{\nabla}^2\bar{R} &= \partial^2\partial_{\mu}\chi^{\mu} - \frac{1}{4}\partial^2\left(\partial_{\lambda} h_{\mu\nu}\partial^{\lambda} h^{\mu\nu}\right) + \frac{1}{2}\partial^2\left(\chi_{\mu}\chi^{\mu}\right) - \partial^2\partial^{\mu}\left(h_{\mu\nu}\chi^{\nu}\right) \\
&\quad - h_{\mu\nu}\partial^{\mu}\partial^{\nu}\partial_{\lambda}\chi^{\lambda} - \chi_{\mu}\partial^{\mu}\partial_{\nu}\chi^{\nu},
\end{aligned}
$$

where $\chi_{\mu} = \partial_{\nu} h^{\nu}{}_{\mu}$. From these, we can see that the $o(h^2)$ term of the Euler density with bar can be written in a total-divergence form in any dimension as

$$
\bar{G}_4 = \partial_{\sigma} L^{\sigma},
$$

where

$$
\begin{aligned}
L^{\sigma} &= \partial_{\lambda} h_{\mu\nu}\partial^{\lambda}\partial^{\sigma} h^{\mu\nu} - \partial^{\sigma} h_{\mu\nu}\partial^2 h^{\mu\nu} - 2\partial_{\lambda} h_{\mu\nu}\partial^{\lambda}\partial^{\mu} h^{\nu\sigma} - 2\partial_{\lambda} h_{\nu}{}^{\sigma}\partial^{\lambda}\chi^{\nu} \\
&\quad + 4\partial^{\sigma} h_{\mu\nu}\partial^{\mu}\chi^{\nu} + \partial_{\lambda} h_{\mu\nu}\partial^{\mu}\partial^{\nu} h^{\lambda\sigma} - \partial_{\lambda} h_{\nu}{}^{\sigma}\partial^{\nu}\chi^{\lambda} - \chi_{\lambda}\partial^{\lambda}\chi^{\sigma} + \chi^{\sigma}\partial_{\lambda}\chi^{\lambda}.
\end{aligned}
$$

Moreover, quantities involving ϕ are expanded up to $o(h^2)$ as follows:

$$\begin{aligned}
\bar{\nabla}^2\phi &= \partial^2\phi - \chi^\mu\partial_\mu\phi - h^{\mu\nu}\partial_\mu\partial_\nu\phi + \frac{1}{2}h^{\mu\lambda}\partial_\mu h^\nu{}_\lambda\partial_\nu\phi + \frac{1}{2}h^{\mu\nu}\chi_\mu\partial_\nu\phi \\
&\quad + \frac{1}{2}h^{\mu\lambda}h^\nu{}_\lambda\partial_\mu\partial_\nu\phi, \\
\bar{\nabla}_\mu\bar{\nabla}_\nu\phi &= \partial_\mu\partial_\nu\phi - \frac{1}{2}\left(\partial_\mu h^\lambda{}_\nu + \partial_\nu h^\lambda{}_\mu - \partial^\lambda h_{\mu\nu}\right)\partial_\lambda\phi \\
&\quad - \frac{1}{4}\Big(h^\sigma{}_\mu\partial_\nu h^\lambda{}_\sigma + h^\sigma{}_\nu\partial_\mu h^\lambda{}_\sigma - h^\lambda{}_\sigma\partial_\mu h^\sigma{}_\nu - h^\lambda{}_\sigma\partial_\nu h^\sigma{}_\mu \\
&\quad - h^\sigma{}_\mu\partial^\lambda h_{\nu\sigma} - h^\sigma{}_\nu\partial^\lambda h_{\mu\sigma} + 2h^{\lambda\sigma}\partial_\sigma h_{\mu\nu}\Big)\partial_\lambda\phi, \\
\bar{\nabla}^4\phi &= \partial^4\phi - \partial^2(h^{\mu\nu}\partial_\mu\partial_\nu\phi + \chi^\mu\partial_\mu\phi) - h^{\mu\nu}\partial_\mu\partial_\nu\partial^2\phi - \chi^\mu\partial_\mu\partial^2\phi \\
&\quad + \partial^2\left(\frac{1}{2}h^{\mu\lambda}h^\nu{}_\lambda\partial_\mu\partial_\nu\phi + \frac{1}{2}h^{\mu\nu}\partial_\mu h^\lambda{}_\nu\partial_\lambda\phi + \frac{1}{2}h^{\mu\nu}\chi_\mu\partial_\nu\phi\right) \\
&\quad + h^{\mu\nu}\Big(\partial_\mu\partial_\nu h^{\lambda\sigma}\partial_\lambda\partial_\sigma\phi + 2\partial_\mu h^{\lambda\sigma}\partial_\nu\partial_\lambda\partial_\sigma\phi + h^{\lambda\sigma}\partial_\mu\partial_\nu\partial_\lambda\partial_\sigma\phi \\
&\quad + \frac{1}{2}\partial_\mu h^\lambda{}_\nu\partial_\lambda\partial^2\phi + \frac{1}{2}\chi_\mu\partial_\nu\partial^2\phi + \partial_\mu\partial_\nu\chi^\lambda\partial_\lambda\phi + 2\partial_\mu\chi^\lambda\partial_\nu\partial_\lambda\phi \\
&\quad + 2\chi^\lambda\partial_\mu\partial_\nu\partial_\lambda\phi\Big) + \chi^\mu\partial_\mu\chi^\nu\partial_\nu\phi + \chi^\mu\chi^\nu\partial_\mu\partial_\nu\phi \\
&\quad + \chi^\mu\partial_\mu h^{\nu\lambda}\partial_\nu\partial_\lambda\phi + \frac{1}{2}h^{\mu\lambda}h^\nu{}_\lambda\partial_\mu\partial_\nu\partial^2\phi,
\end{aligned}$$

and

$$\begin{aligned}
\bar{R}^{\mu\nu}\bar{\nabla}_\mu\bar{\nabla}_\nu\phi &= \partial^\mu\chi^\nu\partial_\mu\partial_\nu\phi - \frac{1}{2}\partial^2 h^{\mu\nu}\partial_\mu\partial_\nu\phi - \frac{1}{2}\partial^\lambda h^{\mu\nu}\partial_\lambda\chi_\mu\partial_\nu\phi \\
&\quad - \frac{1}{2}\partial_\lambda h^{\mu\nu}\partial_\mu\chi^\lambda\partial_\nu\phi + \frac{1}{2}\partial^\lambda h^{\mu\nu}\partial_\mu\chi_\nu\partial_\lambda\phi \\
&\quad + \frac{1}{2}\partial^2 h^{\mu\nu}\partial_\mu h^\lambda{}_\nu\partial_\lambda\phi - \frac{1}{4}\partial^2 h^{\mu\nu}\partial^\lambda h_{\mu\nu}\partial_\lambda\phi \\
&\quad - h^{\mu\nu}\partial^\lambda\chi_\mu\partial_\nu\partial_\lambda\phi - h^{\mu\nu}\partial_\mu\chi^\lambda\partial_\nu\partial_\lambda\phi \\
&\quad + \frac{1}{2}h^{\mu\nu}\partial^2 h^\lambda{}_\mu\partial_\nu\partial_\lambda\phi - \frac{1}{2}\partial_\lambda h^{\mu\nu}\partial_\mu h^{\lambda\sigma}\partial_\nu\partial_\sigma\phi \\
&\quad - \frac{1}{4}\partial^\lambda h^{\mu\nu}\partial^\sigma h_{\mu\nu}\partial_\lambda\partial_\sigma\phi - \frac{1}{2}\partial_\lambda(h^{\lambda\sigma}\partial^\mu h^\nu{}_\sigma)\partial_\mu\partial_\nu\phi \\
&\quad + \frac{1}{2}\partial_\lambda(h^{\mu\sigma}\partial^\nu h^\lambda{}_\sigma)\partial_\mu\partial_\nu\phi + \frac{1}{2}\partial_\lambda(h^{\lambda\sigma}\partial_\sigma h^{\mu\nu})\partial_\mu\partial_\nu\phi,
\end{aligned}$$

$$\begin{aligned}\bar{R}\bar{\nabla}^2\phi &= \partial_\mu\chi^\mu\partial^2\phi - \chi^\mu\left(\partial_\nu\chi^\nu\partial_\mu\phi + \frac{1}{2}\chi_\mu\partial^2\phi\right) \\ &\quad -\frac{1}{4}\partial^\lambda h^{\mu\nu}\partial_\lambda h_{\mu\nu}\partial^2\phi - h^{\mu\nu}\left(\partial_\mu\chi_\nu\partial^2\phi + \partial_\lambda\chi^\lambda\partial_\mu\partial_\nu\phi\right), \\ \bar{\nabla}^\mu\bar{R}\bar{\nabla}_\mu\phi &= \partial^\mu\partial_\nu\chi^\nu\partial_\mu\phi - \frac{1}{2}\partial^\lambda\partial^\sigma h^{\mu\nu}\partial_\lambda h_{\mu\nu}\partial_\sigma\phi - \frac{1}{2}\partial^\mu(\chi^\nu\chi_\nu)\partial_\mu\phi \\ &\quad -\partial^\lambda(h^{\mu\nu}\partial_\mu\chi_\nu)\partial_\lambda\phi - h^{\mu\nu}\partial_\mu\partial_\lambda\chi^\lambda\partial_\nu\phi.\end{aligned}$$

A.3 Expansion Formulas of Gravitational Actions for Dimensional Regularization

Expanding the D-dimensional Weyl action around $D = 4$ using (A.4) yields

$$\begin{aligned}\int d^D x\sqrt{-g}\, C^2_{\mu\nu\lambda\sigma} &= \int d^D x\sqrt{-\hat{g}}\, e^{(D-4)\phi}\, \bar{C}^2_{\mu\nu\lambda\sigma} \\ &= \sum_{n=0}^{\infty}\frac{(D-4)^n}{n!}\int d^D x\sqrt{-\hat{g}}\,\phi^n\,\bar{C}^2_{\mu\nu\lambda\sigma}.\end{aligned}$$

The action for the generalized Euler density $G_D = G_4 + (D-4)\chi(D)H^2$ (5.3) in D dimensions, where $H = R/(D-1)$, is expanded as follows. First, the volume integral of G_4 can be expanded using (A.5) as

$$\begin{aligned}&\int d^D x\sqrt{-g}\, G_4 \\ &= \int d^D x\sqrt{-\bar{g}}\, e^{(D-4)\phi}\left[\bar{G}_4 + (D-3)\bar{\nabla}_\mu J^\mu + (D-3)(D-4)K\right] \\ &= \sum_{n=0}^{\infty}\frac{(D-4)^n}{n!}\int d^D x\sqrt{-\bar{g}}\left\{\phi^n\bar{G}_4 + (D-3)\left(\phi^n\bar{\nabla}_\mu J^\mu + n\phi^{n-1}K\right)\right\} \\ &= \sum_{n=0}^{\infty}\frac{(D-4)^n}{n!}\int d^D x\sqrt{-\bar{g}}\Big\{\phi^n\bar{G}_4 + 4(D-3)\phi^n\bar{R}^{\mu\nu}\bar{\nabla}_\mu\bar{\nabla}_\nu\phi \\ &\quad -2(D-3)\phi^n\bar{R}\bar{\nabla}^2\phi - 2(D-2)(D-3)(D-4)\phi^n\bar{\nabla}^2\phi\bar{\nabla}^\lambda\phi\bar{\nabla}_\lambda\phi \\ &\quad -(D-2)(D-3)^2(D-4)\phi^n(\bar{\nabla}^\lambda\phi\bar{\nabla}_\lambda\phi)^2\Big\}.\end{aligned}$$

Expanding the H squared term multiplied by $D-4$ yields

$$
\begin{aligned}
&(D-4)\int d^Dx\sqrt{-g}\,H^2 \\
&=\sum_{n=0}^{\infty}\frac{(D-4)^n}{n!}\int d^Dx\sqrt{-\hat{g}}\left\{\frac{(D-4)}{(D-1)^2}\phi^n\bar{R}^2-\frac{2(D-6)}{D-1}\phi^n\bar{R}\bar{\nabla}^2\phi\right. \\
&+\frac{2(D-2)}{D-1}\phi^n\bar{\nabla}^\lambda\bar{R}\bar{\nabla}_\lambda\phi+4\phi^n\bar{\nabla}^4\phi+8(D-4)\phi^n\bar{\nabla}^2\phi\bar{\nabla}^\lambda\phi\bar{\nabla}_\lambda\phi \\
&\left.+(D-2)^2(D-4)\phi^n(\bar{\nabla}^\lambda\phi\bar{\nabla}_\lambda\phi)^2\right\}.
\end{aligned}
$$

Substituting these expansion formulas and $\chi(D)=1/2+3(D-4)/4+\chi_3(D-4)^2+\chi_4(D-4)^3+\cdots$ (5.4), the G_D action is expanded as follows[2]:

$$
\begin{aligned}
&\int d^Dx\sqrt{-g}\,G_D=\int d^Dx\sqrt{-g}\left[G_4+(D-4)\chi(D)H^2\right] \\
&=\sum_{n=0}^{\infty}\frac{(D-4)^n}{n!}\int d^Dx\sqrt{-\hat{g}}\left\{\phi^n\bar{G}_4+\frac{D-4}{(D-1)^2}\chi(D)\phi^n\bar{R}^2\right. \\
&+4(D-3)\phi^n\bar{R}^{\mu\nu}\bar{\nabla}_\mu\bar{\nabla}_\nu\phi-2\left[D-3+\frac{D-6}{D-1}\chi(D)\right]\phi^n\bar{R}\bar{\nabla}^2\phi \\
&+\frac{2(D-2)}{D-1}\chi(D)\phi^n\bar{\nabla}^\lambda\bar{R}\bar{\nabla}_\lambda\phi+4\chi(D)\phi^n\bar{\nabla}^4\phi \\
&+2(D-4)\left[-(D-2)(D-3)+4\chi(D)\right]\phi^n\bar{\nabla}^2\phi\bar{\nabla}_\lambda\phi\bar{\nabla}^\lambda\phi \\
&\left.+(D-2)(D-4)\left[-(D-3)^2+(D-2)\chi(D)\right]\phi^n(\bar{\nabla}_\lambda\phi\bar{\nabla}^\lambda\phi)^2\right\} \\
&=\int d^Dx\sqrt{-\hat{g}}\left\{\bar{G}_4+(D-4)\left(2\phi\bar{\Delta}_4\phi+\bar{G}_4\phi-\frac{2}{3}\bar{R}\bar{\nabla}^2\phi+\frac{1}{18}\bar{R}^2\right)\right. \\
&+(D-4)^2\left(\phi^2\bar{\Delta}_4\phi+\frac{1}{2}\bar{G}_4\phi^2+3\phi\bar{\nabla}^4\phi+4\phi\bar{R}^{\mu\nu}\bar{\nabla}_\mu\bar{\nabla}_\nu\phi\right. \\
&\left.-\frac{14}{9}\phi\bar{R}\bar{\nabla}^2\phi+\frac{10}{9}\phi\bar{\nabla}^\lambda\bar{R}\bar{\nabla}_\lambda\phi-\frac{7}{9}\bar{R}\bar{\nabla}^2\phi+\frac{1}{18}\bar{R}^2\phi+\frac{5}{108}\bar{R}^2\right) \\
&+(D-4)^3\left[\frac{1}{3}\phi^3\bar{\Delta}_4\phi+\frac{1}{6}\bar{G}_4\phi^3+\left(4\chi_3-\frac{1}{2}\right)(\bar{\nabla}_\lambda\phi\bar{\nabla}^\lambda\phi)^2\right.
\end{aligned}
$$

---

[2] In two-dimensional quantum gravity, the action in dimensional regularization is given by the volume integral of the D-dimensional scalar curvature, which is expanded around $D=2$ as

$$
\int d^Dx\sqrt{-g}R=\sum_{n=0}^{\infty}\frac{(D-2)^n}{n!}\int d^Dx\sqrt{-\hat{g}}\left[-(D-1)\phi^n\bar{\nabla}^2\phi+\bar{R}\phi^n\right].
$$

The $n=1$ part gives the Liouville action.

$$
\begin{aligned}
&+(8\chi_3-2)\,\bar{\nabla}^2\phi\bar{\nabla}^\lambda\phi\bar{\nabla}_\lambda\phi+\frac{3}{2}\phi^2\bar{\nabla}^4\phi+2\phi^2\bar{R}^{\mu\nu}\bar{\nabla}_\mu\bar{\nabla}_\nu\phi \\
&-\frac{7}{9}\phi^2\bar{R}\bar{\nabla}^2\phi+\frac{5}{9}\phi^2\bar{\nabla}^\lambda\bar{R}\bar{\nabla}_\lambda\phi+\frac{1}{36}\bar{R}^2\phi^2+4\chi_3\phi\bar{\nabla}^4\phi \\
&+\left(\frac{4}{3}\chi_3-\frac{35}{54}\right)\phi\bar{R}\bar{\nabla}^2\phi+\left(\frac{4}{3}\chi_3+\frac{7}{54}\right)\phi\bar{\nabla}^\lambda\bar{R}\bar{\nabla}_\lambda\phi \\
&+\frac{5}{108}\bar{R}^2\phi+\left(-\frac{4}{3}\chi_3+\frac{7}{27}\right)\bar{R}\bar{\nabla}^2\phi+\left(\frac{1}{9}\chi_3-\frac{1}{27}\right)\bar{R}^2\Bigg] \\
&+o((D-4)^4)\Bigg\}, \qquad (A.6)
\end{aligned}
$$

where $\sqrt{-g}\Delta_4$ is the differential operator defined by (2.17) even in D dimensions, which becomes conformally invariant at four dimensions. Note that the coefficients up to $o((D-4)^3)$ do not depend on the value of χ_4.

Appendix B
About Definitions of Fermions in Curved Space

In Minkowski spacetime, the Hermiticity of fermion fields is obvious, and the canonical quantization method can be applied. On the other hand, in Euclidean spacetime, the Hermiticity becomes unclear, and quantization is premised to be performed by the path integral method. Generally, fermions in Euclidean spacetime are defined so that Feynman amplitudes can be derived correctly.

Fermions in Lorentzian manifold Let us define fermions in curved spacetime with locally Lorentzian signature $\eta_{ab} = (-1, 1, 1, 1)$. For brevity, the same symbols as in Chap. 11 are used for vierbein fields, spin connections, curvatures, and so on.

Denoting a gauge transformation that preserves the metric unchanged, a local Lorentz transformation, as $e'^a(x) = \Lambda^a{}_b(x) e^b(x)$, it constitutes the $SO(3,1)$ Lorentz group that satisfies $\eta_{ab} \Lambda^a{}_c(x) \Lambda^b{}_d(x) = \eta_{cd}$. The discussion about definitions of covariant derivatives and curvatures mentioned in Sect. 11.1 can be applied as it is, and the transformation laws are given by replacing $\Phi^a{}_b$ with $\Lambda^a{}_b$. Therefore, the connection 1-form transforms as $\omega'^a{}_b = \Lambda^a{}_c\, \omega^c{}_d\, (\Lambda^{-1})^d{}_b + \Lambda^a{}_c\, d(\Lambda^{-1})^c{}_b$ under the local Lorentz transformation, and the curvature 2-form transforms as $\mathcal{R}'^a{}_b = \Lambda^a{}_c\, \mathcal{R}^c{}_d\, (\Lambda^{-1})^d{}_b$.

The algebra that the $SO(3,1)$ group generator Σ^{ab} satisfies is normalized as

$$\left[\Sigma^{ab}, \Sigma^{cd}\right] = -\eta^{ac}\Sigma^{bd} + \eta^{ad}\Sigma^{bc} + \eta^{bc}\Sigma^{ad} - \eta^{bd}\Sigma^{ac}.$$

The generator acting on vector indices is $(\Sigma^{ab})_{cd} = \delta^a{}_c \delta^b{}_d - \delta^a{}_d \delta^b{}_c$. The elements of the $SO(3,1)$ group with ε^{ab} $(= -\varepsilon^{ba})$ as variables are then given by

$$\Lambda_{cd} = \exp\left(\frac{1}{2}\varepsilon_{ab}\Sigma^{ab}\right)_{cd}.$$

Define the Clifford algebra in Lorentzian space as

$$\{\gamma^a, \gamma^b\} = -2\eta^{ab}. \tag{B.1}$$

K. Hamada, *Trans-Planckian Physics and Inflation*, Fundamental Theories of Physics 223, https://doi.org/10.1007/978-981-96-3476-7

The element of the $Spin(3, 1)$ group composed of this algebra is given by

$$\mathcal{S} = \exp\left(\frac{1}{2}\varepsilon_{ab}\Sigma^{ab}\right), \quad \Sigma^{ab} = -\frac{1}{4}[\gamma^a, \gamma^b]. \tag{B.2}$$

The fermion ψ and its conjugate $\bar{\psi}$ are defined as fields that transform as $\psi' = \mathcal{S}\psi$ and $\bar{\psi}' = \bar{\psi}\mathcal{S}^{-1}$, where if we choose $\gamma^{0\dagger} = \gamma^0, \gamma^{j\dagger} = -\gamma^j$ ($j = 1, 2, 3$), i.e., $\gamma^{a\dagger} = \gamma^0\gamma^a\gamma^0$ as a representation of the gamma matrix, the conjugate fermion satisfying this transformation law can be expressed as $\bar{\psi} = \psi^\dagger\gamma^0$. In fact, using a formula

$$\gamma^0\mathcal{S}^\dagger\gamma^0 = \gamma^0\left(1 + \frac{1}{2}\varepsilon_{ab}\Sigma^{ab\dagger}\right)\gamma^0 = \left(1 - \frac{1}{2}\varepsilon_{ab}\Sigma^{ab}\right) = \mathcal{S}^{-1}$$

derived from $\Sigma^{ab\dagger} = -\gamma^0\Sigma^{ab}\gamma^0$ and $(\gamma^0)^2 = 1$, it can be shown as

$$\bar{\psi}' = \psi'^\dagger\gamma^0 = \psi^\dagger\mathcal{S}^\dagger\gamma^0 = \bar{\psi}\gamma^0\mathcal{S}^\dagger\gamma^0 = \bar{\psi}\mathcal{S}^{-1}.$$

From this, we can see that $\bar{\psi}\psi$ is a local Lorentz invariant scalar as $\bar{\psi}'\psi' = \bar{\psi}\mathcal{S}^{-1}\mathcal{S}\psi = \bar{\psi}\psi$. In addition, the local Lorentz invariance of the kinetic term $\bar{\psi}\gamma^a\nabla_a\psi$ can be shown by using

$$\mathcal{S}^{-1}\gamma^a\mathcal{S} = \Lambda^a{}_b\gamma^b,$$

for $Spin(3, 1)$, as is shown for $Spin(4)$ in Chap. 11. Looking at the Hermitian property, we see that the former is a Hermitian function as $(\bar{\psi}\psi)^\dagger = \psi^\dagger\gamma^{0\dagger}\psi = \psi^\dagger\gamma^0\psi = \bar{\psi}\psi$. On the other hand, since the Hermitian conjugate of $\nabla_a\psi$ is

$$\begin{aligned}
(\nabla_a\psi)^\dagger &= e_a{}^\mu\left(\partial_\mu\psi^\dagger - \frac{1}{2}\omega_{\mu cd}\,\psi^\dagger\gamma^0\Sigma^{cd}\gamma^0\right) \\
&= e_a{}^\mu\left(\partial_\mu\bar{\psi} - \frac{1}{2}\omega_{\mu cd}\,\bar{\psi}\,\Sigma^{cd}\right)\gamma^0 = \nabla_a\bar{\psi}\gamma^0,
\end{aligned}$$

the Hermitian conjugate of the kinetic term is given by

$$\begin{aligned}
(\bar{\psi}\gamma^a\nabla_a\psi)^\dagger &= (\nabla_a\psi)^\dagger\gamma^{a\dagger}\gamma^{0\dagger}\psi \\
&= \nabla_a\bar{\psi}\gamma^0(\gamma^0\gamma^a\gamma^0)\gamma^0\psi = \nabla_a\bar{\psi}\gamma^a\psi.
\end{aligned}$$

The fermion action in Lorentzian space has to be Hermite and be both local Lorentz and diffeomorphism invariant. Therefore, adding i to the kinetic terms and $\sqrt{-g}$ to the whole, it is given by

$$I_\psi^M = \int d^4x\sqrt{-g}\left\{\frac{i}{2}(\bar{\psi}\gamma^a\nabla_a\psi - \nabla_a\bar{\psi}\gamma^a\psi) - m\bar{\psi}\psi\right\}. \tag{B.3}$$

Fermions in Euclidean manifold Since the Hermiticity of the field becomes unclear in Euclidean spacetime, the fermions ψ and $\bar{\psi}$ are treated as independent fields satisfying the transformation law (11.10).

The Clifford algebra $\{\gamma^a, \gamma^b\} = -2\delta^{ab}$ (11.8) that makes up the $Spin(4)$ group is obtained by inverting the time component of (B.1) defined by the Minkowski metric. The action is determined to be invariant under local $SO(4)$ gauge transformation and diffeomorphism. However, the mass term is determined so that the Feynman propagator comes out correctly when Wick-rotated back to Lorentzian space. The action is then given by (11.2), which is the same form as (B.3) (see Footnote 5 in Chap. 11).

Defined in this way, the Feynman amplitude in Lorentzian space can be correctly derived from that in Euclidean space through Wick rotation. This definition was proposed by Osterwalder and Schrader [OS73, OS75].

Conformal invariance The massless fermion action is conformally invariant at any dimension. Under the conformal transformation $\delta_\omega g_{\mu\nu} = 2\omega g_{\mu\nu}$, the vielbein and fermions transform as

$$\delta_\omega e_a^\mu = -\omega e_a^\mu, \qquad \delta_\omega e_{\mu a} = \omega e_{\mu a},$$
$$\delta_\omega \psi = \frac{1-D}{2}\omega\psi, \qquad \delta_\omega \bar{\psi} = \frac{1-D}{2}\omega\bar{\psi},$$

and the spin connection and so on transform as

$$\delta_\omega \omega_{\mu ab} = \left(e_{\mu a} e_b^\lambda - e_{\mu b} e_a^\lambda\right)\partial_\lambda \omega, \quad \delta_\omega\left(\gamma^\mu \nabla_\mu \psi\right) = -\frac{D+1}{2}\omega\gamma^\mu\nabla_\mu\psi,$$

where $\gamma_a \Sigma^{ab} = (D-1)\gamma^b/2$ is used in the second. Thus, it can be shown that the kinetic term of fermions is conformally invariant as

$$\delta_\omega\left(\sqrt{-g}\bar{\psi}\gamma^\mu\nabla_\mu\psi\right) = \left(D\omega + \frac{1-D}{2}\omega - \frac{D+1}{2}\omega\right)\sqrt{-g}\bar{\psi}\gamma^\mu\nabla_\mu\psi = 0$$

in arbitrary D dimensions.

This means that if we redefine the fermion field as $\psi' = e^{(D-1)\phi/2}\psi$ and rewrite the action in terms of ψ', its ϕ-dependence can be removed.

Appendix C
Energy-Momentum Tensor for Each Sector

The energy-momentum tensor $\mathbf{T}_{\mu\nu}$ useful when the metric field is decomposed as (2.3) and (2.4) is presented here. For an action I, it is defined as

$$
\begin{aligned}
\delta I &= \frac{1}{2}\int d^4x\sqrt{-g}T^{\mu\nu}\delta g_{\mu\nu} \\
&= \frac{1}{2}\int d^4x\sqrt{-\bar{g}}\left\{2\bar{T}^{\lambda}_{\ \lambda}\delta\phi + \bar{T}^{\mu\nu}\delta\bar{g}_{\mu\nu}\right\} \\
&= \int d^4x\left\{\mathbf{T}^{\lambda}_{\ \lambda}\delta\phi + \frac{1}{2}\mathbf{T}^{\mu}_{\ \nu}\delta h^{\nu}_{\ \mu}\right\},
\end{aligned}
$$

where the second equality is shown using an variation formula $\delta g_{\mu\nu} = 2e^{2\phi}\bar{g}_{\mu\nu}\delta\phi + e^{2\phi}\delta\bar{g}_{\mu\nu}$ for the metric $g_{\mu\nu} = e^{2\phi}\bar{g}_{\mu\nu}$. The last is an expression when the metric with the bar is further decomposed into $\bar{g}_{\mu\nu} = (\eta e^h)_{\mu\nu}$, where the flat Minkowski metric $\eta_{\mu\nu}$ is employed as the background metric. For each energy-momentum tensor, we should be aware of the metric used to contract the index: the normal $T_{\mu\nu}(g)$ is contracted with the physical metric $g_{\mu\nu}$, $\bar{T}_{\mu\nu}(\phi, \bar{g})$ in the second line is done using $\bar{g}_{\mu\nu}$ excluding the conformal-factor field, and the last $\mathbf{T}_{\mu\nu}(\varphi, h)$ derived below is contracted with $\eta_{\mu\nu}$.

The energy-momentum tensor $\mathbf{T}_{\mu\nu}$ for each sector [HHY06a] used in the text are shown here. The following expressions are full as for the conformal-factor field, in which not only the fluctuation φ but also the homogeneous component $\hat{\phi}$ are included as in $\phi = \hat{\phi} + \varphi$ (10.1). On the other hand, the traceless tensor field $h_{\mu\nu}$ is displayed up to the first order. In addition, $\chi_\mu = \partial_\lambda h^{\lambda}_{\ \mu}$ and $\Box = \partial^\lambda\partial_\lambda = -\partial_\eta^2 + \partial^2$ is the d'Alembertian in flat spacetime. The symmetric product is defined as $a_{(\mu}b_{\nu)} = (a_\mu b_\nu + a_\nu b_\mu)/2$.

Einstein-Hilbert sector From the variation of the Einstein-Hilbert action

$$
\frac{M_{\mathrm{P}}^2}{2}\int d^4x\sqrt{-g}R,
$$

K. Hamada, *Trans-Planckian Physics and Inflation*, Fundamental Theories of Physics 223, https://doi.org/10.1007/978-981-96-3476-7

we obtain the energy-momentum tensor

$$
\begin{aligned}
\mathbf{T}^{\mathrm{EH}}_{\mu\nu} = M_{\mathrm{P}}^2 e^{2\phi}\Big\{ & 2\partial_\mu\partial_\nu\phi - 2\partial_\mu\phi\partial_\nu\phi + \eta_{\mu\nu}(-2\Box\phi - \partial^\lambda\phi\partial_\lambda\phi) - \partial_{(\mu}\chi_{\nu)} \\
& + \frac{1}{2}\Box h_{\mu\nu} - 2h^\lambda_{(\mu}\partial_{\nu)}\partial_\lambda\phi + 2h^\lambda_{(\mu}\partial_{\nu)}\phi\partial_\lambda\phi - 2\partial_{(\mu}h^\lambda_{\nu)}\partial_\lambda\phi + \partial^\lambda h_{\mu\nu}\partial_\lambda\phi \\
& + \eta_{\mu\nu}\left(\frac{1}{2}\partial_\lambda\chi^\lambda + 2h^{\lambda\sigma}\partial_\lambda\partial_\sigma\phi + h^{\lambda\sigma}\partial_\lambda\phi\partial_\sigma\phi + 2\chi^\lambda\partial_\lambda\phi\right)\Big\}. \qquad \text{(C.1)}
\end{aligned}
$$

The trace of this is given by

$$
\begin{aligned}
\mathbf{T}^{\mathrm{EH}\mu}{}_{\mu} = M_{\mathrm{P}}^2 e^{2\phi}\{ & -6\Box\phi - 6\partial^\mu\phi\partial_\mu\phi + \partial_\mu\chi^\mu + 6h^{\mu\nu}\partial_\mu\partial_\nu\phi \\
& + 6\chi^\mu\partial_\mu\phi + 6h^{\mu\nu}\partial_\mu\phi\partial_\nu\phi\}, \qquad \text{(C.2)}
\end{aligned}
$$

which is coincident with the expansion of $\mathbf{T}^{\mathrm{EH}\mu}{}_{\mu} = M_{\mathrm{P}}^2 e^{4\phi} R = M_{\mathrm{P}}^2 e^{2\phi}(\bar{R} - 6\bar{\nabla}^2\phi - 6\bar{\nabla}^\mu\phi\bar{\nabla}_\mu\phi)$ to the first order of $h_{\mu\nu}$.

Riegert sector The energy-momentum tensor of the Riegert sector is derived from the action

$$
\int d^4x\sqrt{-\bar{g}}\left[-\frac{b_1}{(4\pi)^2}\left(2\phi\bar{\Delta}_4\phi + \bar{E}_4\phi + \frac{1}{18}\bar{R}^2\right)\right],
$$

where the last extra term arises when the diffeomorphism-invariant effective action (2.29) is decomposed as in (2.30), and is necessary to obtain the energy-momentum tensor up to the first order of $h_{\mu\nu}$. From the variation of this action, we obtain

$$
\begin{aligned}
\mathbf{T}^{\mathrm{R}}_{\mu\nu} = \frac{b_1}{8\pi^2}\Big\{ & 4\Box\phi\partial_\mu\partial_\nu\phi - 4\partial_{(\mu}\Box\phi\partial_{\nu)}\phi - \frac{8}{3}\partial_\mu\partial_\lambda\phi\partial_\nu\partial^\lambda\phi + \frac{4}{3}\partial_\mu\partial_\nu\partial_\lambda\phi\partial^\lambda\phi \\
& + \frac{2}{3}\partial_\mu\partial_\nu\Box\phi + \eta_{\mu\nu}\left[-\Box\phi\Box\phi + \frac{2}{3}\partial_\lambda\Box\phi\partial^\lambda\phi + \frac{2}{3}\partial_\lambda\partial_\sigma\phi\partial^\lambda\partial^\sigma\phi - \frac{2}{3}\Box^2\phi\right] \\
& - 4\partial_{(\mu}h^\lambda_{\nu)}\Box\phi\partial_\lambda\phi + 2\partial^\lambda h_{\mu\nu}\Box\phi\partial_\lambda\phi - 4h^{\lambda\sigma}\partial_\lambda\partial_\sigma\phi\partial_\mu\partial_\nu\phi - 4\chi^\lambda\partial_\mu\partial_\nu\phi\partial_\lambda\phi \\
& + 4h^{\lambda\sigma}\partial_\lambda\partial_\sigma\partial_{(\mu}\phi\partial_{\nu)}\phi + 4\partial_{(\mu}h^{\lambda\sigma}\partial_\lambda\partial_\sigma\phi\partial_{\nu)}\phi + 4\chi^\lambda\partial_\lambda\partial_{(\mu}\phi\partial_{\nu)}\phi \\
& + \frac{8}{3}h^{\lambda\sigma}\partial_\mu\partial_\lambda\phi\partial_\nu\partial_\sigma\phi + \frac{4}{3}\partial_{(\mu}h^{\lambda\sigma}\partial_{\nu)}\partial_\lambda\phi\partial_\sigma\phi + 4\partial^\lambda h^\sigma_{(\mu}\partial_{\nu)}\partial_\lambda\phi\partial_\sigma\phi \\
& - 4\partial^\lambda h^\sigma_{(\mu}\partial_{\nu)}\partial_\sigma\phi\partial_\lambda\phi - \frac{4}{3}h^{\lambda\sigma}\partial_\mu\partial_\nu\partial_\lambda\phi\partial_\sigma\phi - \frac{4}{3}\partial_{(\mu}h^\lambda_{\nu)}\partial_\lambda\partial_\sigma\phi\partial^\sigma\phi \\
& + \frac{2}{3}\partial^\lambda h_{\mu\nu}\partial_\lambda\partial_\sigma\phi\partial^\sigma\phi + \frac{4}{3}\partial_\mu\partial_\nu h^{\lambda\sigma}\partial_\lambda\phi\partial_\sigma\phi - 4\partial^\lambda\partial_{(\mu}h^\sigma_{\nu)}\partial_\lambda\phi\partial_\sigma\phi \\
& + 2\partial^\lambda\partial^\sigma h_{\mu\nu}\partial_\lambda\phi\partial_\sigma\phi - 4\partial^\lambda\chi_{(\mu}\partial_{\nu)}\phi\partial_\lambda\phi + 4\Box h^\lambda_{(\mu}\partial_{\nu)}\phi\partial_\lambda\phi + \frac{4}{3}\partial_{(\mu}\chi_{\nu)}\partial_\lambda\phi\partial^\lambda\phi \\
& - \frac{2}{3}\Box h_{\mu\nu}\partial_\lambda\phi\partial^\lambda\phi + \frac{4}{3}\partial_\lambda\chi^\lambda\partial_\mu\phi\partial_\nu\phi - 4h^\lambda_{(\mu}\partial_{\nu)}\partial_\lambda\phi\Box\phi + 2h^\lambda_{(\mu}\partial_{\nu)}\Box\phi\partial_\lambda\phi
\end{aligned}
$$

$$
\begin{aligned}
&+2h_{\lambda(\mu}\partial^\lambda\Box\phi\partial_{\nu)}\phi+\frac{8}{3}h^\lambda_{(\mu}\partial_{\nu)}\partial^\sigma\phi\partial_\lambda\partial_\sigma\phi-\frac{4}{3}h^\lambda_{(\mu}\partial_{\nu)}\partial_\lambda\partial_\sigma\phi\partial^\sigma\phi\\
&+\eta_{\mu\nu}\Big[2h^{\lambda\sigma}\partial_\lambda\partial_\sigma\phi\Box\phi+2\chi^\lambda\Box\phi\partial_\lambda\phi-\frac{2}{3}h^{\lambda\sigma}\partial_\lambda\partial_\sigma\partial_\rho\phi\partial^\rho\phi-\frac{2}{3}\chi^\lambda\partial_\lambda\partial_\sigma\phi\partial^\sigma\phi\\
&+2\partial^\lambda\chi^\sigma\partial_\lambda\phi\partial_\sigma\phi-\frac{2}{3}h^{\lambda\sigma}\partial_\lambda\Box\phi\partial_\sigma\phi-\frac{4}{3}h^{\lambda\sigma}\partial_\lambda\partial_\rho\phi\partial_\sigma\partial^\rho\phi-\frac{4}{3}\partial^\rho h^{\lambda\sigma}\partial_\rho\partial_\lambda\phi\partial_\sigma\phi\\
&-\frac{4}{3}\Box h^{\lambda\sigma}\partial_\lambda\phi\partial_\sigma\phi-\frac{2}{3}\partial_\lambda\chi^\lambda\partial_\sigma\phi\partial^\sigma\phi\Big]-\frac{4}{3}\partial_{(\mu}h^{\lambda\sigma}\partial_{\nu)}\partial_\lambda\partial_\sigma\phi-\frac{2}{3}h^{\lambda\sigma}\partial_\lambda\partial_\sigma\partial_\mu\partial_\nu\phi\\
&-\frac{8}{3}\partial_\mu\partial_\nu h^{\lambda\sigma}\partial_\lambda\partial_\sigma\phi-\frac{2}{3}\chi^\lambda\partial_\lambda\partial_\mu\partial_\nu\phi+\frac{8}{3}\partial_{(\mu}\chi^\lambda\partial_{\nu)}\partial_\lambda\phi-\frac{2}{3}\partial_\mu\partial_\nu\chi^\lambda\partial_\lambda\phi\\
&-\frac{2}{3}\partial_{(\mu}h^\lambda_{\nu)}\partial_\lambda\Box\phi+\frac{1}{3}\partial^\lambda h_{\mu\nu}\partial_\lambda\Box\phi+4\partial^\lambda\partial_{(\mu}h^\sigma_{\nu)}\partial_\lambda\partial_\sigma\phi-2\partial^\lambda\partial^\sigma h_{\mu\nu}\partial_\lambda\partial_\sigma\phi\\
&-\frac{14}{3}\partial_{(\mu}\chi_{\nu)}\Box\phi+\frac{7}{3}\Box h_{\mu\nu}\Box\phi-2\partial_\lambda\chi^\lambda\partial_\mu\partial_\nu\phi+4\partial_\lambda\chi_{(\mu}\partial_{\nu)}\partial^\lambda\phi-4\Box h^\lambda_{(\mu}\partial_{\nu)}\partial_\lambda\phi\\
&+\frac{2}{3}\partial_\lambda\partial_{(\mu}\chi^\lambda\partial_{\nu)}\phi-\frac{2}{3}h^\lambda_{(\mu}\partial_{\nu)}\partial_\lambda\Box\phi+\eta_{\mu\nu}\Big[\frac{4}{3}h^{\lambda\sigma}\partial_\lambda\partial_\sigma\Box\phi+\frac{4}{3}\partial^\rho h^{\lambda\sigma}\partial_\rho\partial_\lambda\partial_\sigma\phi\\
&+\frac{4}{3}\chi^\lambda\partial_\lambda\Box\phi+\frac{8}{3}\Box h^{\lambda\sigma}\partial_\lambda\partial_\sigma\phi-\frac{8}{3}\partial^\lambda\chi^\sigma\partial_\lambda\partial_\sigma\phi+\frac{2}{3}\Box\chi^\lambda\partial_\lambda\phi+2\partial_\lambda\chi^\lambda\Box\phi\\
&-\frac{1}{3}\partial^\lambda\partial_\sigma\chi^\sigma\partial_\lambda\phi\Big]-\frac{1}{9}\partial_\mu\partial_\nu\partial_\lambda\chi^\lambda+\frac{1}{9}\eta_{\mu\nu}\Box\partial_\lambda\chi^\lambda\Big\}. \qquad \text{(C.3)}
\end{aligned}
$$

The trace is given by

$$
\begin{aligned}
\mathbf{T}^{\mathrm{R}\mu}{}_{\mu}=\frac{b_1}{8\pi^2}\Big\{&-2\Box^2\phi+4h^{\mu\nu}\Box\partial_\mu\partial_\nu\phi+4\partial^\lambda h^{\mu\nu}\partial_\lambda\partial_\mu\partial_\nu\phi+4\chi^\mu\partial_\mu\Box\phi\\
&+4\Box h^{\mu\nu}\partial_\mu\partial_\nu\phi+2\Box\chi^\mu\partial_\mu\phi+\frac{4}{3}\partial_\mu\chi^\mu\Box\phi-\frac{2}{3}\partial^\mu\partial_\nu\chi^\nu\partial_\mu\phi+\frac{1}{3}\Box\partial_\mu\chi^\mu\Big\},
\end{aligned}
\qquad \text{(C.4)}
$$

which is coincident with the expansion of $\mathbf{T}^{\mathrm{R}\mu}{}_{\mu}=(-b_1/16\pi^2)\times(4\bar{\Delta}_4\phi+\bar{E}_4)$.

Weyl sector The contribution from the Weyl action

$$
\int d^4x\sqrt{-g}\Big[-\frac{1}{t^2}C^2_{\mu\nu\lambda\sigma}\Big]
$$

to the energy-momentum tensor is given by

$$
\mathbf{T}^{\mathrm{W}}_{\mu\nu}=-\frac{2}{t^2}\Big\{\Box^2h_{\mu\nu}-2\Box\partial_{(\mu}\chi_{\nu)}+\frac{2}{3}\partial_\mu\partial_\nu\partial_\lambda\chi^\lambda+\frac{1}{3}\eta_{\mu\nu}\Box\partial_\lambda\chi^\lambda\Big\}, \qquad \text{(C.5)}
$$

and its trace becomes $\mathbf{T}^{\mathrm{W}\mu}{}_{\mu}=0$.

Appendix D
Fundamental Constants

Reduced Planck constant	$\hbar$	=	1.055×10^{-27} cm^2 g s^{-1}
Speed of light	c	=	2.998×10^{10} cm s^{-1}
Newton's constant	G	=	6.672×10^{-8} cm^3 g^{-1} s^{-2}
Planck mass	m_{pl}	=	2.177×10^{-5} g
		=	1.221×10^{19} GeV/c^2
Reduced Planck mass	M_{P}	=	2.436×10^{18} GeV/c^2
Planck length	l_{pl}	=	1.616×10^{-33} cm
Planck time	t_{pl}	=	5.390×10^{-44} s
Boltzmann constant	k_{B}	=	1.381×10^{-16} erg K^{-1}
Megaparsec	1Mpc	=	3.086×10^{24} cm
Hubble constant	H_0	=	$100h$ km s^{-1} Mpc^{-1}
Hubble distance	c/H_0	=	$2998h^{-1}$ Mpc
			(current observation: $h \simeq 0.7$)

Useful constants for converting to natural units ($c = \hbar = k_{\mathrm{B}} = 1$)

1 cm	=	5.068×10^{13} $\hbar$/GeV
1 s	=	1.519×10^{24} $\hbar$/GeV/c
1 g	=	5.608×10^{23} GeV/c^2
1 erg	=	6.242×10^{2} GeV
1 K	=	8.618×10^{-14} GeV/k_{B}

K. Hamada, *Trans-Planckian Physics and Inflation*, Fundamental Theories of Physics 223, https://doi.org/10.1007/978-981-96-3476-7

References

[AB85] I. Avramidy and A. Barvinsky, *Asymptotic Freedom in Higher-Derivative Quantum Gravity*, Phys. Lett. B **159** (1985) 269.

[Abb81] L. Abbott, *The Background Field Method Beyond One Loop*, Nucl. Phys. B **185** (1981) 189.

[Abb+16] B. Abbott, *et al.*, *Observation of Gravitational Waves from a Binary Black Hole Merger*, Phys. Rev. Lett. **116** (2016) 061102.

[ABG48] R. Alpher, H. Bethe, and G. Gamow, *The Origin of Chemical Elements*, Phys. Rev. **73** (1948) 803.

[ACD77] S. Adler, J. Collins, and A. Duncan, *Energy-Momentum-Tensor Trace Anomaly in Spin-1/2 Quantum Electrodynamics*, Phys. Rev. D **15** (1977) 1712.

[ADF85] J. Ambjørn, B. Durhuus, and J. Fröhlich, *Diseases of Triangulated Random Surface Models, and Possible Cures*, Nucl. Phys. B **257** (1985) 433.

[ADM59] R. Arnowitt, S. Deser, and C. Misner, *Dynamical Structure and Definition of Energy in General Relativity*, Phys. Rev. **116** (1959) 1322.

[ADM62] R. Arnowitt, S. Deser, and C. Misner, *An Introduction to Current Research*, in Gravitation, edited by L. Witten (Wiley, NY, 1962) p. 227.

[AI80] S. Avis and C. Isham, *Generalized Spin Structures on Four Dimensional Space-Times*, Commun. Math, Phys. **72** (1980) 72.

[AJ92] J. Ambjørn and J. Jurkiewicz, *Four-Dimensional Simplicial Quantum Gravity*, Phys. Lett. B **278** (1992) 42.

[AM92] I. Antoniadis and E. Mottola, *4D Quantum Gravity in the Conformal Sector*, Phys. Rev. D **45** (1992) 2013.

[AM92a] M. Agishtein and A. Migdal, *Simulations of Four-Dimensional Quantum Gravity as Dynamical Triangulation*, Mod. Phys. Lett. A **7** (1992) 1039.

[AM92b] M. Agishtein and A. Migdal, *Critical Behavior of Dynamically Triangulated Quantum Gravity in Four Dimensions*, Nucl. Phys. B **385** (1992) 395.

[AMM92] I. Antoniadis, P. Mazur, and E. Mottola, *Conformal Symmetry and Central Charges in Four Dimensions*, Nucl. Phys. B **388** (1992) 627.

[AMM97a] I. Antoniadis, P. Mazur, and E. Mottola, *Quantum Diffeomorphisms and Conformal Symmetry*, Phys. Rev. D **55** (1997) 4756.

[AMM97b] I. Antoniadis, P. Mazur, and E. Mottola, *Physical States of the Quantum Conformal Factor*, Phys. Rev. D **55** (1997) 4770.

[AS68] M. Atiyah and I. Singer, *The Index of Elliptic Operators I*, Ann. Math. **87** (1968) 484.

K. Hamada, *Trans-Planckian Physics and Inflation*, Fundamental Theories of Physics 223, https://doi.org/10.1007/978-981-96-3476-7

[AS82] A. Albrecht and P. Steinhardt, *Cosmology for Grand Unified Theories with Radiatively Induced Symmetry Breaking*, Phys. Rev. Lett. **48** (1982) 1220.

[Ash86] A. Ashtekar, *New Variables for Classical and Quantum Gravity*, Phys. Rev. Lett. **57** (1986) 2244.

[AV92] J. Ambjørn and S. Varsted, *Three-Dimensional Simplicial Quantum Gravity*, Nucl. Phys. B **373** (1992) 557.

[Bar80] J. Bardeen, *Gauge-Invariant Cosmological Perturbations*, Phys. Rev. D **22** (1980) 1882.

[BBK+98] S. Bilke, Z. Burda, A. Krzywicki, B. Petersson, J. Tabaczek, and G. Thorleifsson, *4D Simplicial Quantum Gravity Interacting with Gauge Matter Fields*, Phys. Lett. B **418** (1998) 266.

[BC80] L. Brown and J. Collins, *Dimensional Renormalization of Scalar Field Theory in Curved Space-time*, Ann. Phys. **130** (1980) 215.

[BCR83] L. Bonora, P. Cotta-Ramusino, and C. Reina, *Conformal Anomaly and Cohomology*, Phys. Lett. B **126** (1983) 305.

[BD82] N. Birrell and P. Davies, *Quantum Fields in Curved Space* (Cambridge Univ. Press, 1982).

[BFF78] A. Back, P. Freund, and M. Forger, *New Gravitational Instantons and Universal Spin Structures*, Phys. Lett. B **77** (1978) 181.

[BG84] D. Boulware and D. Gross, *Lee-Wick Indefinite Metric Quantization: A Functional Integral Approach*, Nucl. Phys. B **233** (1984) 1.

[BHS83] D. Boulware, G. Horowitz, and A. Strominger, *Zero-Energy Theorem for Scale-Invariant Gravity*, Phys. Rev. Lett. **50** (1983) 1726.

[BIPZ78] E. Brezin, C. Itzykson, G. Parisi, and J. Zuber, *Planar Diagrams*, Commun. Math. Phys. **59** (1978) 35.

[BIZ80] D. Bessis, C. Itzykson, and J. Zuber, *Quantum Field Theory Techniques in Graphical Enumeration*, Adv. Appl. Math. **1** (1980) 109.

[BK87] D. Boulatov and V. Kazakov, *The Ising Model on Random Planar Lattice: The Structure of the Phase Transition and the Exact Critical Exponents*, Phys. Lett. B **186** (1987) 379.

[BK90] E. Brezin and V. Kazakov, *Exactly Solvable Field Theories of Closed Strings*, Phys. Lett. B **236** (1990) 144.

[BKKM86] D. Boulatov, V. Kazakov, I. Kostov, and A. Migdal, *Analytical and Numerical Study of a Model of Dynamically Triangulated Random Surfaces*, Nucl. Phys. B **275** (1986) 641.

[BMP92] P. Bouwknegt, J. McCarthy, and K. Pilch, *BRST Analysis of Physical States for 2D Gravity Coupled to $c<1$ Matter*, Commun. Math. Phys. **145** (1992) 541.

[Bou92] D. Boulatov, *A Model of Three-Dimensional Lattice Gravity*, Mod. Phys. Lett. A **7** (1992) 1629.

[BPST75] A. Belavin, A. Polyakov, A. Schwartz, and Y. Tyupkin, *Pseudoparticle Solutions of the Yang-Mills Equations*, Phys. Lett. B **59** (1975) 85.

[BPZ84] A. Belavin, A. Polyakov, and A. Zamolodchikov, *Infinite Conformal Symmetry in Two-Dimensional Quantum Field Theory*, Nucl. Phys. B **241** (1984) 333.

[BRS75] C. Becchi, A. Rouet, and R. Stora, *Renormalization of the Abelian Higgs-Kibble Model*, Comm. Math. Phys. **42** (1975) 127.

[BRS76] C. Becchi, A. Rouet, and R. Stora, *Renormalization of Gauge Theories*, Ann. Phys. **98** (1976) 287.

[BV85] A. Barvinsky and G. Vilkovisky, *The Generalized Schwinger-DeWitt Technique in Gauge Theories and Quantum Gravity*, Phys. Rep. **119** (1985) 1.

[CD74] D. Capper and M. Duff, *Trace Anomalies in Dimensional Regularization*, Nuovo Cimento A **23** (1974) 173.

[CDG78] C. Callan, R. Dashen, and D. Gross, *Toward a Theory of Strong Interactions*, Phys. Rev. D **17** (1978) 2717.

[CDG79] C. Callan, R. Dashen, and D. Gross, *A Theory of Hadronic Structure*, Phys. Rev. D **19** (1979) 1826.

[CKR94] S. Catteral, J. Kogut, and R. Renken, *Phase Structure of Four-Dimensional Simplicial Quantum Gravity*, Phys. Lett. B **328** (1994) 277.

[COBE] COBE Collaboration, C. Bennett, *et al.*, *Four Year COBE DMR Cosmic Microwave Background Observations: Maps and Basic Results*, Astrophys. J. 464 (1996) L1.

[Col84] J. Collins, *Renormalization* (Cambridge Univ. Press, 1984).

[CS74] S. Chern and J. Simons, *Characteristic Forms and Geometric Invariants*, Ann. Math. **99** (1974) 48.

[CS96] L. Chang and C. Soo, *The Standard Model with Gravity Couplings*, Phys. Rev. D **53** (1996) 5682.

[Dav85] F. David, *Planar Diagrams, Two-Dimensional Lattice Gravity and Surface Models*, Nucl. Phys. B **257** [FS14] (1985) 45.

[Dav88] F. David, *Conformal Field Theories coupled to 2-D Gravity in the Conformal Gauge*, Mod. Phys. Lett. A **3** (1988) 1651.

[DDI76] S. Deser, M. Duff, and C. Isham, *Non-local Conformal Anomalies*, Nucl. Phys. B **111** (1976) 45.

[Des1917] W. De Sitter, *On the Relativity of Inertia: Remarks Concerning Einstein's Latest Hypothesis*, Proc. Kon. Ned. Acad. Wet. **19** (1917) 1217.

[Dew67a] B. DeWitt, *Quantum Theory of Gravity. I. The Canonical Theory*, Phys. Rev. **160** (1967) 1113.

[Dew67b] B. DeWitt, *Quantum Theory of Gravity. II. The Manifestly Covariant Theory*, Phys. Rev. **162** (1967) 1195.

[Dew67c] B. DeWitt, *Quantum Theory of Gravity. III. Applications of the Covariant Theory*, Phys. Rev. **162** (1967) 1239.

[DH80] I. Drummond and S. Hathrell, *Renormalization of the gravitational trace anomaly in QED*, Phys. Rev. D **21** (1980) 958.

[Dir64] P. Dirac, *Lectures on Quantum Mechanics* (Belfer Graduate School of Science, Yeshiva University, New York, 1964).

[DK89] J. Distler and H. Kawai, *Conformal Field Theory and 2D Quantum Gravity*, Nucl. Phys. B **321** (1989) 509.

[DN74] S. Deser and P. Nieuwenhuizen, *Nonrenormalizability of the Quantized Einstein-Maxwell System*, Phys. Rev. Lett. **32** (1974) 245.

[DS90] M. Dougls and S. Shenker, *Strings in Less Than One Dimension*, Nucl. Phys. B **335** (1990) 635.

[DS93] S. Deser and A. Schwimmer, *Geometric Classification of Conformal Anomalies in Arbitrary Dimensions*, Phys. Lett. B **309** (1993) 279.

[Duf77] M. Duff, *Observations on Conformal Anomalies*, Nucl. Phys. B **125** (1977) 334.

[Duf94] M. Duff, *Twenty Years of The Weyl Anomaly*, Class. Quantum Grav. **11** (1994) 1387.

[Dur01] R. Durrer, *The Theory of CMB Anisotropies*, J. Phys. Stud. **5** (2001) 177.

[Dys49a] F. Dyson, *The Radiation Theories of Tomonaga, Schwinger, and Feynman*, Phys. Rev. **75** (1949) 486.

[Dys49b] F. Dyson, *The S Matrix in Quantum Electrodynamics*, Phys. Rev. **75** (1949) 1736.

[EF76] T. Eguchi and P. Freund, *Quantum Gravity and World Topology*, Phys. Rev. Lett. **37** (1976) 1251.

[EGH80] T. Eguchi, P. Gilkey, and A. Hanson, *Gravitation, Gauge Theories and Differential Geometry*, Phys. Rep. **66** (1980) 213.

[EH79] T. Eguchi and A. Hanson, *Self-Dual Solutions to Euclidean Gravity*, Ann. Phys. **120** (1979) 82.

[EHI+97] H. Egawa, T. Hotta, T. Izubuchi, N. Tsuda, and T. Yukawa, *Scaling Behavior in 4D Simplicial Quantum Gravity*, Prog. Theor. Phys. **97** (1997) 539.

[EHY02] H. Egawa, S. Horata, and T. Yukawa, *Clear Evidence of a Continuum Theory of 4D Euclidean Simplicial Quantum Gravity*, Nucl. Phys. B (Proc. Suppl.) **106** (2002) 971.

[Ein1916] A. Einstein, *Die Grundlage der allgemeinen Relativitätstheorie*, Annalen der Phys. **49** (1916) 769.

[Ein1917] A. Einstein, *Kosmologische Betrachtungen zur allgemeinen Relativitätstheorie*, Königlich Preu*ß*ische Akademie der Wissenschaften, Sitzungsberichte, (1917) 142.

[EM96] W. Enright and P. Muir, *Runge-Kutta Software with Defect Control for Boundary Value ODEs*, SIAM J. Sci. Comput. **17** (1996) 479.

[EPP+12] S. El-Showk, M. Paulos, D. Poland, S. Rychkov, D. Simmons-Duffin, and A. Vichi, *Solving the 3D Ising Model with the Conformal Bootstrap*, Phys. Rev. D **86** (2012) 025022.

[Fey48] R. Feynman, *Space-Time Approach to Non-Relativistic Quantum Mechanics*, Rev. of Mod. Phys. **20** (1948) 367.

[Fey49] R. Feynman, *Space-Time Approach to Quantum Electrodynamics*, Phys. Rev. **76** (1949) 769.

[FGG74] S. Ferrara, R. Gatto, and A. Grillo, *Positivity Restriction on Anomalous Dimensions*, Phys. Rev. D **9** (1974) 3564.

[FKN91] M. Fukuma, H. Kawai, and R. Nakayama, *Continuum Schwinger-Dyson Equations and Universal Structures in Two-Dimensional Quantum Gravity*, Int. J. Mod. Phys. A **6** (1991) 1385.

[FMS86] D. Friedan, E. Martinec, and S. Shenker, *Conformal Invariance, Supersymmetry and String Theory*, Nucl. Phys. B **271** (1986) 93.

[FP84] E. Fradkin and M. Palchik, *Conformal Invariance in Quantum Yang-Mills Theory*, Phys. Lett. B **147** (1984) 86.

[FQS84] D. Friedan, Z. Qiu, and S. Shenker, *Conformal Invariance, Unitarity, and Critical Exponents in Two Dimensions*, Phys. Rev. Lett. **52** (1984) 1575.

[Fre84] M. Freeman, *The Renormalization of Nonabelian Gauge Theories in Curved Space-Time*, Ann. Phys. **153** (1984) 339.

[Fri1922] A. Friedmann, *Über die Krümmung des Raumes*, Z. Phys. **10** (1922) 377.

[FT82a] E. Fradkin and A. Tseytlin, *Renormalizale Asymptotically Free Quantum Theory of Gravity*, Nucl. Phys. B **201** (1982) 469.

[FT82b] E. Fradkin and A. Tseytlin, *Asymptotic Freedom in Extended Conformal Supergravities*, Phys. Lett. B **110** (1982) 117.

[Fub1904] G. Fubini, *Sulle metriche definite da una forma Hermitiana*, Atti del Reale Istituto Veneto di Scienze, Lettere ed Arti, **63** (1904) 501.

[Gam48] G. Gamow, *The Origin of Elements and the Separation of Galaxies*, Phys. Rev. **74** (1948) 505.

[Gil09] L. Gilder, *The Age of Entanglement: When Quantum Physics Was Reborn* (Vintage Books, 2009).

[GL91] M. Goulian and M. Li, *Correlation Functions in Liouville Theory*, Phys. Rev. Lett. **66** (1991) 2051.

[Gla61] S. Glashow, *Partial Symmetries of Weak Interactions*, Nucl. Phys. **22** (1961) 579.

[GM90] D. Gross and A. Migdal, *Nonperturbative Two-Dimensional Quantum Gravity*, Phys. Rev. Lett. **64** (1990) 127.

[GP78] G. Gibbons and C. Pope, $\mathbb{CP}^2$ *as a Gravitational Instantons*, Commun. Math. Phys. **61** (1978) 239.

[GSW87] M. Green, J. Schwarz, and E. Witten, *String Theory I, II* (Cambridge Univ. Press, 1987).

[Gut81] A. Guth, *The Inflationary Universe: A Possible Solution to the Horizon and Flatness Problems*, Phys. Rev. D **23** (1981) 347.

[GV92] M. Gross and S. Varsted, *Elementary Moves and Ergodicity in D-Dimensional Simplicial Quantum Gravity*, Nucl. Phys. B **378** (1992) 367.

[GW73] D. Gross and F. Wilczek, *Ultraviolet Behavior of Nonabelian Gauge Theories*, Phys. Rev. Lett. **30** (1973) 1343.

[Ham94] K. Hamada, *Ward Identities of* W_∞ *Symmetry in Liouville Theory coupled to* $c_M < 1$ *Matter*, Phys. Lett. B **324** (1994) 141.

[Ham02] K. Hamada, *Resummation and Higher Order Renormalization in 4D Quantum Gravity*, Prog. Theor. Phys. **108** (2002) 399.

[Ham05] K. Hamada, *Building Blocks of Physical States in a Non-Critical 3-Brane on $R \times S^3$*, Int. J. Mod. Phys. A **20** (2005) 5353.

[Ham09] K. Hamada, *Conformal Field Theory on $R \times S^3$ from Quantized Gravity*, Int. J. Mod. Phys. A **24** (2009) 3073.

[Ham12a] K. Hamada, *Background-Free Quantum Gravity based on Conformal Gravity and Conformal Field Theory on M^4*, Phys. Rev. D **85** (2012) 024028.

[Ham12b] K. Hamada, *BRST Invariant Higher Derivative Operators in 4D Quantum Gravity based on CFT*, Phys. Rev. D **85** (2012) 124036.

[Ham12c] K. Hamada, *BRST Analysis of Physical Fields and States for 4D Quantum Gravity on $R \times S^3$*, Phys. Rev. D **86** (2012) 124006.

[Ham14a] K. Hamada, *Determination of Gravitational Counterterms Near Four Dimensions from Renormalization Group Equations*, Phys. Rev. D **89** (2014) 104063.

[Ham14b] K. Hamada, *Renormalization Group Analysis for Quantum Gravity with a Single Dimensionless Coupling*, Phys. Rev. D **90** (2014) 084038.

[Ham18] K. Hamada, *Quantum Gravity and Cosmology Based on Conformal Field Theory* (Cambridge Scholar Publishing, Newcastle, 2018)

[Ham20a] K. Hamada, *Localized Massive Excitation of Quantum Gravity as a Dark Particle*, Phys. Rev. D **102** (2020) 026024.

[Ham20b] K. Hamada, *Diffeomorphism Invariance Demands Conformal Anomalies*, Phys. Rev. D **102** (2020) 125005.

[Ham22] K. Hamada, *Revealing A Trans-Planckian World Solves The Cosmological Constant Problem*, Prog. Theor. Exp. Phys. **2022** (2022) 103E02.

[Ham24a] K. Hamada, *Study of Nonlinear Evolution of Spacetime Fluctuations in Quantum Gravity Inflation for Deriving Primordial Spectrum*, Universe **10** (2024) 33.

[Ham24b] K. Hamada, *Quantum Gravity Effective Action Provides Entropy of the Universe*, Particles **7** (2024) 465.

[Har70] E. Harrison, *Fluctuations at the Threshold of Classical Cosmology*, Phys. Rev. D **1** (1970) 2726.

[Hat82a] S. Hathrell, *Trace Anomalies and $\lambda\phi^4$ Theory in Curved Space*, Ann. Phys. **139** (1982) 136.

[Hat82b] S. Hathrell, *Trace Anomalies and QED in Curved Space*, Ann. Phys. **142** (1982) 34.

[Haw74] S. Hawking, *Black hole explosions?*, Nature. **248** (1974) 30.

[Haw75] S. Hawking, *Particle Creation by Black Holes*, Commun. Math. Phys. **43** (1975) 199.

[HETY01] S. Horata, H. Egawa, N. Tsuda, and T. Yukawa, *Phase Structure of Four-Dimensional Simplicial Quantum Gravity with a $U(1)$ Gauge Field*, Prog. Theor. Phys. **106** (2001) 1037.

[HEY03] S. Horata, H. Egawa, and T. Yukawa, *Grand Canonical Simulation of 4D Simplicial Quantum Gravity*, Nucl. Phys. B (Proc. Suppl.) **119** (2003) 921.

[HH03] K. Hamada and S. Horata, *Conformal Algebra and Physical States in a Non-critical 3-brane on $R \times S^3$*, Prog. Theor. Phys. **110** (2003) 1169.

[HHY06a] K. Hamada, S. Horata, and T. Yukawa, *Space-time Evolution and CMB Anisotropies from Quantum Gravity*, Phys. Rev. D **74** (2006) 123502.

[HHY06b] K. Hamada, S. Horata, and T. Yukawa, *Background Free Quantum Gravity and Cosmology* in Focus on Quantum Gravity Research edited by D. Moore (Nova Science Publisher, NY, 2006), Chapter 1.

[HHY10] K. Hamada, S. Horata, and T. Yukawa, *From Conformal Field Theory Spectra to CMB Multipoles in Quantum Gravity Cosmology*, Phys. Rev. D **81** (2010) 083533.

[HM16] K. Hamada and M. Matsuda, *Two-Loop Quantum Gravity Corrections to The Cosmological Constant in Landau Gauge*, Phys. Rev. D **93** (2016) 064051.

[HM17] K. Hamada and M. Matsuda, *Physical Cosmological Constant in Asymptotically Background-Free Quantum Gravity*, Phys. Rev. D **96** (2017) 026010.

[HP78] S. Hawking and C. Pope, *Generalized Spin Structures in Quantum Gravity*, Phys. Lett. B **73** (1978) 42.
[HS95] W. Hu and N. Sugiyama, *Toward Understanding CMB Anisotropies and their Implications*, Phys. Rev. D **51** (1995) 2599.
[HS99] K. Hamada and F. Sugino, *Background-Metric Independent Formulation of 4D Quantum Gravity*, Nucl. Phys. B **553** (1999) 283.
[HTY98] H. Hagura, N. Tsuda, and T. Yukawa, *Phase and Fractal Structures of Three-Dimensional Simplicial Gravity*, Phys. Lett. B **418** (1998) 273.
[Hub1929] E. Hubble, *A Relation between Distance and Radial Velocity among Extra-Galactic Nebulae*, Proceedings of the National Academy of Sciences. **15** (1929) 168.
[HW70] K. Huang and S. Weinberg, *Ultimate Temperature and the Early Universe*, Phys. Rev. Lett. **25** (1970) 895.
[HY05] K. Hamada and T. Yukawa, *CMB Anisotropies Reveal Quantized Gravity*, Mod. Phys. Lett. A **20** (2005) 509.
[IP82] C. Isham and C. Pope, *A Spinor Field Representation of the Stiefel-Whitney Class*, Phys. Lett. B **114** (1982) 137.
[IZ80] C. Itzykson and J. Zuber, *Quantum Field Theory* (McGraw-Hill Inc. 1980).
[Kaj+98] T. Kajita, *et al.*, *Evidence for Oscillation of Atmospheric Neutrinos*, Phys. Rev. Lett. **81** (1998) 1562.
[Kaz86] V. Kazakov, *Ising Model on a Dynamical Planar Random Lattice: Exact Solution*, Phys. Lett. A **119** (1986) 140.
[KKM85] V. Kazakov, I. Kostov, and A. Migdal, *Critical Properties of Randomly Triangulated Planar Random Surfaces*, Phys. Lett. B **157** (1985) 295.
[Kie12] C. Kiefer, *Quantum Gravity* Third Edition (Oxford Univ. Press, 2012).
[KO79] T. Kugo and I. Ojima, *Local Covariant Operator Formalism of Non-Abelian Gauge Theories and Quark Confinement Problem*, Prog. Theor. Phys. Suppl. **66** (1979) 1.
[KO83] M. Kato and K. Ogawa, *Covariant Quantization of String based on BRST Invariance*, Nucl. Phys. B **212** (1983) 443.
[KPZ88] K. Knizhnik, A. Polyakov, and A. Zamolodchikov, *Fractal Structure of 2D-Quantum Gravity* , Mod. Phys. Lett. A **3** (1988) 819.
[KS84] H. Kodama and M. Sasaki, *Cosmological Perturbation Theory*, Prog. Theor. Phys. Suppl. **78** (1984) 1.
[KS01] E. Komatsu and D. Spergel, *Acoustic Signatures in the Primary Microwave Background Bispectrum*, Phys. Rev. D **63** (2001) 063002.
[KT90] E. Kolb and M. Turner, *The Early Univrse* (Westview Press,1990).
[Lan81] P. Langacker, *Grand Unified Theories and Proton Decay*, Phys. Rep. **72** (1981) 185.
[Lem1927] G. Lemaître, *Un Univers homogéne de masse constante et de rayon croissant rendant compte de la vitesse radiale des nébuleuses extra-galactiques*, Annales de la Société Scientifique de Bruxelles **A47** (1927) 49.
[Lin82] A. Linde, *A New Inflationary Universe Scenario: A Possible Solution of the Horizon, Flatness, Homogeneity, Isotropy and Primordial Monopole Problems*, Phys. Lett. B **108** (1982) 389.
[LW69] T. Lee and G. Wick, *Negative Metric and Unitarity of the S-Matrix*, Nucl. Phys. B **9** (1969) 209.
[LZ91] B. Lian and G. Zuckermann, *2D Gravity with $c = 1$ Matter*, Phys. Lett. B **254** (1991) 417.
[Mac77] G. Mack, *All Unitary Ray Representations of the Conformal Group $SU(2,2)$ with Positive Energy*, Commun. Math. Phys. **55** (1977) 1.
[MC81] V. Mukhanov and G. Chibisov, *Quantum Fluctuations and a Nonsingular Universe*, Pis'ma Zh. Eksp. Teor. Fiz. **33** (1981) 549 [JETP Lett. **33** (1981) 532].
[Meh67] M. Mehta, *Random Matrices* (Academic Press, 1967; Elsevier Academic Press, 2004, third edition).
[Mor94] T. Morris, *The Exact Renormalization Group and Approximate Solutions*, Int. J. Mod. Phys. A **9** (1994) 2411.

[Mut87] T. Muta, *Foundation of Quantum Chromodynamics* (World Scientific, 1987).
[Nak72] N. Nakanishi, *Indefinite-Metric Quantum Field Theory*, Prog. Theor. Phys. Suppl. **51** (1972) 1.
[Nak90] M. Nakahara, *Geometry, Topology and Physics* (IOP Publishing, 1990; Taylor & Francis Incorporated, 2021, third edition).
[Nie81] P. Nieuwenhuizen, *Supergravity*, Phys. Rep. **68** (1981) 189.
[Oog93] H. Ooguri, *Schwinger-Dyson Equation in Three-Dimensional Simplicial Quantum Gravity*, Prog. Theor. Phys. **89** (1993) 1.
[OS73] K. Osterwalder and R. Schrader, *Axioms for Euclidean Green's Functions I.*, Commun. Math. Phys. **31** (1973) 83.
[OS75] K. Osterwalder and R. Schrader, *Axioms for Euclidean Green's Functions II.*, Commun. Math. Phys. **42** (1975) 281.
[Ost1850] M. Ostrogradsky, *Mémoires sur les équations différentielles, relatives au problème des isopérimètres*, Mem. Acad. St. Petersbourg **VI 4** (1850) 385.
[OTY98] S. Oda, N. Tsuda, and T. Yukawa, *Grand-Canonical Simulation of Two-Dimensional Simplicial Gravity*, Prog. Theor. Phys. **99** (1998) 875.
[Pac91] U. Pachner, *P.L. Homeomorphic Manifolds are Equivalent by Elementary Shellings*, Europ. J. Combinatorics **12** (1991) 129.
[Pan83] S. Paneitz, *A Quartic Conformally Covariant Differential Operator for Arbitrary Pseudo-Riemannian Manifolds*, Preprint 1983, SIGMA **4** (2008) 036.
[Pee93] P. Peebles, *Principles of Physical Cosmology* (Princeton Univ. Press, 1993).
[Per+99] S. Perlmutter, *et al.*, *Measurements of* Ω *and* Λ *from 42 High-Redshift Supernovae*, Astrophys. J. **157** (1999) 565.
[Pla1899] M. Planck, *Über irreversible Strahlungsvorgänge*, Sitzungsberichte der
[PLANCK] Planck Collaboration, Ade, P., *et al.*, Astron. Astrophys., https://doi.org/10.1051/0004-6361/201321591(2014).
[Pol98] J. Polchinski, *String Theory I, II* (Cambridge Univ. Press, 1998).
[Poli73] H. Politzer, *Reliable Perturbative Results for Strong Interactions?*, Phys. Rev. Lett. **30** (1973) 1346.
[Poly81] A. Polyakov, *Quantum Geometry of Bosonic Strings*, Phys. Lett. B **103** (1981) 207.
[PR68] G. Ponzano and T. Regge, *Semiclassical Limit of Racah Coefficients* in Spectroscopic and group theoretical methods in physics, eds. F. Bloch, S. G. Cohen, A. De-Shalit, S. Sambursky, and I. Talmi (North-Holland Publishing, Amsterdam, 1968) p.1.
[PS95] M. Peskin and D. Schroeder, *An Introduction to Quantum Field Theory* (Westview Press, 1995). Königlich Preuβischen Akademie der Wissenschaften **5** (1899) 440.
[PT09] L. Parker and D. Toms, *Quantum Field Theory in Curved Spacetime* (Cambridge Univ. Press, 2009).
[PU50] A. Pais and G. Uhlenbeck, *On Field Theories with Non-Localized Action*, Phys. Rev. **79** (1950) 145.
[PW65] A. Penzias and R. Wilson, *A Measurement of Excess Antenna Temperature at 4080-Mc/s*, Astrophys. J. **142** (1965) 419.
[Reg61] T. Regge, *General Relativity without Coordinates*, Nuovo Cimento **19** (1961) 558.
[Reu98] M. Reuter, *Nonperturbative Evolution Equation for Quantum Gravity*, Phys. Rev. D **57** (1998) 971.
[Rie84] R. Riegert, *A Non-Local Action for the Trace Anomaly*, Phys. Lett. B **134** (1984) 56.
[Rie+21] A. Riess, S. Casertano, W. Yuan, J. Bowers, L. Macri, J. Zinn, and D. Scolnic, *Cosmic Distances Calibrated to 1% Precision with Gaia EDR3 Parallaxes and Hubble Space Telescope Photometry of 75 Milky Way Cepheids Confirm Tension with* Λ*CDM*, Astrophys. J. Lett. 908 (2021) L6.
[Rie+98] A. Riess, B. Schmidt, *et al.*, *Observational Evidence from Supernovae for an Accelerating Universe and a Cosmological Constant*, Astron. J. **116** (1998) 1009.
[RO84] M. Rubin and C. Ordóñez, *Eigenvalues and Degeneracies for n-Dimensional Tensor Spherical Harmonics*, J. Math. Phys. **25** (1984) 2888.

[RS18] M. Reuter and F. Saueressig, *Quantum Gravity and the Functional Renormalization Group* (Cambridge Univ. Press, 2018).

[SW67] R. Sachs and A. Wolfe, *Perturbations of a Cosmological Model and Angular Variations of the Microwave Background*, Astrophys. J. **147** (1967) 73.

[Sal68] A. Salam, *Weak and Electromagnetic Interactions*, Conf. Proc. **C 680519** (1968) 367.

[Sat81] K. Sato, *First Order Phase Transition of a Vacuum and Expansion of the Universe*, Mon. Not. R. Astron. Soc. **195** (1981) 467.

[Sch1916] K. Schwarzschild, *Über das Gravitationsfeld eines Massenpunktes nach der Einsteinschen Theorie*, Sitzungsberichte der Königlich Preuβischen Akademie der Wissenschaften **7** (1916) 189.

[Sch48] J. Schwinger, *On Quantum-Electrodynamics and the Magnetic Moment of the Electron*, Phys. Rev. **73** (1948) 416.

[Sch51] J. Schwinger, *On Green's functions of quantized fields I + II*. PNAS. **37** (1951) 452.

[SDSS] SDSS Collaboration, C. Ahn, *et al.*, *The Tenth Data Release of the Sloan Digital Sky Survey: First Spectroscopic Data from the SDSS-III Apache Point Observatory Galactic Evolution Experiment*, Astrophys. J. Suppl. **211** (2014) 17.

[Sei90] N. Seiberg, *Notes on Quantum Liouville Theory and Quantum Gravity*, Prog. Theor. Phys. Suppl. **102** (1990) 319.

[SKDE] Super-Kamiokande Collaboration, K. Abe, *et al.*, *Search for Proton Decay via* $p \to e^+\pi^0$ *and* $p \to \mu^+\pi^0$ *in* 0.31 *Megaton · Years Exposure of the Super-Kamiokande Water Cherenkov Detector* Phys. Rev. D **95** (2017) 012004.

[Sta80] A. Starobinsky, *A New Type of Isotropic Cosmological Models Without Singularity*, Phys. Lett. B **91** (1980) 99.

[Ste77] K. Stelle, *Renormalization of Higher-Derivative Quantum Gravity*, Phys. Rev. D **16** (1977) 953.

[Stu1905] E. Study, *Kürzeste Wege im komplexen Gebiet*, Mathematische Annalen **60** (1905) (3).

[SVZ80] M. Shifman, A. Vanishtein, and V. Zakharov, *Instanton Density in a Theory with Massless Quarks*, Nucl. Phys. B **163** (1980) 46.

[tHo74] G. 't Hooft, *A planar diagram theory for strong interactions*, Nucl. Phys. B **72** (1974) 461.

[tHo76] G. 't Hooft, *Computation of the Quantum Effects due to a Four-Dimensional Pseudoparticle*, Phys. Rev. D **14** (1976) 3432.

[Tho99] G. Thorleifsson, *Three-Dimensional Simplicial Gravity and Degenerate Triangulations*, Nucl. Phys. B **538** (1999) 278.

[tHV72] G. 't Hooft and M. Veltman, *Regularization and Renormalization of Gauge Fields*, Nucl. Phys. B **44** (1972) 189.

[tHV74] G. 't Hooft and M. Veltman, *One-Loop Divergencies in the Theory of Gravitation*, Ann. Inst. Henri Poincare **XX** (1974) 69.

[Tom46] S. Tomonaga, *On a Relativistically Invariant Formulation of the Quantum Theory of Wave Fields*, Prog. Theor. Phys. **1** (1946) 27.

[Tom77] E. Tomboulis, *1/N Expansion and Renormalization in Quantum Gravity*, Phys. Lett. B **70** (1977) 361.

[Tom80] E. Tomboulis, *Renormalizability and Asymptotic Freedom in Quantum Gravity*, Phys. Lett. B **97** (1980) 77.

[TV92] V. Turaev and O. Viro, *State Sum Invariants of 3-Manifolds and Quantum 6j-Symbols*, Topology **31** (1992) 865.

[Tyu75] I. Tyutin, Lebedev preprint FIAN, 1975.

[UD62] R. Utiyama and B. DeWitt, *Renormalization of a Classical Gravitational Field Interacting with Quantized Matter Fields*, J. Math. Phys. **3** (1962) 608.

[Uti56] R. Utiyama, *Invariant Theoretical Interpretation of Interaction*, Phys. Rev. **101** (1956) 1597.

[Vel75] M. Veltman, *Quantum Theory of Gravitation* in Methods in Field Theory, Les Houches 1975.

[VMK88] D. Varshalovich, A. Moskalev, and V. Khersonskii, *Quantum Theory of Angular Momentum* (World Scientific, Singapore, 1988).
[VZNS82] A. Vanishtein, V. Zakharov, V. Novikov, and M. Shifman, *ABC of Instantons*, Sov. Phys. Usp. **25** (1982) 195.
[WB83] J. Wess and J. Bagger, *Supersymmetry and Supergravity*, (Princeton Univ. Press, Princeton, 1983).
[Wei67] S. Weinberg, *A Model of Leptons*, Phys. Rev. Lett. **19** (1967) 1264.
[Wei79] S. Weinberg, *Ultraviolet Divergences in Quantum Theories of Gravitation* in General Relativity, an Einstein Centenary Survey, eds. S. Hawking and W. Israel (Cambridge Univ. Press, Cambridge, 1979) p. 790.
[Wei89] S. Weinberg, *The Cosmological Constant Problem*, Rev. Mod. Phys. **61** (1989) 1.
[Weig82] D. Weingarten, *Euclidean Quantum Gravity on a Lattice*, Nucl. Phys. B **210** [FS6] (1982) 229.
[Wet93] C. Wetterich, *Exact Evolution Equation for the Effective Potential*, Phys. Lett. B **301** (1993) 90.
[Wey1918] H. Weyl, *Reine Infinitesimalgeometrie*, Mathematische Zeitschrift **2** (1918) 384.
[Whe64] J. Wheeler, *Geometrodynamics and the Issue of the Final State* in Relativity, Groups and Topology, eds B. S. DeWitt and C. M. DeWitt (Gordon and Breach, NY, 1964) p. 317.
[Wic54] G. Wick, *Properties of Bethe-Salpeter Wave Functions*. Phys. Rev. **96** (1954) 1124.
[Wil75] K. Wilson, *The Renormalization Group: Critical Phenomena and the Kondo Problem*, Rev. of Mod. Phys. **47** (1975) 773.
[Wit88] E. Witten, $2+1$-*Dimensional Gravity as an Exactly Soluble System*, Nucl. Phys. B **311** (1988) 46.
[Wit92] E. Witten, *Ground Ring of Two-Dimensional String Theory*, Nucl. Phys. B **373** (1992) 187.
[WMAP] WMAP Collaboration, C. Bennett, *et al.*, *Nine-Year Wilkinson Microwave Anisotropy Probe (WMAP) Observations: Final Maps and Results*, Astrophys. J. Suppl. Ser. **208** (2013) 20.
[Woo15] R. Woodard, *Ostrogradsky's Theorem on Hamiltonian Instability*, Scholarpedia. **10** (2015) 32243, arXiv1506.02210 [hep-th].
[WWW19] J. Wang, X. Wen, and E. Witten, *A New* $SU(2)$ *Anomaly*, J. Math. Phys. **60** (2019) 052301.
[WZ71] J. Wess and B. Zumino, *Consequences of Anomalous Ward Identities*, Phys. Lett. B **37** (1971) 95.
[YM54] C. Yang and R. Mills, *Conservation of Isotopic Spin and Isotopic Gauge Invariance*, Phys. Rev. **96** (1954) 191.
[Zel67] Ya. Zel'dovich, *Cosmological Constant and Elementary Particles*, Pis'ma Zh. Eksp. Theor. Fiz. **6** (1967) 883 [JETP letters **6** (1967) 316].
[Zel72] Ya. Zel'dovich, *A Hypothesis, Unifying the Structure and the Entropy of the Universe*, Mon. Not. R. Astron. Soc. **160** (1972) P1.

Index

K. Hamada, *Trans-Planckian Physics and Inflation*, Fundamental Theories of Physics 223, https://doi.org/10.1007/978-981-96-3476-7

The manufacturer's authorised representative in the EU is Springer Nature Customer Service Centre GmbH, Europaplatz 3, 69115 Heidelberg, Germany. If you have any concerns regarding our products, please contact ProductSafety@springernature.com

Printed and bound by CPI Group (UK) Ltd, Croydon, CR0 4YY

01/10/2026

02225928-0002